SPRINGER PROTOCOLS HANDBOOKS

For further volumes:
http://www.springer.com/series/7651

Springer Protocols Handbooks collects a diverse range of step-by-step laboratory methods and protocols from across the life and biomedical sciences. Each protocol is provided in the Springer Protocol format: readily-reproducible in a step-by-step fashion. Each protocol opens with an introductory overview, a list of the materials and reagents needed to complete the experiment, and is followed by a detailed procedure supported by a helpful notes section offering tips and tricks of the trade as well as troubleshooting advice. With a focus on large comprehensive protocol collections and an international authorship, Springer Protocols Handbooks are a valuable addition to the laboratory.

Phytoconstituent Analysis

Qualitative to Molecular Profiling

Edited by

Dhruti Amin

C. G. Bhakta Institute of Biotechnology, Uka Tarsadia University, Bardoli, Surat, Gujarat, India

Natarajan Amaresan

Saveetha Institute of Basic Medical Sciences, Saveetha Institute of Medical and Technical Sciences (SIMATS), Saveetha University, Chennai, Tamil Nadu, India

Sanket Ray

Microbiology Department, Naran Lala College of Professional and Applied Sciences, Navsari, Gujarat, India

Humana Press

Editors
Dhruti Amin
C. G. Bhakta Institute of Biotechnology
Uka Tarsadia University, Bardoli
Surat, Gujarat, India

Sanket Ray
Microbiology Department
Naran Lala College of Professional and Applied Sciences
Navsari, Gujarat, India

Natarajan Amaresan
Saveetha Institute of Basic Medical Sciences,
Saveetha Institute of Medical and Technical Sciences (SIMATS)
Saveetha University
Chennai, Tamil Nadu, India

ISSN 1949-2448 ISSN 1949-2456 (electronic)
Springer Protocols Handbooks
ISBN 978-1-0716-5399-9 ISBN 978-1-0716-5400-2 (eBook)
https://doi.org/10.1007/978-1-0716-5400-2

This Humana imprint is published by the registered company Springer Science+Business Media, LLC, part of Springer Nature.
The registered company address is: 1 New York Plaza, New York, NY 10004, U.S.A.

Foreword

Plants are a rich and enduring source of bioactive secondary metabolites that form the foundation of traditional medicine, modern drug discovery, nutraceuticals, and functional foods. With rapid advances in analytical chemistry, molecular biology, and computational sciences, phytoconstituent research has evolved from classical chemical tests to integrated, high-resolution, and data-driven approaches. This volume *Phytoconstituent Analysis: Qualitative to Molecular Profiling* aptly captures this progression and presents it in a systematic and scientifically rigorous manner.

This book is structured into four well-defined parts. *Part I* focuses on standardized qualitative and quantitative protocols for major classes of phytoconstituents, including alkaloids, terpenoids, phenolic acids, glycosides, polyketides, non-ribosomal peptides, steroids, flavonoids, saponins, coumarins, and tannins. *Part II* advances into chromatographic and metabolic profiling techniques, addressing the growing demand for precise, reproducible, and high-throughput analytical methodologies. *Part III* integrates molecular and genetic approaches to elucidate biosynthetic pathways and regulatory mechanisms of secondary metabolites, reflecting current trends in functional and systems biology. *Part IV* highlights emerging applications of artificial intelligence, machine learning, and big data analytics in phytoconstituent profiling and biosynthesis prediction.

The editors, Dr. Dhruti Amin, Dr. Natraj Amaresan, and Dr. Sanket Ray, are to be commended for assembling a comprehensive and cohesive protocol-based reference that bridges classical phytochemical analysis with modern molecular and computational tools. The contributions from experts ensure both scientific depth and practical applicability.

This book will serve as a valuable resource for students, researchers, academicians, and professionals in phytochemistry, biotechnology, pharmacognosy, and allied life sciences, supporting reproducible research and methodological advancement in phytoconstituent analysis.

Prof. Ujjval Trivedi
Post Graduate Department of Biosciences
Sardar Patel University
Vallabh Vidyanagar, Gujarat, India

Preface

Phytoconstituents are a diverse group of plant-produced secondary metabolites vital for plant defense, adaptation, and ecological roles. They are also key sources of bioactive compounds with applications in human health and medicine, driving research in plant sciences, pharmacognosy, biotechnology, natural products chemistry, and drug discovery. Accurate identification, quantification, and characterization of these compounds depend on reliable, standardized analytical methods. Without such tools, progressing from basic screening to mechanistic studies is difficult and inconsistent.

This protocol book, titled *Phytoconstituent Analysis: Qualitative to Molecular Profiling*, offers a comprehensive collection of validated protocols. It covers the full analytical spectrum, from classical phytochemical assays to modern molecular biology techniques and emerging computational methods, providing researchers a unified resource for advancing investigations.

The book is organized into four parts, each focused on a different aspect of phytoconstituent analysis: *Part I*: Protocols for chemical detection and quantification of secondary metabolites, emphasizing reagent prep, assay execution, calibration, and result interpretation for reliable screening. *Part II*: Advanced chromatographic and high-resolution techniques like HPLC, GC-MS, LC-MS/MS, and NMR, with guidance on sample prep, instrument settings, method development, and validation for accurate metabolomic analysis. *Part III*: Molecular methods linking phytochemistry and molecular biology, with protocols for exploring genetic, transcriptomic, and biosynthetic regulation of secondary metabolites, aiding in genetic engineering. *Part IV*: AI in phytoconstituent analysis, using machine learning, deep learning, and big-data analytics for data integration, pattern recognition, property prediction, pathway inference, and virtual screening, enhancing experimental workflows and insights.

To maximize usability and ensure reproducibility, every chapter adheres to a standardized, reader-friendly protocol format. This includes exhaustive lists of required materials and reagents, clear step-by-step procedural instructions, highlighted critical notes on key considerations, potential pitfalls, and comprehensive troubleshooting guidance for common issues.

This volume is designed to function as both a practical hands-on laboratory manual and a reliable long-term reference resource. It is intended for a broad audience, including undergraduate and postgraduate students, early-career researchers, seasoned academics, and industry professionals actively engaged in phytochemical screening, advanced metabolomics, molecular biology of plant secondary metabolism, and computational phytochemistry.

By promoting the adoption of standardized, high-quality analytical practices across these interconnected disciplines, we hope this book will foster greater consistency, comparability, and rigor in phytoconstituent research worldwide, ultimately contributing to more reproducible, impactful, and innovative scientific outcomes.

Last but not least, we extend our heartfelt gratitude to the distinguished researchers, expert academicians, and leading scientists who have generously contributed as authors and peer reviewers. Their expertise, dedication, and insightful contributions have greatly enriched the content and scientific integrity of this volume. We are deeply grateful to our family members for their unwavering patience, understanding, and unconditional support throughout the demanding process of compiling and editing this scholarly work.

We sincerely appreciate Springer Nature for entrusting us with this opportunity and for the exceptional professionalism of the editorial, production, and support teams, whose relentless efforts and guidance were instrumental in bringing this project to fruition. Finally, we express profound thanks to our respective institutions for providing the necessary research facilities, resources, encouragement, and conducive environment that made this endeavor possible. We are grateful from the bottom of our hearts for all the support received along this journey.

Bardoli, India *Dhruti Amin*
Chennai, India *Natarajan Amaresan*
Navsari, India *Sanket Ray*

Contents

Part II Advanced Metabolic Profiling Techniques for Phytoconstituent Analysis

Part III Molecular Methods for Phytoconstituent Analysis

Part IV Artificial Intelligence (AI) in Phytoconstituent Analysis

Contributors

KRUTIKA ABHYANKAR • *Division of Biomedical and Life Sciences, School of Science, Navrachana University, Vadodara, Gujarat, India*

NATARAJAN AMARESAN • *Saveetha Institute of Basic Medical Sciences, Saveetha Institute of Medical and Technical Sciences (SIMATS), Saveetha University, Chennai, Tamil Nadu, India*

DHRUTI AMIN • *C. G. Bhakta Institute of Biotechnology, Uka Tarsadia University, Bardoli, Surat, Gujarat, India*

RUKHSAR ANSARI • *Microbiology Department, Naran Lala College of Professional and Applied Sciences, Navsari, Gujarat, India*

ANKIT AVASTHI • *C. G. Bhakta Institute of Biotechnology, Uka Tarsadia University, Bardoli, Surat, Gujarat, India*

JUHI BAROT • *Microbiology Department, Naran Lala College of Professional and Applied Sciences, Bhagwati Sankul, Navsari, Gujarat, India*

AHMET BEYATLI • *School of Agriculture, Food and Ecosystem Science, Faculty of Science, The University of Melbourne, Melbourne, Victoria, Australia*

PREETI BHATT • *Babu Madhav Institute of Information Technology, Uka Tarsadia University, Maliba Campus, Gopal Vidyanagar, Bardoli-Mahuva Road, Surat, Gujarat, India*

SAPNA CHANDWANI • *C. G. Bhakta Institute of Biotechnology, Uka Tarsadia University, Bardoli, Surat, Gujarat, India*

ANKIT CHAUDHARY • *Kishorbhai Institute of Agriculture Sciences and Research Centre, Uka Tarsadia University, Bardoli, Surat, Gujarat, India*

ANJALI DESAI • *Microbiology Department, Naran Lala College of Professional and Applied Sciences, Navsari, Gujarat, India*

KRUNAL DHOLIYA • *Framtix Holdings AB, Malmo, Sweden; Arts, Science and Commerce College, Kholwad, Surat, Gujarat, India*

AMIT GAMIT • *C. G. Bhakta Institute of Biotechnology, Uka Tarsadia University, Maliba Campus, Bardoli, Surat, Gujarat, India*

PRIYANG KANOJIYA • *Microbiology Department, Naran Lala College of Professional and Applied Sciences, Navsari, Gujarat, India*

HITARTH KANSARA • *Division of Biomedical and Life Sciences, School of Science, Navrachana University, Vadodara, Gujarat, India*

ABDULLAH KHAN • *C. G. Bhakta Institute of Biotechnology, Uka Tarsadia University, Maliba Campus, Bardoli, Surat, Gujarat, India*

MANDAR KULKARNI • *P. D. Patel Institute of Applied Sciences, Charotar University of Science and Technology, CHARUSAT Campus, Changa, Anand, Gujarat, India*

SARIKA KUMARI • *C.G. Bhakta Institute of Biotechnology, Uka Tarsadia University, Maliba Campus, Bardoli, Surat, Gujarat, India*

QUANG LE DANG • *Institute of Materials Science, Vietnam Academy of Science and Technology (VAST), 18 Hoang Quoc Viet, Hanoi, Vietnam; Graduate University of Science and Technology, Vietnam Academy of Science and Technology (VAST), 18 Hoang Quoc Viet, Hanoi, Vietnam*

MAYUR MARATHE • *C. G. Bhakta Institute of Biotechnology, Uka Tarsadia University, Bardoli, Surat, Gujarat, India*

MARIANA MARQUES • *i4HB – Associate Laboratory Institute for Health and Bioeconomy, Faculty of Pharmacy, University of Porto, Rua de Jorge Viterbo Ferreira, 228, 4050-313, Porto, Portugal, Porto, Portugal; UCIBIO – Applied Molecular Biosciences Unit, MedTech-Laboratory of Pharmaceutical Technology, Faculty of Pharmacy, University of Porto, Rua de Jorge Viterbo Ferreira, 228, 4050-313, Porto, Portugal, Porto, Portugal*

CRESSIDA MEHTA • *C. G. Bhakta Institute of Biotechnology, Uka Tarsadia University, Bardoli, Surat, Gujarat, India*

MANSI P. MEHTA • *Department of Biotechnology, VNSGU, Surat, Gujarat, India*

SNAA MISTRY • *Kishorbhai Institute of Agriculture Sciences and Research Centre, Uka Tarsadia University, Bardoli, Surat, Gujarat, India*

BHIKHU S. MORE • *SDJ International College, Palsana, Surat, Gujarat, India*

AMI NAIK • *C. G. Bhakta Institute of Biotechnology, Uka Tarsadia University, Maliba Campus, Bardoli, Surat, Gujarat, India*

HETVI NAIK • *C. G. Bhakta Institute of Biotechnology, Uka Tarsadia University, Maliba Campus, Bardoli, Surat, Gujarat, India*

CUONG QUOC NGUYEN • *Faculty of Health Sciences, College of Natural Sciences, Can Tho University, Can Tho, Vietnam*

NHU Y. NGUYEN-THI • *Faculty of Biological, Chemical, and Food Technology, Can Tho University of Technology, Can Tho, Vietnam*

KETANKUMAR PANCHAL • *P. D. Patel Institute of Applied Sciences, Charotar University of Science and Technology, CHARUSAT Campus, Changa, Anand, Gujarat, India*

NILESHKUMAR PANDYA • *Arts, Science and Commerce College, Kholwad, Surat, Gujarat, India*

ANAND PATEL • *Tarsadia Institute of Chemical Science, Uka Tarsadia University, Maliba Campus, Gopal Vidyanagar, Bardoli-Mahuva Road, Surat, Gujarat, India*

BHAVIN PATEL • *Tarsadia Institute of Chemical Science, Uka Tarsadia University, Maliba Campus, Gopal Vidyanagar, Bardoli-Mahuva Road, Surat, Gujarat, India*

DHARA PATEL • *Department of Biotechnology and Bioengineering, Institute of Advanced Research, Gandhinagar, IndiaSurat,*

HARSHVADAN PATEL • *Department of Microbiology, School of Science and Technology, Vanita Vishram Women's University, Surat, Gujarat, India*

NAFISA PATEL • *yMicrobiology Department, Naran Lala College of Professional and Applied Sciences, Bhagwati Sankul, Navsari, Gujarat, India*

ROSHANI PATEL • *Department of Biotechnology and Bioengineering, Institute of Advanced Research, Gandhinagar, IndiaSurat, ; School of Biotechnology and Bioengineering, Institute of Advanced Research, Gandhinagar, Gujarat, Gujarat, India*

SHUBHAM PATEL • *C. G. Bhakta Institute of Biotechnology, Uka Tarsadia University, Bardoli, Surat, Gujarat, India*

KAJAL PATIL • *Babu Madhav Institute of Information Technology, Uka Tarsadia University, Maliba Campus, Gopal Vidyanagar, Bardoli-Mahuva Road, Surat, Gujarat, India*

DIANA PINTO • *REQUIMTE/LAQV, ISEP, Polytechnic of Porto, Rua Dr. António Bernardino de Almeida, Porto, Portugal*

ANIL PRAJAPATI • *M.B. Patel Science College, Charotar Education Society, Near Sardar Gunj, Anand, Gujarat, India*

SANKET RAY • *Microbiology Department, Naran Lala College of Professional and Applied Sciences, Navsari, Gujarat, India*

FRANCISCA RODRIGUES • *i4HB – Associate Laboratory Institute for Health and Bioeconomy, Faculty of Pharmacy, University of Porto, Rua de Jorge Viterbo Ferreira, 228, 4050-313, Porto, Portugal, Porto, Portugal; UCIBIO – Applied Molecular Biosciences Unit, MedTech-Laboratory of Pharmaceutical Technology, Faculty of Pharmacy, University of Porto, Rua de Jorge Viterbo Ferreira, 228, 4050-313, Porto, Portugal, Porto, Portugal*

SHRUTI SATASHIA • *Atmanand Saraswati Science College, Surat, Gujarat, India*

RAKESH SAVANT • *Babu Madhav Institute of Information Technology, Uka Tarsadia University, Maliba Campus, Gopal Vidyanagar, Bardoli-Mahuva Road, Surat, Gujarat, India*

GAURAV S. SHAH • *Department of Biotechnology, VNSGU, Surat, Gujarat, India*

ABHISHEK SHARMA • *School of Biotechnology and Bioengineering, Institute of Advanced Research, Gandhinagar, Gujarat, Gujarat, India*

GLORY SIMON • *Department of Animal Genetics and Breeding, College of Veterinary Science and Animal Husbandry, Kamdhenu University, Anand, Gujarat, India*

VISHWAM SINGH • *Division of Biomedical and Life Sciences, School of Science, Navrachana University, Vadodara, Gujarat, India*

HAFIZ SULERIA • *School of Agriculture, Food and Ecosystem Science, Faculty of Science, The University of Melbourne, Melbourne, Victoria, Australia*

FILIPA TEIXEIRA • *REQUIMTE/LAQV, ISEP, Polytechnic of Porto, Rua Dr. António Bernardino de Almeida, Porto, Portugal; ICBAS - School of Medicine and Biomedical Sciences, University of Porto, Rua Jorge Ferreira Viterbo 228, 4050-313 Porto, Portugal, Porto, Portugal*

QUANG DE TRAN • *Faculty of Health Sciences, College of Natural Sciences, Can Tho University, Can Tho, Vietnam*

RISHITA TRIVEDI • *C. G. Bhakta Institute of Biotechnology, Uka Tarsadia University, Maliba Campus, Bardoli, Surat, Gujarat, India*

KEVIN VAIWALA • *C. G. Bhakta Institute of Biotechnology, Uka Tarsadia University, Bardoli, Surat, Gujarat, India*

HIRANGI VANSIA • *C. G. Bhakta Institute of Biotechnology, Uka Tarsadia University, Maliba Campus, Bardoli, Surat, Gujarat, India*

Part I

Qualitative and Quantitative Protocols for Phytoconstituent Analysis

Chapter 1

Qualitative and Quantitative Analysis of Alkaloids

Cressida Mehta, Mayur Marathe, Kevin Vaiwala, Shubham Patel and Dhruti Amin

Abstract

Alkaloids are a diverse group of nitrogen-containing secondary metabolites that play significant roles in pharmacology, toxicology, and chemotaxonomy. This chapter discusses alkaloid extraction, classical tests such as Dragendorff, Mayer, Wagner, and Hager, as well as chromatographic and spectroscopic confirmation techniques. It also covers quantitative methods ranging from gravimetric and UV-Visible analysis to advanced HPLC. It also covers practical steps such as reagent preparation, instrument calibration, method validation, and recent advances that enhance sensitivity, throughput, and sustainability.

Keywords Microwave-assisted extraction, Ultrasound-assisted extraction, Supercritical fluid extraction, Bromocresol green assay, High-performance liquid chromatography

1 Introduction

Alkaloids are a diverse group of naturally occurring nitrogen-containing organic compounds, most of which have heterocyclic structures [1]. They include key structural classes such as pyridine, isoquinoline, indole, quinoline, and tropane alkaloids, each exhibiting diverse biological activities, including analgesic, antimalarial, anticancer, cardiotonic, neuroactive, and toxic effects [2]. Because of their pharmacological properties, alkaloids are among the most important groups of phytochemicals in drug discovery, pharmacognosy, and the quality control of herbal medicines [3]. Reliable detection and accurate quantification of alkaloids are essential for phytochemical standardization, toxicological evaluation, and pharmaceutical applications [4]. Classical qualitative tests remain valuable for rapid preliminary screening, whereas modern quantitative techniques, such as UV–visible spectrophotometry, high-performance liquid chromatography (HPLC), and mass spectrometry, provide precise and reproducible measurements [5]. Recent advances in alkaloid research emphasize the use of environmentally friendly, rapid extraction techniques such

Dhruti Amin et al. (eds.), *Phytoconstituent Analysis*, Springer Protocols Handbooks,
https://doi.org/10.1007/978-1-0716-5400-2_1,

as microwave-assisted extraction (MAE) and ultrasound-assisted extraction (UAE), along with highly sensitive multi-analyte HPLC methods capable of simultaneously quantifying multiple alkaloids in complex plant matrices [6, 7]. These developments enhance analytical efficiency, reduce solvent use, and enable high-throughput processing of samples, thereby improving the overall quality and sustainability of alkaloid analysis [8].

2 Materials

2.1 Sample Collection

1. Fresh plant materials (leaves, bark, roots, seeds).
2. Airtight glass containers (amber), Shade-drying trays, 40–60 mesh sieves.
3. Hot air oven (40–45°C), Mechanical grinder, Refrigerator (4°C).

2.2 Extraction Methods

2.2.1 Acid–Base Extraction

1. Powdered plant material, Methanol/70% ethanol, Hydrochloric acid (1%), Dichloromethane (DCM), Chloroform ($CHCl_3$)/ Ethyl acetate, Ammonium hydroxide (NH_3OH), Anhydrous sodium sulfate.
2. Conical flasks (250 mL), Glass beakers, funnels, and filters.
3. Separatory funnel, Rotary evaporator/nitrogen evaporator.

2.2.2 Microwave-Assisted Extraction (MAE) and Ultrasound-Assisted Extraction (UAE)

1. Methanol/acidified methanol (0.1% HCl)
2. Ultrasonic bath (sonicator), Microwave extraction unit, Centrifuge
3. Centrifuge tubes, Filter papers

2.2.3 Supercritical Fluid Extraction

1. CO_2 gas cylinder,
2. SFE apparatus
3. Ethanol/Methanol

2.3 Qualitative Screening Tests

Dragendorff's Reagent(Potassium Bismuth Iodide Reagent)

1. Solution A: Dissolve 0.85 g bismuth subnitrate in 10 mL glacial acetic acid and 40 mL distilled water.
2. Solution B: Dissolve 8 g potassium iodide (KI) in 20 mL distilled water.
3. Mix Solution A and Solution B and dilute to 100 mL with distilled water.

Mayer's Reagent (Potassium Mercuric Iodide Reagent)

1. Solution A: Dissolve 1.36 g mercuric chloride ($HgCl_2$) in 60 mL distilled water.
2. Solution B: Dissolve 5 g potassium iodide (KI) in 10 mL distilled water.
3. Mix both solutions and make up the volume to 100 mL with distilled water.

Wagner's Reagent—(Iodine–Potassium Iodide Reagent): Dissolve 1.27 g Iodine and 2 g Potassium Iodide (KI) in 100 mL DistilledWater

Hager's Reagent (Saturated Picric Acid Solution): 1 g Picric Acid in 100 mL Distilled Water

2.4 Thin Layer Chromatography (TLC)

1. Dragendorff spray reagent.
2. Standard alkaloids (atropine/berberine/harmine).
3. Silica gel TLC plates (10 × 10 cm), Capillary spotting tubes.
4. TLC developing chamber, UV lamp (254/365 nm).
5. Mobile phases: a. Chloroform, methanol, and ammonium hydroxide. b. Toluene, ethyl acetate, and diethylamine.

2.5 Quantitative Analysis

2.5.1 Precipitation Gravimetric Method

1. 1% HCl solution, Dragendorff reagent.
2. Pre-weighed filter papers, Glass funnels.
3. Hot air oven (60–70°C), Desiccator, Analytical balance.

2.5.2 UV-Visible Spectrophotometric Method

1. Standard alkaloids (Atropine/berberine/harmine).
2. Bromocresol green (BCG) reagent.
3. Phosphate buffer (pH 4.7).
4. 0.1 N HCl.
5. Chloroform.
6. Spectrophotometer (420 nm).
7. Volumetric flasks, cuvettes.

2.5.3 HPLC-Based Quantitative Analysis

1. C18 column (150 × 4.6 mm, 5 μm).
2. Methanol, acetonitrile (HPLC grade), Formic acid/glacial acetic acid, Standard alkaloids for calibration, Water (Milli-Q/HPLC grade).
3. Syringe filters (0.22 μm), HPLC vials.
4. HPLC system with UV/PDA detector.

5. Sonicator.
6. Centrifuge.

3 Methods

3.1 Sample Collection [9]

1. Collect plant material (leaves, bark, roots, seeds) and record voucher specimen details. Air-dry in shade or oven-dry at ≤ 40–45°C to constant weight (*see* **Note 1**).
2. Grind dried material to a fine powder and pass through a 40–60 mesh sieve. Store in airtight containers at 4°C, protected from light and moisture.

3.2 Extraction Methods

3.2.1 Acid-Base Extraction [10]

1. Weigh 5–10 g of powdered plant material into a 250 mL conical flask.
2. Add 100 mL of methanol or 70% ethanol and wait for 24 h (or sonicate 30–60 min for better recovery). Filter and concentrate the solvent under reduced pressure to obtain the crude extract.
3. Dissolve the residue in 1% HCl (enough to make pH ~2–3). Transfer to separatory funnel and extract non-alkaloid neutral components with DCM (2 × 50 mL). Discard organic (non-alkaloid) layers.
4. Basify the aqueous acidic layer with NH_4OH (pH 9–10) to free the alkaloids and extract with DCM/$CHCl_3$/ethyl acetate (3 × 50 mL). Mix well organic extracts, dry over anhydrous Na_2SO_4, filter, and evaporate solvent to yield the alkaloid fraction.
5. Store residue at −20°C or proceed to fractionation or analysis (*see* **Note 2**).

3.2.2 Microwave-Assisted Extraction (MAE) and Ultrasound-Assisted Extraction (UAE)

1. Microwave-assisted extraction (MAE) [11]: Use methanol or acidified methanol (0.1% HCl) with controlled power and time (e.g., 100–300 W for 2–10 min depending on matrix) followed by filtration and neutralization. MAE reduces solvent and time.
2. Ultrasound-assisted extraction (UAE) [12]: Sonicate powdered sample with solvent (methanol or acidified solvent) for 20–40 min at ambient temperature; centrifuge and collect supernatant. UAE improves yield and is simple to implement.

3.2.3 Supercritical Fluid Extraction [13]

1. SFE (CO_2 with co-solvent) can be applied for selective nonpolar alkaloids and to reduce solvent consumption; it requires specialized equipment. Optimization is sample-dependent.

3.3 Qualitative Screening Tests

3.3.1 Dragendorff's Test [14]

1. To 1 mL of extract (in acidic medium), add one to two drops of Dragendorff's reagent. Appearance of orange-brown precipitate indicates alkaloids present (*see* **Note 3**).

3.3.2 Mayer's Test [14]

1. Add one to two drops of Mayer's reagent to dilute the extract cream precipitate, which indicates alkaloids.

3.3.3 Wagner's Test [15]

1. Add a few drops to extract—positive yields brown–red precipitate.

3.3.4 Hager's Test [15]

1. Add Hager's reagent; yellow precipitate indicates the presence of alkaloids (see **Notes 4 and 5**).

3.3.5 Additional Color Tests [14]

1. Dragendorff on TLC: Spray TLC plate with Dragendorff's reagent to visualize alkaloid spots.

3.4 Thin Layer Chromatography (TLC) [16]

1. Prepare silica gel TLC plates (10×10 cm). Spot sample (alkaloid fraction) and reference standards
2. Mobile phases (on the basis of alkaloid class):
   a. Chloroform: methanol: ammonium hydroxide (90:10:1)—for many basic alkaloids.
   b. Toluene: ethyl acetate: diethylamine (7:2:1)
3. Dry the plate and visualize under UV 254/365 nm, then spray with Dragendorff's reagent (spots turn orange/brown). Compare Rf with standards (*see* **Note 6**).

3.5 Quantitative Analysis

3.5.1 Precipitation Gravimetric Method [17]

1. Extract plant powder with acidic aqueous solution (e.g., 1% HCl) and filter.
2. For free alkaloids and some gravimetric variants, directly add Dragendorff reagent to the acidic extract to precipitate alkaloid-bismuth complexes.
3. Filter the precipitate through pre-weighed filter paper, wash thoroughly with water, and dry at 60–70°C to constant weight, cool in a desiccator, and weigh.
4. Calculate % total alkaloids = (mass of dried precipitate × correction factor/mass of sample) × 100.

3.5.2 UV-Visible Spectrophotometric Method [18]

1. Prepare reagent: bromocresol green solution and phosphate buffer pH 4.7.
2. Dissolve alkaloid extract in acid (0.1 N HCl), adjust to pH ~4.7. Add a measured volume of BCG reagent to form a yellow complex.

3. Extract the complex into chloroform (or other organic solvent), separate the organic layer, and measure absorbance at 420 nm.
4. Quantify using calibration curve from a suitable standard (atropine/berberine/harmine) and unit used mg alkaloid/g dry weight.
5. Validation: linearity, LOD/LOQ, precision (RSD), accuracy (recovery). Typical linear ranges vary (2–10 μg/mL for some BCG methods).

3.5.3 HPLC-Based Quantitative Analysis [19]

1. Weigh sample of plant extract equivalent to (X) mg dry material.
2. Dissolve alkaloid fraction in methanol to make a convenient stock (e.g., 1–2 mg/mL). Vortex briefly.
3. Sonicate 10–15 min to ensure complete dissolution.
4. Dry eluate under gentle nitrogen or rotary evaporation and reconstitute in mobile-phase A/B mix or methanol to the desired concentration (example target: 1.0 mg/L).
5. Centrifuge (5 min, 3000–5000 × g) if particulate matter remains.
6. Filter final solution through 0.22 μm syringe filter into HPLC vial. Label and keep at 4°C until analysis.
7. Column: C18, 150 × 4.6 mm, 5 μm (alternative: 100–150 mm × 2.1–4.6 mm, 1.7–5 μm for UHPLC).
8. Mobile phase A: 0.1% formic acid in water.
9. Mobile phase B: Acetonitrile.
10. Gradient (example): 0–1 min, 5% B (equilibration); 1–12 min, linear to 60% B; 12–14 min, ramp to 95% B (wash); 14–18 min, re-equilibrate to 5% B.
11. Flow rate: 1.0 mL/min (for 4.6 mm ID), adjust to 0.3–0.4 mL/min for 2.1 mm ID.
12. Column temperature: 30°C.
13. Detection: UV at 345 nm (or λmax of target); PDA scan recommended.
14. Injection volume: 10 μL (reduce for narrow-bore columns).
15. System suitability: inject standard before run (check theoretical plates >2000, tailing factor ≤1.5, %RSD area ≤2% for replicate injections).
16. Prepare all samples in duplicate and filter them through 0.2 μ m membrane filter before injection.

17. Use a Phenomenex Kinetex C18 column (150 mm × 4.6 mm; 5 μm) for separation.
18. Prepare Mobile Phase:

    A : 99:1 water/99.8% glacial acetic acid

    B : 67:32:1 water/acetonitrile/99.8% glacial acetic acid.

19. Set instrument conditions: flow rate 0.700 mL/min (approx. 54 bar), detection wavelength 275 nm, column temperature 25°C, and injection volume 50 μL.
20. Apply the external standard method for quantification.
21. Run the gradient program as follows:
    - 0–10 min: 10 → 20% B
    - 10–16 min: 20 → 40% B
    - 16–20 min: 40 → 50% B
    - 20–25 min: 50 → 70% B
    - 25–30 min: hold at 70% B
    - 30–40 min: 70 → 10% B
    - 40–45 min: hold at 10% B for re-equilibration
22. Maintain a total run time of 45 min and process chromatographic data to quantify analytes using the external calibration standards (*see* **Note 7**).

4 Notes

1. Use a low drying temperature for heat-sensitive alkaloids.
2. Choose suitable extraction solvents like acidified methanol/ethanol, or use acid–base extraction
3. Always use fresh Dragendorff and Mayer reagents for accurate results.
4. Handle picrate reagents carefully as dry picrates can be explosive.
5. Never allow picric acid (Hager's reagent) to dry to avoid danger.
6. Confirm positive tests using TLC, HPLC, or MS.
7. Validate methods by checking linearity, LOD/LOQ, precision, recovery, and stability.

References

1. Roberts MF (Ed) (2013) Alkaloids: biochemistry, ecology, and medicinal applications. Springer Science & Business Media
2. Dewick PM (2002) Medicinal natural products: a biosynthetic approach. John Wiley & Sons
3. Hostettmann K, Marston A, Ndjoko K, Wolfender JL (2000) The potential of African plants as a source of drugs. Curr Organic Chem 4(10):973–1010
4. Sasidharan S, Chen Y, Saravanan D, Sundram KM, Latha LY (2011) Extraction, isolation and characterization of bioactive compounds from plants' extracts. Afr J Tradit Complement Altern Med 8(1):1–10
5. Marston A, Hostettmann K (2006) Developments in the application of counter-current chromatography to plant analysis. J Chromatogr A 1112(1-2):181–194
6. Chemat F, Rombaut N, Sicaire AG, Meullemiestre A, Fabiano-Tixier AS, Abert-Vian M (2017) Ultrasound assisted extraction of food and natural products. Mechanisms, techniques, combinations, protocols and applications. A review. Ultrason Sonochem 34:540–560
7. Azwanida NN (2015) A review on the extraction methods use in medicinal plants, principle, strength and limitation. Med Aromat Plants 4(196):2167–0412
8. Wolfender JL, Marti G, Thomas A, Bertrand S (2015) Current approaches and challenges for the metabolite profiling of complex natural extracts. Jf Chromatogr A 1382:136–164
9. Wallis TE. Text book of pharmacognosy Delhi, India: CBS Publishers and Distributors; 1989. p. 356-549
10. Houghton P, Raman A (2012) Laboratory handbook for the fractionation of natural extracts. Springer Science & Business Media
11. Ganzler K, Salgó A, Valkó K (1986) Microwave extraction: a novel sample preparation method for chromatography. J Chromatogr A 371:299–306
12. Vinatoru M (2001) An overview of the ultrasonically assisted extraction of bioactive principles from herbs. Ultrason Sonochem 8(3):303–313
13. King JW (2002) Supercritical fluid extraction: present status and prospects. Grasas Aceites 53(1):8–21
14. Harborne AJ (1998) Phytochemical methods a guide to modern techniques of plant analysis. springer Science & Business Media.
15. Wagner H, Bladt S (1996) Plant drug analysis: a thin layer chromatography atlas. Springer Berlin Heidelberg, Berlin, Heidelberg
16. Stahl E (1967). Thin-layer chromatography: a laboratory handbook. Berlin: Heidelberg: New York: Springer Verlag.xx + 979 pp
17. Obadoni BO, Ochuko PO (2002) Phytochemical studies and comparative efficacy of the crude extracts of some haemostatic plants in Edo and Delta States of Nigeria. Glob J Pure Appl Sci 8(2):203–208
18. Shamsa F, Monsef H, Ghamooshi R, Verdian-Rizi M (2008) Spectrophotometric determination of total alkaloids in some Iranian medicinal plants. Thai J Pharm Sci 32(1):17–20
19. He XG (2000) On-line identification of phytochemical constituents in botanical extracts by combined high-performance liquid chromatographic–diode array detection–mass spectrometric techniques. J Chromatogr A 880(1-2):203–232

Chapter 2

Detection and Measurement of Terpenoids

Amit Gamit, Ankit Avasthi and Dhruti Amin

Abstract

Terpenoids are one of the most diverse groups of plant secondary metabolites, synthesized through the mevalonate (MVA) and methylerythritol phosphate (MEP) pathways and widely distributed in aromatic, medicinal, and economically important plants. Their structural complexity and broad bioactivity make them valuable in pharmaceuticals, nutraceuticals, cosmetics, and agricultural formulations. Accurate detection and measurement of terpenoids are essential for assessing plant quality, validating biological activities, and supporting metabolic studies. This chapter outlines an integrated workflow for the extraction, processing, and analytical evaluation of terpenoids from plant materials. It describes commonly used extraction approaches suitable for both volatile and non-volatile terpenoids, followed by qualitative and quantitative methods for determining terpenoid presence and concentration. Spectroscopic and chromatographic tools used for structural confirmation and profiling are also highlighted. Overall, this chapter provides a streamlined and practical guide for researchers aiming to characterize terpenoids effectively in plant-based samples.

Keywords Terpenoids extraction, Chromatography, Phytochemical screening, Structural characterization

1 Introduction

Terpenoids, or isoprenoids, represent the largest and most structurally diverse class of natural compounds, with more than 80,000 identified molecules occurring across a wide range of biological systems, including plants, microorganisms, fungi, marine organisms, and some animals [1]. They are biosynthesized from repeating five-carbon isoprene units (C_5H_8), making them unique in both their origin and structural diversity [2]. Terpenoids are classified according to the number of isoprene units they contain, forming groups such as hemiterpenes (C_5), monoterpenes (C_{10}), sesquiterpenes (C_{15}), diterpenes (C_{20}), sesterterpenes (C_{25}), triterpenes (C_{30}), tetraterpenes (C_{40}), and polyterpenes (C_{40}+) [3, 4]. Each class demonstrates wide-ranging physicochemical characteristics and biological activities, enabling their roles in essential oils, plant

Dhruti Amin et al. (eds.), *Phytoconstituent Analysis*, Springer Protocols Handbooks,
https://doi.org/10.1007/978-1-0716-5400-2_2,

defense, pigmentation, and structural components, such as natural rubber [5–7]. Terpenoids are abundantly found in medicinal and aromatic plants, such as Mentha species, Eucalyptus, Citrus, *Artemisia annua*, *Azadirachta indica*, Lavandula, *Ocimum*, *Pinus* species, and *Panax ginseng*, as well as in fruits and vegetables rich in carotenoids like carrots, tomatoes, sweet potatoes, and leafy greens. Their distribution varies across leaves, flowers, seeds, bark, roots, and specialized structures, such as glandular trichomes, resin ducts, and latex-containing tissues [8–12].

Terpenoids are synthesized from the universal precursors IPP and DMAPP, produced through the cytosolic MVA and plastidic MEP pathways. The MVA route forms sesquiterpenes and triterpenes, while the MEP pathway yields monoterpenes and diterpenes. Enzymes like terpene synthases and cytochrome P450s diversify terpenoid structures [1, 7]. Terpenoid extraction varies with polarity and volatility: distillation isolates volatile compounds, while solvents and Soxhlet are suited for less volatile terpenoids (Fig. 1). Green methods, such as supercritical CO_2, ultrasound, and microwaves, enhance yield and protect heat-sensitive metabolites, whereas Solid-phasemicroextraction (SPME) enables rapid, non-destructive analysis of volatiles [6, 7].

Terpenoid detection relies on complementary qualitative, chromatographic, spectrophotometric, and spectroscopic techniques. Classical color reactions like Salkowski and Liebermann–Burchard rapidly confirm triterpenoids and steroids, while spectrophotometric assays, such as vanillin–acetic acid, quantify total terpenoids. Chromatographic methods remain central: TLC/HPTLC for profiling, GC–MS for volatile terpenes, and HPLC/UHPLC or LC–MS/MS for non-volatile and complex mixtures. Structural elucidation depends on FTIR, NMR, and HR-MS, which reveal functional groups and molecular composition (Fig. 1) [1, 6, 7].

Terpenoids have numerous applications across multiple fields, including medicine and pharmaceuticals. They serve as potent bioactive compounds with antimicrobial, antiviral, anticancer, antimalarial, antidiabetic, anti-inflammatory, and antioxidant properties. Notable examples include artemisinin (antimalarial), taxol (anticancer), menthol (analgesic), ginsenosides (immunomodulatory), and carotenoids (antioxidant and provitamin A activity) [13–17]. In agriculture, terpenoids function as natural pesticides, plant growth regulators, such as gibberellins, and allelochemicals involved in plant–plant and plant–insect interactions [14]. The food, fragrance, and cosmetic industries rely extensively on terpenoids for flavor, aroma, preservation, and natural pigmentation [6]. Additionally, metabolically engineered bacteria and plants are increasingly used to produce industrially important terpenoids through advanced biotechnological approaches [7]. Because of their ecological significance, terpenoids also contribute to plant

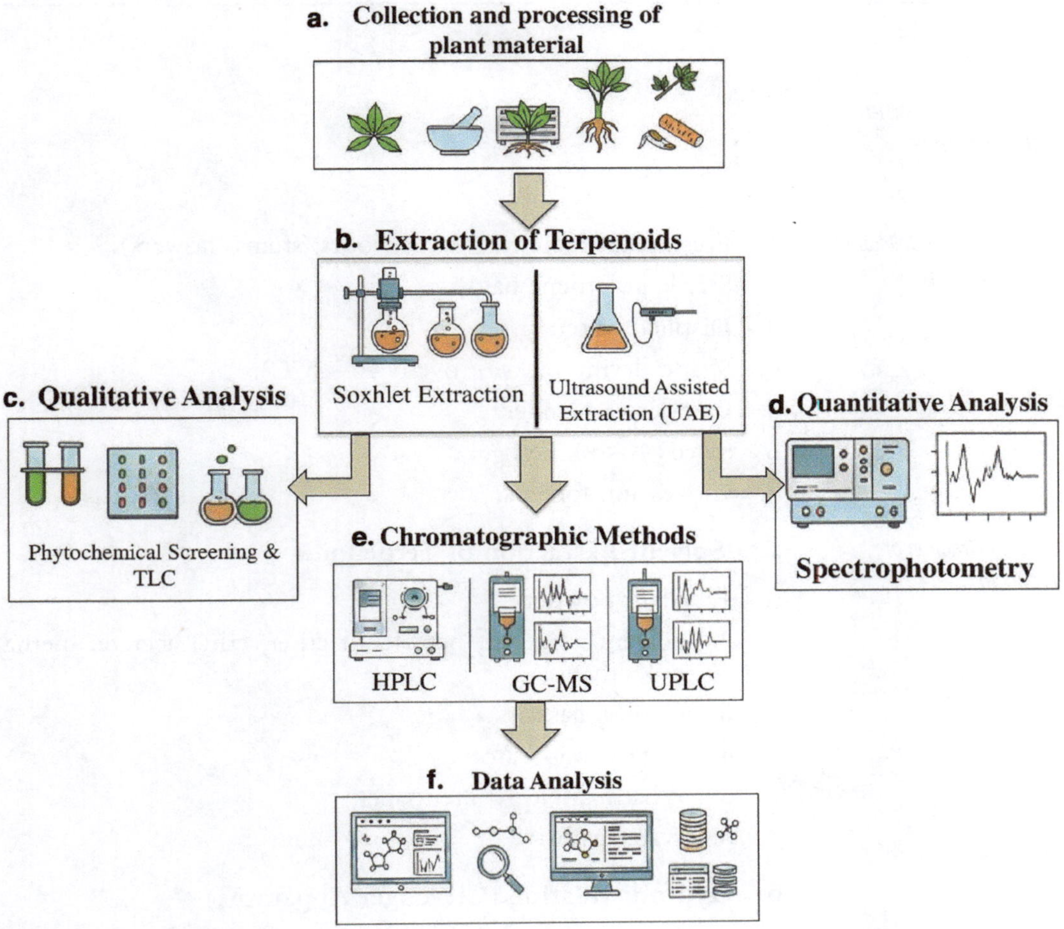

Fig. 1 Workflow of Terpenoids detection showing (**a**) collection and processing of plant material, (**b**) extraction, (**c**) qualitative, (**d**) quantitative analysis, (**e**) chromatographic techniques, and (**f**) data analysis

resilience against environmental stresses, including pathogens, herbivores, ultraviolet radiation, and temperature fluctuations [5].

Given these diverse functions and widespread applications, accurate detection and precise quantification of terpenoids are essential for standardizing herbal formulations, evaluating plant chemotypes, ensuring the quality of essential oils, advancing drug discovery, informing metabolic engineering, and elucidating plant biochemical pathways. A clear understanding of terpenoid extraction and analytical techniques is therefore essential for researchers in phytochemistry, natural product research, pharmacognosy, biotechnology, and plant sciences. This chapter focuses on the principles, methodologies, and detailed protocols for detecting, measuring, and analyzing terpenoids, serving as a comprehensive guide for laboratory-based investigations.

2 Materials

2.1 Plant Sample Collection and Extraction

2.1.1 Sample Collection and Processing

1. Fresh plant material (leaves/roots/stems/flowers).
2. Sterile polythene bags.
3. Distilled water.
4. Shade-drying racks or oven (40–45°C).
5. Grinder or blender.
6. Sieve (40–60 mesh).
7. Gloves and forceps.

2.1.2 Extraction of Terpenoid

a. **Solvent Extraction of Terpenoids**

1. Plant powder.
2. Solvents: hexane, petroleum ether, ethyl acetate, methanol, ethanol.
3. Conical flasks.
4. Rotary evaporator.
5. Whatman No. 1 filter paper.
6. Water bath.

b. **Hydrodistillation (Clevenger Apparatus)**

1. Fresh or dried plant material (50–200 g).
2. Clevenger-type hydrodistillation apparatus.
3. Distilled water.
4. Heating mantle.
5. Anhydrous sodium sulphate.

c. **Ultrasound-Assisted Extraction (UAE)**

1. Plant powder.
2. Solvent (ethanol, methanol, hexane, or mixtures).
3. Ultrasonic bath (20–40 kHz) or probe sonicator.
4. Conical flask.
5. Water bath (optional for temperature control).

d. **Soxhlet Extraction**

1. Soxhlet apparatus.
2. Plant powder.

3. Solvent (hexane, ethanol, methanol).
4. Heating mantle.
5. Round-bottom flask.

2.2 Detection and Measurement of Terpenoids

2.2.1 Qualitative Detection

Preliminary Phytochemical Analysis

1. **Salkowski Test**
   1. Plant extract.
   2. Chloroform.
   3. Concentrated sulfuric acid (H_2SO_4).
   4. Test tubes.
   5. Micropipettes or measuring cylinders.
2. **Liebermann–Burchard test**
   1. Plant extract.
   2. Acetic anhydride solution.
   3. Concentrated sulfuric acid (H_2SO_4).
   4. Test tubes.
   5. Micropipettes or dropper.

2.2.2 Qualitative Detection of Total Terpenoid Content

a. **Vanillin–Sulfuric Acid Colorimetric Assay**
   1. Plant sample extract.
   2. Rotary evaporator + evaporation flask.
   3. 8% vanillin ethanol solution.
   4. 80% sulfuric acid.
   5. Test tubes or reaction tubes.
   6. Water bath (set at 60°C).
   7. Ice bath.
   8. Micropipettes and tips.
   9. UV–Vis spectrophotometer.
   10. Cuvettes.
   11. Calibration standards for terpenoids.
   12. Distilled water.

2.2.3 Chromatographic Techniques

1. **TLC Profiling of Terpenoids**
   1. Merck 60 F silica gel TLC plates (10 × 2 cm).

2. Plant extract.
3. Capillary tubes or micropipettes.
4. Developing chamber.
5. Mobile phase solvents: *n*-hexane, acetone.
6. Ceric sulfate reagent.
7. Glassware (beakers, flasks).
8. Heat source (hot plate or oven for gentle heating).
9. Ruler (for measuring distances to calculate Rf).

2. **HPLC Method for Terpenoid Quantification**
   1. Waters 2695 HPLC system.
   2. Diode Array Detector (DAD).
   3. Sphere Clone ODS(2) C18 column (5 μm, 4.6 × 250 mm).
   4. Column oven (to maintain 35°C).
   5. Analytical balance.
   6. Ultrasonic bath (optional for sample prep).
   7. Filtration unit (0.22 or 0.45 μm filters).
   8. Microcentrifuge tubes or HPLC vials.
   9. Micropipettes and tips.
   10. Water with 0.1% phosphoric acid (Mobile Phase A).
   11. Acetonitrile, HPLC grade (Mobile Phase B).
   12. Methanol, HPLC grade.
   13. Terpene reference standards (individual standards).
   14. 50% aqueous methanol (for dilutions).
   15. Sample extracts.
   16. HPLC vials with caps.
   17. Syringe filters (0.22 or 0.45 μm PTFE or nylon).
   18. Glassware for mobile phase preparation.

3. **GC–MS Analysis of Terpenoids**
   1. Gas Chromatograph (GC).
   2. Mass Spectrometer (MS).
   3. GC Injector unit.
   4. FS-Supreme-5 ms Plus column (5% Phenylpolysilphenylene, 30 m × 0.25 mm, 0.25 μm film).
   5. Temperature-controlled oven.
   6. Data acquisition and GC–MS analysis software.
   7. SPME fiber assembly.
   8. Analytical balance.

9. Micropipettes and tips.
10. GC vials with septa.
11. Hexane (HPLC/GC grade).
12. Terpenoid standards (monoterpenes, sesquiterpenes, etc.).
13. Sample extracts.
14. High-purity hydrogen gas (carrier gas).
15. High-purity gases for MS (helium or nitrogen if needed for makeup gas).
16. GC vials (1.5 mL).
17. Septum-sealed vial caps.
18. SPME fiber (suitable for volatile terpenes).

2.2.4 Spectroscopic Technique

1. **FTIR of Terpenoids**

1. Terpenoid extract or purified terpenoid sample.
2. FTIR spectrometer (with ATR or transmission mode).
3. ATR crystal (e.g., diamond or ZnSe).
4. Spectroscopic-grade KBr powder.
5. Pellet die and hydraulic press.
6. Mortar and pestle (for mixing sample with KBr).
7. Desiccator.
8. Clean glass slides or vials for sample handling.
9. Ethanol or methanol (for cleaning ATR crystal).
10. Spatula and forceps.

3 Methods

3.1 Sample Collection and Extraction

3.1.1 Sample Collection and Preprocessing [18–20]

1. Collect healthy, disease-free plant parts during peak metabolite synthesis (usually early morning).
2. Place samples in clean sterile polythene bags and transport to the laboratory in an ice box to prevent degradation.
3. Wash samples gently with distilled water to remove dust and surface impurities.
4. Shade-dry the plant material at room temperature for 7–14 days to preserve heat-sensitive terpenoids.
5. Alternatively, dry in a hot-air oven at 40–45°C for faster processing

6. Grind the dried samples to fine powder using a grinder.
7. Pass through a 40–60 mesh sieve to obtain uniform particle size.
8. Store powdered samples in airtight amber containers at 4°C until extraction.

3.1.2 Extraction of Terpenoid [21]

a. **Solvent Extraction of Terpenoids** [18, 19]
   1. Take 50 g of dried plant powder in a conical flask.
   2. Add 500 mL of solvent based on polarity of target terpenoids (hexane for non-polar, methanol/ethanol for semi-polar).
   3. Macerate for 24–72 h at room temperature with occasional shaking.
   4. Filter using Whatman No. 1 filter paper.
   5. Repeat extraction two additional times to maximize yield.
   6. Combine filtrates and concentrate using a rotary evaporator at 40–45°C.
   7. Remove residual solvent using a water bath or nitrogen stream.
   8. Store crude extract at 4°C in amber vials.

b. **Hydrodistillation** [22, 23]
   1. Weigh 50 g of plant material and place into a round-bottom flask.
   2. Add 500 mL distilled water (ratio 1:10 w/v).
   3. Assemble Clevenger apparatus and heat the mixture to a boil.
   4. Continue distillation for 6 h until no more essential oil is collected.
   5. Collect the distillate and separate oil from aqueous phase.
   6. Dry the collected oil using anhydrous sodium sulfate to remove moisture.
   7. Store essential oil in amber vials at 4°C until analysis.

c. **Ultrasound-Assisted Extraction (UAE)** [24] (*see* **Note 1**)
   1. Mix 5–20 g plant powder with solvent at 1:10–1:20 ratio (w/v).
   2. Place flask in ultrasonic bath set at 20–40 kHz frequency.
   3. Sonicate for 20–60 min at 30–40°C, maintaining temperature using a water bath to avoid terpenoid degradation.
   4. Filter extract using Whatman filter paper.
   5. Concentrate filtrate using a rotary evaporator.

6. Store UAE extract at 4°C for further analysis.

d. **Soxhlet Extraction** [18, 19]

1. Place 10–20 g of plant powder inside a thimble.
2. Fill the Soxhlet flask with 150–250 mL of solvent.
3. Heat the system gently to allow solvent to evaporate, condense, and siphon back.
4. Continue extraction for 6–8 h 10–12 cycles depending on plant type.
5. Collect solvent extract from the flask and concentrate using a rotary evaporator.
6. Remove remaining solvent using a water bath.
7. Store extract at 4°C in airtight amber vials.

3.2 Detection and Measurement of Terpenoids [25, 26]

3.2.1 Qualitative Detection

Preliminery Phytochemical Analysis [27]

1. **Salkowski Test**
   1. Take 1 mL of each plant extract into a clean test tube.
   2. Add 5 mL of chloroform to the extract and mix gently.
   3. Carefully add 1 mL of concentrated H_2SO_4 along the inner wall of the test tube to form a separate lower layer (do not shake).
   4. Allow the mixture to stand undisturbed for phase separation.
   5. Observe the lower chloroform layer for color development.
   6. Formation of a scarlet red color in the lower layer indicates the presence of sterols/terpenoids (Fig. 2).

2. **Liebermann–Burchard test**
   1. Take 1 mL of each plant extract into a clean test tube.
   2. Add a few drops of acetic anhydride solution to the extract.
   3. Carefully add a few drops of concentrated H_2SO_4 along the inner wall of the test tube (do not mix vigorously).
   4. Allow the mixture to stand undisturbed for color development.
   5. Observe the formation of color at the interface.
   6. A red-brown ring indicates the presence of triterpenoids.
   7. A green color indicates the presence of sterols.

Fig. 2 Salkowski Test; positive test for Terpenoids

3.2.2 Qualitative Detection of Total Terpenoid Content

a. **Vanillin–Sulfuric Acid Colorimetric Assay** [28]

1. Transfer the prepared sample extract into a rotary evaporator flask.
2. Evaporate the solvent completely using a rotary evaporator.
3. Allow the residue to cool to room temperature.
4. Add 0.5 mL of 8% vanillin ethanol solution to the cooled residue and mix gently.
5. Add 5 mL of 80% sulfuric acid carefully to the mixture.
6. Place the reaction tube in a 60°C water bath for 15 min.
7. Immediately transfer the tube to an ice bath for 15 min to stop the reaction.
8. Prepare detection and blank controls following the same procedure.

9. Measure the absorbance of the reaction mixture at the appropriate wavelength (543 nm for hawthorn leaves and 558 nm for the liquorice system)
10. Calculate total terpenoid content using the absorbance values and calibration curve.
11. Plot absorbance (y-axis) versus standard concentration (x-axis, μg/mL). Fit a linear regression curve:

$$A = a + bC$$

Where, A = absorbance, C = concentration, a = y-intercept, b = slope.

Calculations:

Using the regression equation of the standard curve, calculate terpenoid concentration in the diluted sample:

$$C_{\text{sample (diluted)}} = \frac{A_{\text{sample}} - a}{b}$$

Calculate the original terpenoid concentration in the undiluted extract:

$$C_{\text{sample (original)}} = C_{\text{sample (diluted)}} \times \text{Dilution Factor}$$

Express results as:

$$\text{Total terpenoid content}\left(\text{mg/g extract}\right) = \frac{C_{\text{sample}} \times V_{\text{extract}} \times D.F.}{M_{\text{extract}}}$$

where:

C_{sample} = concentration from calibration curve (μg/mL)

V_{extract} = volume of extract used (mL)

D.F. = dilution factor

M_{extract} = mass of extract used (mg)

3.2.3 Chromatographic Techniques

1. **TLC Profiling of Terpenoids** [29]
   1. Prepare the TLC setup using Merck 60 F_{254} silica gel as the stationary phase coated on the TLC plate.
   2. Use a TLC plate measuring 10 × 2 cm for the analysis.

3. Prepare the mobile phase using *n*-hexane: acetone in different solvent ratios:
   - 90:10 (v/v).
   - 80:20 (v/v).
   - 70:30 (v/v).
   - 60:40 (v/v).
4. Apply the dry plant extract onto the TLC plate using a capillary tube or micro-pipette.
5. Place the TLC plate in the developing chamber containing the mobile phase and allow the solvent front to rise.
6. After testing different mobile phase ratios, select the best eluent: *n*-hexane: acetone (80:20 v/v) for total terpenoid analysis.
7. Develop the TLC plate using the selected eluent and allow the solvent front to reach an appropriate height.
8. Remove the TLC plate and air-dry it completely.
9. Spray the dried TLC plate with ceric sulfate reagent.
10. Heat the sprayed plate gently to develop color.
11. Observe the appearance of red-colored spots, indicating the presence of terpenoids and steroids.
12. Document the results and compare the Rf values with standards or literature.
13. Rf calculation: After plate dries, calculate Rf for each spot:

$$R_f = \frac{\text{Distance of compound from baseline(cm)}}{\text{Distance of solvent front from baseline(cm)}}$$

2. **HPLC Method for Terpenoid Quantification** [30]

1. Use a Waters 2695 HPLC system equipped with a diode array detector (DAD).
2. Install a Sphere Clone ODS(2) C18 column, 5 μm particle size, 4.6 × 250 mm (Phenomenex, USA).
3. Maintain the column temperature at 35°C throughout the analysis.
4. Prepare the mobile phases:
   - Solvent A: Water containing 0.1% phosphoric acid.
   - Solvent B: 100% acetonitrile.
5. Set up the linear gradient program as follows:
   - 0–20 min: Increase B from 5% → 20%.
   - 20–70 min: Increase B from 20% → 95%.

- 70–75 min: Hold at 95% B.

1. Set the flow rate to 0.75 mL/min.
2. Monitor the chromatogram at 210 nm using the diode array detector.
3. Use an injection volume of 25 µL for each sample.
4. Prepare reference standard stock solutions at 1.00 mg/mL in methanol.
5. Prepare calibration solutions by diluting the stock solutions with 50% aqueous methanol to obtain a concentration range of 6.25–100 µg/mL.
6. Inject calibration solutions to prepare external calibration curves.
7. Confirm that peak area vs. concentration shows a linear relationship over the range 6.25–100 µg/mL.
8. Inject sample extracts and determine the concentration of each terpene using the calibration curve.

3. **GC–MS Analysis of Terpenoids** [31]
   1. Inject 1 µL of hexane extract, or alternatively insert the SPME fiber directly into the GC injector.
   2. Set the injector temperature to 230°C.
   3. Use the column: FS-Supreme-5 ms Plus.
      - 5% Phenylpolysilphenylene.
      - 30 m × 0.25 mm, film thickness 0.25 µm.
   4. Select Hydrogen as the carrier gas at a flow rate of 1 mL/min.
   5. Set up the temperature program as follows:
      - Hold at 50°C for 3 min.
      - Ramp at 5°C/min to 170°C.
      - Ramp at 100°C/min to 300°C.
      - Hold at 300°C for 2 min.
   6. Set the MS operation mode to Electron Impact (EI) at 70 eV.
   7. Use a scan range of 46–350 m/z.
   8. Set the transfer line temperature to 250°C.
   9. Set the ion source temperature to 220°C.

Identification of Terpenoid Products

1. Analyze the obtained GC chromatograms using the associated GC–MS software.

2. Compare the mass spectra of sample peaks with library databases included in the software.
3. Identify tentative terpene matches based on the best library hits.
4. Observe that hexane extract and SPME analysis yield similar terpene profiles but differ in relative abundance of compounds.
5. Use diagnostic ion masses to classify terpenes:
   - Monoterpenes (acyclic): m/z 69, 93 (e.g., linalool).
   - Monoterpenes (cyclic): m/z 93, 121 (e.g., γ-terpinene).
   - Aromatic monoterpenes: *p*-cymene (m/z 119), thymol (m/z 135).
   - Sesquiterpenes: m/z 69, 93 (e.g., *β*-caryophyllene).
6. Confirm compound identity using authentic standards, comparing both:
   - Retention time.
   - Mass spectral fragments under the same GC–MS conditions.

3.2.4 Spectroscopic Technique

1. **FTIR of Terpenoids** [32]
   1. Ensure the terpenoid extract or purified sample is completely dry and free from residual solvents or moisture.
   2. For ATR mode, place one to two drops of liquid extract or 2–5 mg of dried extract directly on the ATR crystal; for KBr pellet mode, dry the sample in a desiccator.
   3. If using KBr, grind 1–2 mg sample with 100–150 mg spectroscopic-grade KBr to form a fine, homogeneous mixture.
   4. Compress the KBr mixture in a pellet die using a hydraulic press to obtain a transparent pellet.
   5. Turn on the FTIR spectrometer and allow it to stabilize for about 10–15 min.
   6. Run a background scan with an empty ATR crystal or a blank KBr pellet.
   7. Set spectral parameters: range 4000–400 cm^{-1}, resolution 4 cm^{-1}, and 16–32 scans for optimal signal.
   8. Load the sample: either place the ATR arm onto the sample or insert the KBr pellet into the sample holder.
   9. Begin the scan and allow the instrument to record the IR absorption spectrum.
   10. Save and label the obtained spectrum for interpretation.
   11. Identify key terpenoid peaks such as C–H stretching (2800–3000 cm^{-1}) and C=C stretching (~1640 cm^{-1}).

12. Look for C–O stretching bands in oxygenated terpenoids around 1050–1150 cm^{-1}.
13. Examine the fingerprint region (600–1400 cm^{-1}) for structural features unique to specific terpenoids.
14. Compare the spectrum with literature values, standard terpenoid spectra, or spectral libraries for confirmation.

4 Note

1. UAE increases extraction efficiency, reduces time, and preserves heat-sensitive monoterpenes [33].

References

1. Christianson DW (2017) Structural and chemical biology of terpenoid cyclases. Chem Rev 117:11570–11648. https://doi.org/10.1021/acs.chemrev.7b00287
2. Hyldgaard M, Mygind T, Meyer RL (2012) Essential oils in food preservation: mode of action, synergies, and interactions with food matrix components. Front Microbiol 3:12
3. Mabou FD, Belinda I, Yossa N (2021) Terpenes: structural classification and biological activities. IOSR J Pharm Biol Sci 16:12–40. https://doi.org/10.9790/3008-1603012540
4. Alvarez-Martínez FJ, Barrajon-Catalan E, Herranz-Lopez M, Micol V (2021) Antibacterial plant compounds, extracts, and essential oils: an updated review on their effects and putative mechanisms of action. Phytomedicine 90:153626
5. Hillier SG, Lathe R (2019) Terpenes, hormones and life: isoprene rule revisited. J Endocrinol 242:R9–R22. https://doi.org/10.1530/JOE-19-0171
6. Câmara JS, Perestrelo R, Ferreira R, Berenguer CV, Pereira JA, Castilho PC (2024) Plant-derived terpenoids: a plethora of bioactive compounds with several health functions and industrial applications—a comprehensive overview. Molecules 29:3861. https://doi.org/10.3390/molecules29133861
7. Nagegowda DA, Gupta P (2020) Advances in biosynthesis, regulation, and metabolic engineering of plant specialized terpenoids. Plant Sci 294:110457. https://doi.org/10.1016/j.plantsci.2019.110457
8. Mahato SB, Garai S (1998) Triterpenoid saponins. Prog Chem Org Nat Prod 74:1–196
9. Lacalla-Dubois M, Wagner H (2000) Bioactive saponins from plants: an update. Stud Nat Prod Chem 21:633–686
10. Francis G, Kerem Z, Makkar HP, Becker K (2002) The biological action of saponins in animal systems: a review. Br J Nutr 88:587–605
11. Vincken JP, Heng L, de Groot A, Gruppen H (2007) Saponins: classification and occurrence in the plant kingdom. Phytochemistry 68:275–297
12. Dinda B, Debnath S, Mohanta BC, Harigaya Y (2010) Naturally occurring triterpenoid saponins. Chem Biodivers 7:2327–2580
13. Fan M, Yuan S, Li L, Zheng J, Zhao D, Wang C, Wang H, Liu X, Liu J (2023) Application of terpenoid compounds in food and pharmaceutical products. Fermentation 9(2):119. https://doi.org/10.3390/fermentation9020119
14. Ninkuu V, Zhang L, Yan J, Fu Z, Yang T, Zeng H (2021) Biochemistry of terpenes and recent advances in plant protection. Int J Mol Sci 22(11):5710. https://doi.org/10.3390/ijms22115710
15. Sharifi-Rad J, Kumar NVA, Zucca P, Varoni EM, Dini L, Panzarini E, Rajkovic J, Tsouh Fokou PV, Azzini E, Peluso I, Mishra AP, Cho WC (2021) Paclitaxel: application in modern oncology. J Oncol 2021:3687700. https://doi.org/10.1155/2021/3687700
16. Lee BJ, Weyers M, Haynes RK, van der Kooy F (2023) Discovery of artemisinin in *Artemisia annua*, its current production, and relevance to sub-Saharan Africa. S Afr J Bot 153:21–27. https://doi.org/10.1016/j.sajb.2022.12.017
17. Guimarães AC et al. (2019) Antibacterial activity of terpenes and terpenoids present in essential oils. Molecules 24(13):2471. https://doi.org/10.3390/molecules24132471
18. Harborne JB (1998) Phytochemical methods: A guide to modern techniques of plant analysis, 3rd ed. Chapman & Hall

19. Azwanida NN (2015) A review on the extraction methods used in medicinal plants: Principle, strength and limitation. Med Aromat Plants 4(3):1–6
20. Sasidharan S, Chen Y, Saravanan D, Sundram KM, Latha LY (2011) Extraction, isolation, and characterization of bioactive compounds from plant extracts. Afr J Tradit Complement Altern Med 8(1):1–10
21. Isidore E, Karim H, Ioannou I (2021) Extraction of phenolic compounds and terpenes from *Cannabis sativa* L. by-products: from conventional to intensified processes. Antioxidants 10:942
22. European Pharmacopoeia (2004) Council of Europe. European Directorate for the Quality of Medicines, Strasbourg
23. Guenther E (1972) The essential oils, vols 1–6. Robert E. Krieger Publishing
24. Tian Y, Xu Z, Zheng B, Lo YM (2013) Optimization of ultrasonic-assisted extraction using response surface methodology. Food Chem 150:376–382
25. Rodríguez-Llorente D, Cañada-Barcala A, Álvarez-Torrellas S, Águeda VI, García J, Larriba M (2020) Use of eutectic solvents, terpenes and terpenoids in liquid–liquid extraction processes: a review. Processes 8:1220
26. Uwineza PA, Waśkiewicz A (2020) Supercritical fluid extraction of bioactive compounds from plant materials: recent advances. Molecules 25:3847
27. Guédiri I, Boubekri C, Smara O (2020) Preliminary phytochemical screening from different extracts of *Solanum nigrum* plant growing in south Algeria. J Fundam Applied Sci 12(2):624–633
28. Guo Y, Han Z, Zhang J, Lu Y, Li C, Liu G (2024) Development of a high-speed and ultrasensitive UV/Vis-CM for detecting total triterpenes in traditional Chinese medicine and its application. Heliyon 10(11):e32239
29. Prates L, Perez A (2024) The evidence for human agency in the late Pleistocene megafaunal extinctions. https://doi.org/10.1016/B978-0-12
30. Lü H, Lee RP, Huang J, Chen J, Go VLW, Li Z, Lu QY (2020) A new HPLC–UV method for quantification of terpenoids and antioxidant activity of commercial loquat leaf tea. J Food Meas Charact 14(2):1085–1091
31. Choi YH, Erol O, Irmisch S (eds) (2025) Natural product isolation and identification: methods and protocols, 1st edn. Springer. https://doi.org/10.1007/978-1-0716-4350-1
32. Singh PK, Singh J, Medhi T, Kumar A (2022) Phytochemical screening, quantification, FT-IR and in silico characterization of bioactives in polyherbal formulation. ACS Omega 7(37):33067–33078. https://doi.org/10.1021/acsomega.2c03616
33. Vinatoru M (2001) Overview of ultrasonically assisted extraction of bioactive principles from herbs. Ultrason Sonochem 8(3):303–313

Check for updates

Chapter 3

Qualitative and Quantitative Assessment of Phenolic Acids

Cressida Mehta, Mayur Marathe, Kevin Vaiwala, Shubham Patel and Dhruti Amin

Abstract

Phenolic acids are important secondary metabolites widely distributed in plants and known for their antioxidant, antimicrobial, anti-inflammatory, and nutritional properties. The analysis of phenolic acids is essential for understanding plant biochemistry, quality control of herbal products, and phytopharmaceutical research. This chapter provides practical methods for qualitative screening using colorimetric tests and TLC, and for quantitative estimation using the Folin–Ciocalteu method and HPLC analysis. Detailed procedures for sample preparation, reagent preparation, chromatographic workflow, instrument settings, and data interpretation are included. These methods are suitable for analyzing phenolic acids in leaves, stems, fruits, seeds, roots, or medicinal plant extracts.

Keywords Phytochemical screening, Thin-layer chromatograph, Folin–Ciocalteu assay, High-performance liquid chromatography

1 Introduction

Phenolic acids are a major class of plant-derived secondary metabolites within the polyphenol group [1]. They are primarily classified into two types based on their carbon structure: hydroxybenzoic acids (such as gallic acid, protocatechuic acid, and vanillic acid) and hydroxycinnamic acids (such as caffeic acid, ferulic acid, p-coumaric acid, and sinapic acid) [2, 3]. These compounds contribute to plant defense, pigmentation, and growth regulation and provide significant health benefits due to their strong antioxidant activity [4].

In plant sciences, assessing phenolic acids is essential for phytochemical analysis, chemotaxonomy, food quality testing, and pharmacognostic standardization [5]. Qualitative tests provide a preliminary indication of phenolic presence, while quantitative methods accurately determine their concentration [6]. Modern chromatographic techniques, such as High-Performance Liquid Chromatography (HPLC), provide precise identification and quantification of individual phenolic acids [7].

Dhruti Amin et al. (eds.), *Phytoconstituent Analysis*, Springer Protocols Handbooks,
https://doi.org/10.1007/978-1-0716-5400-2_3,

2 Materials

2.1 Plant Sample Preparation

1. Fresh/dried plant material (1–5 g).
2. Solvents: Methanol, Ethanol, Acetone, Distilled water.
3. Whatman No. 1 filter paper, Centrifuge tubes (15 mL, 50 mL).
4. Water bath (40–60°C), Mortar and pestle/grinder, Sonicator.

2.2 Qualitative Tests

1. Ferric chloride ($FeCl_3$) solution (5%).
2. Lead acetate (10%).
3. Gelatin solution (1%).
4. Hydrochloric acid (HCl).
5. Sodium hydroxide (NaOH).

2.3 Thin Layer Chromatography (TLC)

1. Mobile phase:
   A. Toluene: Ethyl acetate: Formic acid (5:4:1)
   B. Methanol:Water: Acetic acid (8:1:1)
2. Anisaldehyde-sulphuric acid reagent/ferric chloride spray.
3. Standards: Gallic acid, Caffeic acid, Ferulic acid, p-coumaric acid.
4. UV Chamber (254, 366 nm).
5. Capillary tubes, TLC sprayers, Silica gel TLC plates (0.25 mm).

2.4 UV–Visible Spectrophotometric Method

1. Folin–Ciocalteu reagent, Sodium carbonate (7.5% w/v), For standard: Gallic acid (1 mg/mL stock).
2. Quartz cuvettes, Test tubes and rack.
3. UV–Vis spectrophotometer.

2.5 High-Performance Liquid Chromatography (HPLC)

1. HPLC system with UV/Vis detector.
2. C18 analytical column (250 mm × 4.6 mm, 5 μm).
3. Mobile phases: A: Water + 0.1% formic acid or B: Methanol or Acetonitrile.
4. Standards of phenolic acids (HPLC-grade).
5. Syringe filters (0.22 μm).
6. Autosampler vials.

3 Methods

3.1 Plant Sample Preparation [8]

1. Dry the plant material and grind it to a fine powder using a mechanical grinder.
2. Weigh accurately 1 g of the powdered sample and transfer it to a conical flask.

3. Add 10 mL of 80% methanol and mix thoroughly.
4. Heat the mixture in a water bath at 50–60°C for 30 min or sonicate for 20 min.
5. Allow the extract to cool to room temperature.
6. Centrifuge at 5000 rpm for 10 min.
7. Filter the supernatant through Whatman No. 1 filter paper.
8. Collect the filtrate and store at 4°C until further analysis (*see* **Notes 1 and 2**).

3.2 Qualitative Tests

3.2.1 Ferric Chloride Test [9]

1. Pipette 2 mL of plant extract into a test tube.
2. Add three to four drops of 5% $FeCl_3$ solution.
3. Observe for the development of blue, green, or black coloration, indicating the presence of phenolic compounds.

3.2.2 Lead Acetate Test [9]

1. Take 2 mL of plant extract in a test tube.
2. Add 1 mL of 10% lead acetate solution.
3. Observe for the formation of a bulky white or yellow precipitate, confirming the presence of phenolic compounds.

3.2.3 Gelatin Test [9]

1. Mix 2 mL of plant extract with 1 mL of 1% gelatin solution containing sodium chloride.
2. Observe for the formation of a white precipitate, indicating the presence of phenolic compounds.

3.2.4 Sodium Hydroxide Test [9]

1. Take 1 mL of plant extract in a test tube.
2. Add two to three drops of sodium hydroxide (NaOH) solution.
3. Observe the development of an intense yellow color.
4. Add dilute hydrochloric acid (HCl) and note the disappearance or reduction of the yellow color, confirming the presence of phenolic acids.

3.3 Thin Layer Chromatography (TLC) [10]

1. Apply the plant extract and standard phenolic acids onto a precoated TLC plate using capillary tubes.
2. Develop the plate in a suitable mobile phase up to a distance of 8–9 cm from the point of application.
3. Remove the plate and air-dry.
4. Examine under UV light at 254 and 366 nm.
5. Spray the plate with anisaldehyde–sulphuric acid reagent or $FeCl_3$ reagent for visualization.
6. Calculate the Rf values and compare with those of standard phenolic acids (*see* **Note 3**).

3.4 UV–Visible Spectrophotometric Method [11]

1. Pipette 0.1 mL of plant extract into a test tube.
2. Add 0.5 mL of distilled water.
3. Add 0.25 mL of Folin–Ciocalteu reagent and allow to stand for 5 min.
4. Add 1.25 mL of 7.5% Nal_2CO_3 solution.
5. Mix thoroughly and incubate at room temperature for 30 min.
6. Measure the absorbance at 765 nm using a UV–Visible spectrophotometer against reagent blank.
7. Prepare a calibration curve using gallic acid (10–100 μg/mL) and express the results as gallic acid equivalents (GAE) (*see* **Note 4**).

3.5 High-Performance Liquid Chromatography (HPLC) [12]

3.5.1 Sample Preparation

1. Filter the extract through a 0.22 μm syringe filter.
2. Transfer the filtrate into an HPLC vial.

3.5.2 Chromatographic Conditions for Analysis

1. Column: C18 reverse-phase column.
2. Flow rate: 1.0 mL/min
3. Injection volume: 20 μL
4. Detection wavelength: 280 nm (320 nm for cinnamic acid derivatives).
5. Apply gradient elution as follows:
   - 0–5 min: 95% A
   - 5–20 min: 70% A
   - 20–30 min: 50% A
   - 30–35 min: return to initial conditions

3.5.3 Analysis

1. Inject standard phenolic compounds to determine retention times.
2. Inject sample extracts under identical chromatographic conditions.
3. Identify peaks by comparing retention times with standards.
4. Quantify compounds using calibration curves constructed from peak area versus concentration (*see* **Note 5**).

4 Notes

1. Plant extraction solvent can vary (methanol, ethanol, acetone). Methanol generally gives higher phenolic content.
2. Avoid strong light exposure during extraction and analysis because phenolics degrade easily.
3. TCL plates must be activated at 110°C for 10 min.
4. Folin–Ciocalteu reagent reacts with all reducing compounds; therefore, values represent total phenolics, not individual acids.
5. For precise identification of each phenolic acid, HPLC or LC–MS is recommended.

References

1. Bravo L (1998) Polyphenols: chemistry, dietary sources, metabolism, and nutritional significance. Nutr Rev 56(11):317–333
2. Crozier A, Jaganath IB, Clifford MN (2009) Phenolics: chemistry, bioavailability and effects on health. Nat Prod Rep 26(8):1001–1043
3. Dai J, Mumper RJ (2010) Plant phenolics: extraction, analysis and their antioxidant and anticancer properties. Molecules 15(10):7313–7352
4. Cheynier V (2012) Phenolic compounds: from plants to foods. Phytochem Rev 11(2):153–177
5. Khoddami A, Wilkes MA, Roberts TH (2013) Techniques for analysis of plant phenolic compounds. Molecules 18(2):2328–2375
6. Michalak A (2006) Phenolic compounds and their antioxidant activity in plants growing under heavy metal stress. Pol J Environ Stud 15(4):523–530
7. Shahidi F, Ambigaipalan P (2015) Phenolics and polyphenolics in foods, beverages and spices: antioxidant activity and health effects–a review. J Funct Foods 18:820–897
8. Naczk M, Shahidi F (2004) Extraction and analysis of phenolics in food. J Chromatogr A 1054(1-2):95–111
9. Farnsworth NR (1966) Biological and phytochemical screening of plants. J Pharm Sci 55(3):225–276
10. Wagner H, Bladt S (1996) Plant drug analysis: a thin layer chromatography atlas. Springer Berlin Heidelberg, Berlin, Heidelberg
11. Singleton V L, Orthofer R, Lamuela-Raventós RM (1999) [14] Analysis of total phenols and other oxidation substrates and antioxidants by means of folin-ciocalteu reagent. In: Methods in enzymology, vol 299. Academic press, pp 152–178
12. Stalikas CD (2007) Extraction, separation, and detection methods for phenolic acids and flavonoids. J Sep Sci 30(18):3268–3295

Check for updates

Chapter 4

Testing and Quantification Techniques of Glycosides

Anjali Desai, Rukhsar Ansari, Priyang Kanojiya and Sanket Ray

Abstract

This chapter presents analytical methods for the testing and quantification of glycosides, including UV-Vis colorimetric methods for preliminary screening, capillary electrophoresis with UV/MS detection (CE-UV/MS) for versatile separation, ultra-high-performance liquid chromatography with tandem mass spectrometry (UHPLC-MS/MS) for high-resolution analysis, and high-performance liquid chromatography with dual photodiode array and evaporative light scattering detection (HPLC-PDA/ELSD) for universal compound detection.

Keywords Glycosides, UHPLC-MS/MS, Capillary electrophoresis, UV-Vis colorimetric, HPLC-PDA/ELSD

1 Introduction

Glycosides are among the most structurally diverse and biologically important classes of secondary metabolites found in plant kingdom, serving essential roles in plant defense mechanisms and cellular signaling pathways. These complex molecules consist of a carbohydrate moiety (glycone) covalently linked to a non-sugar component (aglycone) through glycosidic bonds. Glycosides include steroidal saponins, phenolic glycosides such as salicin and arbutin, flavonoid glycosides such as quercetin derivatives, rutin, and isoquercitrin, triterpenoid and cardiac glycosides such as digitoxin and digoxin, and anthraquinone glycosides such as sennosides and barbaloin [1].

Many glycosides are difficult to detect using conventional UV-based analytical methods because having poor or completely absent UV chromophoric properties, while some glycosides show structural similarity between closely related compounds that can complicate identification and quantification. Many glycosides demonstrate inherent stability issues during extraction and analysis procedures, potentially leading to degradation or transformation products that affect quantitative accuracy. To address these multifaceted analytical challenges, this protocol presents four complementary and

Dhruti Amin et al. (eds.), *Phytoconstituent Analysis*, Springer Protocols Handbooks,
https://doi.org/10.1007/978-1-0716-5400-2_4,

strategically selected analytical approaches, each specifically designed to target different aspects of glycoside analysis.

Capillary electrophoresis coupled with UV or mass spectrometric detection (CE-UV/MS) provides an effective and versatile separation technique that combines electrophoretic separation mechanisms with either UV detection for chromophoric compounds or mass spectrometric detection for enhanced sensitivity and specificity. CE-UV variant proves particularly valuable for analyzing UV-active flavonoid glycosides, while CE-MS enables the detection and quantification of UV-inactive compounds [2].

Ultra-high-performance liquid chromatography coupled with tandem mass spectrometry (UHPLC-MS/MS) represents the current procedure in glycoside analysis, combining high-resolution chromatographic separation achieved through 2 μm particle columns under high pressure with highly selective and sensitive mass spectrometric detection using triple quadrupole instruments capable of multiple reaction monitoring (MRM) for targeted quantification [3].

The fourth analytical approach employs high-performance liquid chromatography with dual detection systems (HPLC-PDA/ELSD), representing a widespread strategy that utilizes both photodiode array detection for UV-active compounds and evaporative light scattering detection for universal compound detection regardless of chromophoric properties. The photodiode array detector enables monitoring at multiple wavelengths, including 196 nm for non-chromophoric glycosides and 254 nm and 280 nm for aromatic compounds, while providing complete UV spectra for peak purity confirmation and compound identification [4].

The selection of the most appropriate analytical method depends on several critical factors, including the nature of target compounds with consideration of their UV activity and structural characteristics, sample complexity ranging from simple purified extracts to highly complex plant matrices, sensitivity requirements that may vary from major component quantification to trace-level detection, and available instrumentation capabilities from basic UV-Vis spectrophotometers to sophisticated LC-MS systems. A strategic method involves initial use of colorimetric or UV-based methods for screening and preliminary quantification, followed by confirmation and detailed analysis using mass spectrometry-based technique.

2 Materials

2.1 UV-Vis Colorimetric

1. Methanol (HPLC grade), ethanol, deionized water, concentrated H_2SO_4.
2. p-anisaldehyde solution, vanillin solution.
3. Reference standard pure glycoside.

4. Ultrasonic bath or shaker, water bath, ice bath, centrifuge, rotary evaporator, capped reaction tubes.
5. UV-Vis spectrophotometer (scan 200–800 nm).
6. PPE and fume hood for acids.

2.2 CE-UV/MS

1. Standards: rutin, isoquercitrin, ginsenoside Rg1, salicin.
2. Uncoated 50 μm fused silica.
3. HPLC/MS-grade methanol, acetonitrile, ultrapure water, formic acid, sodium tetraborate or ammonium acetate.
4. Syringe filters 0.22 μm.

2.3 UHPLC-MS/MS Method

1. C18 column (1.7 μm, 50 × 2.1 mm).
2. Data processing and instrument control using MassLynx software (version 4.1).
3. UHPLC-MS/MS System: UHPLC coupled with Xevo TQ-S triple quadrupole mass spectrometer.
4. Evaporator: Labconco RapidVap evaporator or equivalent.
5. Derivatization reagent: L-cysteine ethyl ester hydrochloride (for cyanide derivatization).
6. Solid-phase extraction (SPE) cartridges: Oasis® HLB 1cc (30 mg) cartridges.

2.4 HPLC-PDA/ELSD

1. Water and acetonitrile.
2. HPLC-grade methanol (80%, v/v).
3. High-purity nitrogen gas (>99.5%).

3 Method

3.1 UV-VIS Colorimetric

1. Prepare a stock solution of the standard glycoside in methanol (5 mg/mL) and dilute to working standards covering 20–230 μg/mL (*see* **Note 1**).
2. Dry and powder the plant material (100–500 mg), then extract with 50%–80% methanol (*see* **Note 2**) by sonication for 10–20 min or shake 1 h at room temperature; centrifuge suspension at 10,000 × *g* for 10 min.
3. Collect the supernatant, repeat extraction, and combine supernatants. Evaporate collected supernatant to dryness, reconstitute in methanol, and centrifuge before analysis.
4. Add 100 μL sample or standard solution, 100 μL chromogenic reagent (*see* **Note 3**), and 2.0–2.5 mL aqueous acid to a capped tube. Mix and incubate at 60°C for 10–20 min. Stop reaction

by place in ice for 10 min. If phase separation is present, use clear supernatant for absorbance measurement.

5. Scan a standard reaction (200–800 nm) to confirm λ_{max} for specific glycosides. For p-anisaldehyde/H_2SO_4 + saponins, $\lambda_{max} \approx$ 600 nm; for steroidal sapogenins, $\lambda_{max} \approx$ 425 nm.
6. Measure absorbance of standards and samples at the selected λ. Prepare a blank using solvent + reagents (no analyte). Run standards and samples in the same reaction batch to avoid drift [5].

3.2 CE-UV/MS

1. Weigh 50–100 mg plant sample and extract it with 1 mL of 80% methanol (or +0.1% formic acid for MS) (*see* **Note 4**).
2. Sonicate suspension for 20–30 min, then centrifuge at 10,000 × *g* for 10 min. Filter supernatant through a 0.22 μm filter and add internal standard to all standards and samples.
3. For UV-active glycosides, use 25 mM sodium tetraborate buffer (pH 9.0) at 25°C, +25 kV, and hydrodynamic injection. Detect flavonoids at 360 nm, phenolic glycosides at 280 nm, and anthraquinone glycosides at 450 nm.
4. For UV-inactive or trace glycosides, CE-MS is performed using a volatile 10–20 mM ammonium acetate buffer, pH 6–8 with a sheath-flow ESI interface and sheath liquid of 50:50 methanol:water containing 0.1% formic acid positive mode or 5 mM ammonium acetate negative mode (*see* **Note 5**).
5. Targeted analysis uses SIM or MRM/SRM transitions, while untargeted analysis employs HRMS full-scan with extracted ion chromatograms, detecting ions as $[M+H]^+$ or $[M+Na]^+$ in positive mode and $[M–H]^-$ in negative mode.
6. Calibration curve is prepared out by preparing various concentration of standards in either running buffer or a blank plant extract, injecting each level in triplicate.
7. In CE-UV, quantification is based on the analyte peak area, whereas for CE-MS it uses the peak area ratio of analyte to internal standard.
8. A calibration curve is plotted to determine the limit of detection (LOD, signal-to-noise ratio of 3) and limit of quantification (LOQ, signal-to-noise ratio of 10), and final glycoside concentrations are reported in μg/g dry weight.
9. To ensure the method's accuracy and reliability, validation should demonstrate linearity with R^2 values of at least 0.990, precision with intra- and interday RSDs not exceeding 5% for UV detection or 10% for MS detection, recovery within the range of 80%–120%, and specificity verified either by UV spectral matching for UV-active glycosides or by exact mass identification when using MS detection [6].

3.3 UHPLC-MS/MS Method

1. Using a C18 column and a UHPLC connected to a triple quadrupole mass spectrometer (UHPLC-MS/MS), cyanogenic glycosides were separated (*see* **Note 6**).
2. Phase A and Phase B of the mobile phase consist of 0.1% formic acid in water and 0.1% formic acid in acetonitrile, respectively. Chromatographic gradient employed for UHPLC–MS/MS analysis is mentioned in Table 1.
3. The column's flow rate stays at 200 μL/min. The requirements for the MS/MS electrospray ionization (ESI) source set are as follows: nebulizer gas 500 L/h N_2, capillary voltage 3.07 kV, cone voltage 21, temperature 150°C, and desolvation temperature 350°C.
4. Optimize each analyte's collision energies, which should vary from 17 to 30 eV (*see* **Note 7**).
5. The temperatures of the sample and column are set to 10°C and 40°C, respectively.
6. MassLynx program is used for data processing and instrument control, and concentrations of standard solutions range from 1 ng/mL to 10 μg/mL.
7. Every analysis is carried out in triplicate, including a blank; method's intra- and interday precisions had a CV% below 5%.
8. Excel is used for statistical analysis, and the mean ± standard error of the mean (SEM) is the outcome.
9. Coefficient of variation for six consecutive injections is used to verify candidate concentrations for verification of the lower limit of detection (LLOD) and lower limit of quantification (LLOQ) (*see* **Note 8**) [7].

3.4 HPLC-PDA/ELSD

HPLC systemHPLC System: Shimadzu Prominence-i LC-2030C 3D Plus or equivalent dual-detector system, photodiode array (PDA) detector, evaporative light scattering detector (ELSD)—Sedex 85 LT-ELSD or equivalent, Phenomenex Luna C18 (150 mm × 4.6 mm, 5 μm).

1. Mobile phase composition:

Table 1
UHPLC Gradient Elution Program of Glycosides for UHPLC–MS/MS Analysis

Time (min)	%A (water)	%B (acetonitrile)
0.0–1.0	95	5
1.0–3.0	95–80	20
3.0–7.0	80–40	60
7.0–8.0	0	100
8.1–10.0	95	5 (re-equilibration)

Mobile Phase A: HPLC-grade water.
Mobile Phase B: HPLC-grade acetonitrile.

2. Optimized analytical conditions for HPLC–PDA/ELSD are describe in Table 2.
Two gradient elution programs were optimized based on the analytical requirement:
Extended gradient program (86 min): used for comprehensive profiling of glycosides. Rapid (*see* **Note 10**) gradient program (10 min): applied for quick screening purposes. Gradient elution programs for rapid glycoside analysis are provided in Tables 3 and 4 respectively.
3. PDA detector is set at 196 nm as the primary wavelength for detecting non-chromophoric glycosides, with additional detection at 254 nm and 280 nm for aromatic glycosides. Bandwidth: ±4 nm; data collection rate: 5 Hz; time constant: 0.1 s.
4. ELSD drift tube temperature was maintained at 40°C with nitrogen gas pressure set at 340 kPa. Gain was auto-optimized and filter was set to 1 Hz.

Table 2
Chromatographic Conditions of Glycosidesfor HPLC–PDA/ELSD Analysis

Parameter	Specification
Flow rate	0.8 mL/min
Injection volume	25 μL
Column temperature	30°C (*see***Note 9**)
Run time	86 min (extended analysis) or 10 min (rapid screening)

Table 3
Extended Gradient Elution Program of Glycosides for HPLC–PDA/ELSD Analysis

Time (min)	% A (water)	% B (acetonitrile)
0–25	77	23
25–35	77→68	23→32
35–50	68	32
50–68	68→65	32→35
68–75	65→5	35→95
75–76	5→77	95→23
76–86	77	23

Table 4
Rapid Gradient Elution Program of Glycosides forHPLC–PDA/ELSD Analysis

Time (min)	% A (water)	% B (acetonitrile)
0.0–1.0	95	5
1.0–3.0	95→80	5→20
3.0–7.0	80→40	20→60
7.0–8.0	40→0	60→100
8.0–8.1	0→95	100→5
8.1–10.0	95	5 (re-equilibration)

5. Approximately 100 mg of finely powdered plant material (< 1 mm) was extracted with 5 mL of 80% methanol using ultrasonic-assisted extraction for 20 min at room temperature; the process was repeated twice.
6. Supernatants were combined after centrifugation at 4000 rpm for 5 min, evaporated to dryness, and reconstituted in 6 mL of distilled water.
7. SPE-C18 cartridges were conditioned with 6 mL methanol (*see* **Note 11**) and 12 mL water; samples were then loaded, washed with 6 mL water, and glycosides were eluted with 4 mL methanol.
8. Eluted samples were brought to volume in a 5 mL volumetric flask and filtered through a 0.45 μm PTFE syringe filter before HPLC injection.
9. Reference standards were prepared in 80% methanol at concentrations of 0.1–2.5 mg/mL and diluted serially (0.75 factor) to obtain 5–6 calibration levels (*see* **Note 12**).
10. All standard solutions were stored at 4°C and used within 7 days to ensure stability and accuracy.
11. Method met system suitability criteria (resolution >1.5, tailing factor <2.0, RSD <2%, RT RSD <1%) and was validated for specificity, linearity (R^2 > 0.999), precision, accuracy (80%–120% recovery), and sensitivity (LOD/LOQ at 3:1 and 10:1 S/N ratios).
12. Quantification was performed using PDA for UV-absorbing glycosides and ELSD for non-UV active ones; QAMS was applied where necessary, with peak purity confirmed via PDA spectra and integration checked manually (*see* **Note 13**) [8].

4 Notes

1. Store glycoside stock solutions in dark conditions at 4°C to prevent photo degradation.
2. The choice between 50%–80% methanol depends on the glycoside type. Saponins typically extract better with 70%–80% methanol, while phenolic glycosides prefer 50%–60% methanol.
3. Maintain consistent temperature (±2°C) during the color development reaction. Higher rates cause degradation. Lower temperatures result in incomplete color development.
4. New capillaries require conditioning with 1 M NaOH (30 min), water (10 min), and running buffer (20 min). Between runs, rinse with 0.1 M NaOH (2 min), water (2 min), and running buffer (3 min) to maintain reproducible migration times and peak areas.
5. Optimize formic acid concentration (0.05%–0.2%) to balance sensitivity and peak shape.
6. C18 columns with smaller particle sizes (1.7–2.1 μm) provide superior resolution but require higher pressures (up to 15,000 psi). For thermally labile glycosides, maintain column temperature at 30°C or lower.
7. For each glycoside, optimize at least two transitions: one for quantification (highest intensity) and one for confirmation. Collision energies typically range from 15 to 35 eV for glycosides.
8. Filter all samples through 0.22 μm filters to prevent column contamination. For complex matrices, consider solid-phase extraction (SPE) using HLB cartridges to remove salts and polar interferences.
9. Lower temperatures (35–40°C) are suitable for thermally labile compounds but may cause peak tailing. Higher temperatures (50–60°C) improve peak shape but may decompose sensitive glycosides.
10. The extended gradient (86 min) is necessary for complex saponin mixtures where structural isomers require baseline separation.
11. Methanol elution should be performed slowly (1 drop/s) for maximum recovery.
12. When reference standards are unavailable, use chemically similar compounds as external standards.
13. Use PDA for chromophoric compounds (flavonoid and phenolic glycosides) and ELSD for non-chromophoric compounds (saponins, cardiac glycosides).

14. When reference standards are unavailable, use chemically similar compounds as external standards.

References

1. Bhatla SC, Lal MA (2023) Secondary metabolites. In: Plant physiology, development and metabolism. Springer, pp 765–808
2. Stolz A et al. (2019) Recent advances in capillary electrophoresis-mass spectrometry: instrumentation, methodology and applications. Electrophoresis 40(1):79–112
3. Motilva M-J, Serra A, Macià A (2013) Analysis of food polyphenols by ultra high-performance liquid chromatography coupled to mass spectrometry: an overview. J Chromatogr A 1292:66–82
4. Lalchandani DS, Chenkual L, Porwal PK (2025) Stability of Herbal Formulations: Emerging Trends and Techniques.In: Singh A, Kulhari H, Saharan VA, (eds). Formulating Pharma-, Nutra-, and Cosmeceutical Products from HerbalSubstances: Dosage Forms and Delivery Systems. John Wiley and Sons, Inc, NJ, USA, pp 195-217
5. Le Bot M et al. (2022) An accurate, cost-effective and simple colorimetric method for the quantification of total triterpenoid and steroidal saponins from plant materials. Food Chem 383:132597
6. Arráez-Román D et al. (2007) Identification of phenolic compounds from pollen extracts using capillary electrophoresis–electrospray time-of-flight mass spectrometry. Anal Bioanal Chem 389(6):1909–1917
7. Cui Y et al (2019) A UHPLC-Q-Exactive-Orbitrap-MS method for simultaneous determination of three flavonoids from parasitic Loranthus and their pharmacokinetics in rat plasma. Trop J Pharm Res 18(10):2175–2182
8. Ngo T-M-D et al. (2025) Simultaneous quantification of main saponins in Panax vietnamensis by HPLC-PDA/ELSD using the quantitative analysis of multi-components using the single-marker method. Metabolites 15(7):419

Chapter 5

Analytical Methods for Plant Polyketides

Ami Naik

Abstract

Plant polyketides, notably flavonoids, stilbenes, and chalcones, constitute a structurally diverse group of secondary metabolites crucial for plant defense, pigmentation, and adaptation to abiotic and biotic stresses. These compounds are chiefly synthesized by Type III polyketide synthases (PKSs), homodimeric enzymes that orchestrate starter loading, iterative chain elongation, and intramolecular cyclization within a single active-site cavity. This chapter presents a comprehensive, stepwise methodological framework for the investigation of plant polyketides, from sample collection to advanced analytical characterization. It details standardized protocols for sample preparation, including controlled washing, lyophilization or shade-drying, and fine grinding to promote efficient solvent penetration while preserving metabolite stability. Extraction workflows emphasize repeated methanol or ethanol extraction assisted by sonication to maximize yield. Rapid qualitative assessment is achieved through classical phytochemical tests (such as ferric chloride and aluminium chloride assays) and thin-layer chromatography. For accurate quantification and structural elucidation, the chapter compares UV-Visible spectrophotometry and high-performance liquid chromatography (HPLC), while highlighting liquid chromatography–tandem mass spectrometry (LC–MS/MS) as the method of choice for high sensitivity and specificity in complex plant matrices.

Keywords Phytochemicals, Flavonoids, Qualitative, Quantitative

1 Introduction

Plant polyketides constitute one of the most chemically versatile classes of plant secondary metabolites, contributing substantially to plant physiology, survival, and adaptation. These compounds include a wide range of structurally distinct molecules, such as flavonoids, stilbenes, chalcones, anthraquinones, and xanthones, each characterized by variations in carbon backbone length, degree of oxidation, and cyclization patterns. Through these structural features, plant polyketides perform a variety of ecological functions, such as visual pigments that attract pollinators, antioxidants that protect photosynthetic tissues from radiation, and antimicrobivore agents that strengthen plant defense mechanisms [1].

Dhruti Amin et al. (eds.), *Phytoconstituent Analysis*, Springer Protocols Handbooks,
https://doi.org/10.1007/978-1-0716-5400-2_5,

The biosynthesis of plant polyketides is primarily orchestrated by Type III polyketide synthases (PKSs), a unique class of enzymes that differ markedly from the large, multidomain Type I PKSs found in bacteria and fungi. Type III PKSs are relatively small (~40–45 kDa), homodimeric cytosolic enzymes that utilize simple acyl-CoA substrates, such as malonyl-CoA, to catalyze sequential decarboxylative condensations. Remarkably, all catalytic steps—including chain initiation, elongation, cyclization, and aromatization—occur within a single active-site cavity. Subtle differences in active-site architecture and amino acid composition enable Type III PKSs to generate an extraordinary diversity of products from similar substrates, explaining the extensive structural diversity observed among plant polyketides despite the apparent enzymatic simplicity [2].

This chapter presents a comprehensive and systematic methodological framework for the investigation of plant polyketides, combining classical phytochemical techniques with advanced chromatographic and spectrometric methodologies. Emphasis is placed on standardized sample preparation procedures, including controlled washing to remove surface contaminants, appropriate drying methods to prevent enzymatic degradation, fine grinding to enhance solvent penetration, and optimized solvent extraction strategies to ensure maximum recovery of target metabolites. Careful attention to these preparatory steps is essential for achieving reproducible results and minimizing variability arising from plant matrix complexity.

Qualitative characterization of plant polyketides is addressed through a series of well-established phytochemical assays that allow rapid screening and preliminary classification of metabolite subclasses. Color-based reactions, including ferric chloride tests for phenolic compounds, aluminium chloride assays for flavonoids, and vanillin–sulfuric acid reactions for broader polyketide detection, provide quick and cost-effective confirmation of compound presence [2]. These assays are complemented by thin-layer chromatography (TLC), which enables simultaneous separation and visualization of multiple components, offering valuable insights into the chemical diversity and relative abundance of phenolics, flavonoids, and related polyketide derivatives within complex plant extracts.

2 Materials

2.1 Plant Material

1. Fresh or dried leaves, stems, roots, or seeds.
2. Dry plant material should be finely powdered (~0.5–1 mm).

2.2 Solvents

1. Methanol, Ethanol—extraction.
2. Acetonitrile—HPLC mobile phase.
3. Ultrapure water (18.2 MΩ).
4. Ethyl acetate—partitioning.

2.3 Reagents for Colorimetric Screening

1. Ferric chloride (1% w/v in water)—phenolic detection.
2. Aluminium chloride (10% w/v in methanol)—flavonoid detection.
3. Vanillin–sulfuric acid reagent—triterpenoids/polyketides.

2.4 Chromatography and Detection

1. Thin Layer Chromatography (TLC) plates (Silica gel 60 F254).
2. HPLC system.
3. LC–MS/MS system (advanced quantification).
4. Standard polyketides.

2.5 Laboratory Equipment

1. Centrifuge (4000–5000 rpm).
2. Sonicator (ultrasonic bath or probe).
3. Rotary evaporator.
4. Lyophilizer.
5. Micropipettes, volumetric flasks.
6. UV-Vis spectrophotometer.
7. Amber storage containers.

3 Methods

3.1 Sample Preparation

Washing of Plant Material

1. Fresh plant parts (leaves, stems, roots, or seeds) are washed thoroughly with distilled water to remove adhering dust, soil, and debris.
2. A brief rinse with 0.1% (w/v) sodium hypochlorite solution is performed for surface sterilization, followed by multiple washes with distilled water to remove residual disinfectant.

Drying

3. The washed material is shade-dried under adequate airflow, avoiding direct sunlight to prevent photodegradation and oxidative loss of polyketides.
4. For heat-sensitive or delicate tissues, freeze-drying (lyophilization) is employed until the material becomes completely dry and brittle.

Grinding

5. The dried plant material is ground into a fine, uniform powder (particle size 0.5–1 mm) using a mortar and pestle or a mechanical grinder.
6. Excessive heat generation during grinding is avoided to preserve thermolabile and oxidation-sensitive metabolites.

Storage

7. The powdered material is transferred to airtight, properly labeled containers and stored at −20°C until further extraction and analysis.

Rationale

8. Proper washing, controlled drying, and uniform grinding enhance extraction efficiency, analytical reproducibility, and metabolite stability, thereby preserving the chemical integrity of polyketides.

3.2 Extraction of Polyketides [3]

Weighing of Sample

1. Accurately weigh 1–5 g of finely powdered plant material and transfer it into a clean, dry conical flask (*see* **Note 1**).

Solvent Addition

2. Add 10–20 mL of extraction solvent per gram of plant powder.
3. Use methanol or 70% (v/v) ethanol for general polyketide extraction.
4. For relatively hydrophobic polyketides, employ a methanol: ethyl acetate (1:1, v/v) solvent system.

Sonication-Assisted Extraction

5. Place the conical flask in a sonication bath and sonicate for 20–30 min at room temperature.
6. Apply intermittent sonication cycles to prevent overheating and degradation of thermolabile metabolites.

Centrifugation

7. Centrifuge the extract at 4000 rpm for 10 min to separate the solvent phase from plant debris.

Repeated Extraction

8. Repeat the solvent addition, sonication, and centrifugation Steps 2–3 times to ensure maximum recovery of polyketides (*see* **Note 2**).

Solvent Evaporation

9. Pool all collected supernatants and concentrate the extract by evaporation under reduced pressure using a rotary evaporator at ≤40°C.
10. The resulting crude polyketide-rich extract may be used for subsequent qualitative and quantitative analyses.

3.3 Qualitative Analysis [4]

3.3.1 Colorimetric Phytochemical Screening of Polyketides

Determine the presence of polyketides and related secondary metabolites using simple color reactions.

Phenolics (Ferric Chloride Test)

1. Pipette 0.5 mL of the plant extract into a clean, dry test tube (*see* **Note 3**).
2. Add 2 mL of freshly prepared 1% (w/v) ferric chloride ($FeCl_3$) solution.
3. Mix gently and allow the reaction to proceed at room temperature.
4. Observe the color change in the reaction mixture.
5. The appearance of a blue–black or green coloration confirms the presence of phenolic polyketides (*see* **Note 4**).

Flavonoids (Aluminium Chloride Test)

1. Mix 0.5 mL of the plant extract with 1 mL of 10% (w/v) aluminium chloride ($AlCl_3$) in methanol in a clean test tube (*see* **Note 5**).
2. Mix the contents gently and allow the reaction to proceed at room temperature.
3. Observe the development of color in the reaction mixture.
4. The appearance of a yellow coloration indicates a positive reaction for flavonoid polyketides.
5. The intensity of the color may be quantified spectrophotometrically using UV–Vis analysis at an appropriate wavelength.

Triterpenoids/Polyketides (Vanillin–Sulfuric Acid Test)

1. Mix 0.2 mL of the plant extract with 0.5 mL of vanillin–sulfuric acid reagent in a clean test tube.
2. Heat the reaction mixture gently in a water bath for 2–5 min.
3. Allow the mixture to cool to room temperature.
4. Observe the development of color.
5. The appearance of a pink-red or violet coloration indicates the presence of triterpenoids or related polyketides, resulting from condensation of vanillin with hydroxyl or aldehyde functional groups in the polyketide structure (*see* **Note 6**).

3.3.2 Thin Layer Chromatography (TLC) for Polyketides [5]

To qualitatively separate and detect polyketide compounds in plant extracts and compare their chromatographic behavior with reference standards.

Plate Preparation

1. Use precoated silica gel TLC plates (layer thickness 0.25–0.5 mm).

2. Draw a baseline approximately 1 cm from the lower edge of the plate using a pencil.

Sample Application

3. Apply 5–10 μL of the plant extract onto the baseline using a capillary tube or micropipette.
4. Spot a standard polyketide compound alongside the sample for comparison.
5. Allow the applied spots to dry completely to prevent diffusion during development.

Plate Development

6. Place the TLC plate vertically in a presaturated TLC chamber containing the mobile phase.
7. Example mobile phase for polyphenolic polyketides: Chloroform: Methanol: Water (8:2:0.2, v/v/v).
8. Ensure that the solvent level remains below the sample spots.
9. Allow the solvent front to migrate upward until it reaches approximately 0.5 cm below the top edge of the plate.

Visualization

10. Remove the plate from the chamber and allow it to air-dry.
11. Visualize separated compounds under:
    - UV light at 254 and 366 nm, and/or
    - Spray reagents such as anisaldehyde–sulfuric acid, vanillin–sulfuric acid, or ferric chloride for color development.
12. Record the number of spots, color, shape, and intensity of each band.

Calculation of Rf Values

13. Measure the distance travelled by each compound and the solvent front from the baseline.
14. Calculate the retardation factor (Rf) using the formula:

$$Rf = \frac{\text{Distance travelled by the compound}}{\text{Distance travelled by the solvent front}}$$

15. Compare the Rf values and color reactions of sample spots with those of the standard polyketide to confirm identity.

Rationale:

16. TLC provides a simple, rapid method for qualitative detection, separation, and comparison of polyketides.
17. Spot patterns and Rf values help in assessing the presence of known compounds and detecting possible novel metabolites.

3.4 Quantitative Analysis

3.4.1 UV-Visible Spectrophotometry for Polyketides [6]

To determine the concentration of polyketides in plant extracts using UV–Visible absorbance measurements.

Preparation of Standard Solutions

1. Prepare a series of standard polyketide solutions in methanol over a concentration range of 5–100 μg/mL.
2. Mix thoroughly to ensure complete dissolution and homogeneity.

Measurement of Standards

3. Measure the absorbance of each standard solution at its maximum absorption wavelength (λ_{max}) using a UV–Vis spectrophotometer (*see* **Note 7**).
4. Use methanol as the reagent blank to zero the instrument.

Sample Preparation and Measurement

5. Dilute the plant extract in methanol to ensure the absorbance falls within the linear range of the standard curve.
6. Measure the absorbance of the diluted sample at the same λ_{max} used for the standards.

Construction of Calibration Curve

7. Plot absorbance versus concentration for the standard solutions.
8. Perform linear regression analysis to obtain the calibration equation and the slope of the standard curve, in accordance with the Beer–Lambert law.

Quantification of Polyketides (see Note 8)

9. Calculate the concentration of polyketides in the sample using the following equation:

$$C_{\text{sample}}\left(\text{mg/mL}\right) = \frac{A_{\text{sample}} - A_{\text{blank}}}{\text{slope of standard curve}}$$

10. Express results as mg polyketide equivalents per mL of extract, or convert to mg/g dry weight where applicable.

3.4.2 High Performance Liquid Chromatography (HPLC) for Quantification of Polyketides [5, 7]

To accurately quantify polyketide compounds in plant extracts using HPLC based on peak area comparison with authenticated standards.

Sample Preparation

1. Dissolve the crude plant extract in the HPLC mobile phase to obtain a final concentration of 1 mg/mL.

2. Prepare both standard and sample solutions in the same solvent system to minimize solvent-related chromatographic variations.
3. Filter each solution through a 0.22 µm PTFE syringe filter to remove particulate matter.
4. Transfer the filtrate into low-binding HPLC vials to prevent adsorption of analytes.

Chromatographic Conditions

5. Column: Reverse-phase C18 column (150 × 4.6 mm, 5 µm particle size).
6. Mobile Phase: Solvent A: Water containing 0.1% (v/v) formic acid; Solvent B: Acetonitrile containing 0.1% (v/v) formic acid.
7. Gradient program
    - 0–30 min: Linear gradient from 5% B to 95% B
    - 30–35 min: Hold at 95% B
    - 35–42 min: Re-equilibration to 5% B
    - Flow Rate: 0.5–1.0 mL/min
8. Detection: UV absorbance at 254–280 nm, selected based on the λ_{max} of the target polyketide.

Standard Calibration

9. Prepare standard polyketide solutions in the concentration range of 5–100 µg/mL using the mobile phase.
10. Inject each standard under identical chromatographic conditions.
11. Plot peak area versus concentration and perform linear regression analysis to generate the calibration curve.

Sample Analysis and Quantification

12. Inject the prepared plant extract under the same chromatographic conditions as the standards.
13. Identify the target polyketide peak based on retention time and UV spectrum comparison with the standard (*see* **Note 9**).
14. Record the corresponding peak area.
15. Calculate the polyketide content using the following equation:

$$\text{Polyketide content}\left(\%\text{w/w}\right) = \frac{\text{Peak area of sample}}{\text{Peak area of standard}} \times \text{Concentration of standard} \times 100$$

3.4.3 LC–MS/MS Quantification of Polyketides [8] (Advanced Quantification)

To achieve high-sensitivity and high-specificity quantification of polyketides in plant extracts using liquid chromatography coupled with tandem mass spectrometry (LC–MS/MS).

Sample Preparation

1. Dissolve the crude plant extract in methanol to a final concentration of 0.1–1 mg/mL (*see* **Note 9**).
2. Filter the solution through a 0.22 µm PTFE syringe filter to remove particulate matter.
3. Transfer the filtrate into low-binding LC–MS vials to minimize analyte adsorption and loss.

Chromatographic Separation

4. Perform separation using a reversed-phase C18 LC column (100–150 mm length, 2.6–5 µm particle size).
5. Use the following mobile phases:
   - Solvent A: Water with 0.1% (v/v) formic acid
   - Solvent B: Acetonitrile with 0.1% (v/v) formic acid
6. Apply a gradient elution program, e.g.:
   - 5%–95% B over 20–30 min, followed by a hold at high organic content and re-equilibration to initial conditions.
7. Maintain a flow rate of 0.2–0.5 mL/min, compatible with MS detection.

Mass Spectrometric Detection

8. Employ electrospray ionization (ESI) in positive ion mode, which is suitable for polyhydroxylated polyketides.
9. Detect analytes using Multiple Reaction Monitoring (MRM) by monitoring specific parent → fragment ion transitions.
10. High-resolution MS (HR-MS), when employed, enables discrimination between structural isomers and detection of common adduct ions such as $[M+Na]^+$, frequently observed for polyketide compounds.

Calibration and Quantification

11. Prepare standard polyketide solutions at known concentrations in methanol or mobile phase.
12. Inject standards under identical LC–MS/MS conditions and record MRM signal intensities (*see* **Note 8**).
13. Construct a calibration curve by plotting MRM signal intensity versus concentration and apply linear regression analysis.
14. Inject the prepared plant extract and record corresponding MRM signals.

15. Calculate the absolute concentration of polyketides using the equation:

$$C_{\text{sample}}\left(\mu\text{g/mL}\right) = \frac{\text{Signal intensity of sample}}{\text{Slope of calibration curve}}$$

16. Express results as µg/mL of extract or convert to µmol/g of dry plant material, as required.

4 Notes

1. Freeze-drying (lyophilization) of plant material prior to extraction enhances polyketide yield by disrupting cellular architecture and improving solvent penetration.
2. Multiple extraction cycles combined with careful solvent handling maximize polyketide recovery, while minimizing oxidative or thermal degradation.
3. Freshly prepared samples should be used for qualitative tests to prevent oxidation-related artifacts that may interfere with color reactions.
4. The ferric chloride reaction is based on the formation of ferric–phenolate complexes through coordination of Fe^{3+} ions with phenolic hydroxyl groups.
5. In the aluminium chloride assay, Al^{3+} ions form stable chelate complexes with hydroxyl groups of flavonoids, resulting in the characteristic yellow coloration.
6. The vanillin–sulfuric acid reaction involves condensation of vanillin with hydroxyl or aldehyde functional groups present in polyketide structures, producing colored derivatives.
7. All glassware and cuvettes must be thoroughly cleaned and free from scratches to avoid light scattering and absorbance errors during spectrophotometric analysis.
8. Replicate injections ($n \geq 3$) are recommended to improve analytical accuracy, precision, and reproducibility in chromatographic and spectrometric analyses.
9. Standards and samples should be prepared in the same solvent system to minimize matrix effects and ion suppression, particularly in LC–MS/MS analysis.

5 Precautions

1. Handle plant extracts under dark and cooled conditions (<10°C) to minimize thermal and photodegradation of polyketides.
2. Use freshly prepared samples for UV–Vis and HPLCHPLC analyses to ensure reliable and reproducible signal responses.
3. Avoid excessive concentrations of formic acid (>0.1%), as higher acid levels may suppress ionization efficiency, particularly in negative-ion mode mass spectrometry.
4. TLC and UV–Vis spectrophotometry are suitable for preliminary screening and semiquantitative estimation, whereas HPLC and LC–MS/MS provide accurate and definitive quantitative results.
5. Always include blank samples and triplicate analyses to ensure analytical reproducibility and data reliability.
6. Store purified polyketide extracts in amber, airtight containers at –20°C under an inert atmosphere (e.g., nitrogen) to prevent oxidative degradation.

References

1. Bisht R et al (2021) An overview of the medicinally important plant type III PKS derived polyketides. Front Plant Sci 12:746908
2. Yu Z-Q et al (2024) Targeted isolation, identification, and antioxidant evaluation of aromatic polyketides from a plant-derived fungus *Ophiobolus cirsii* LZU-1509. Fitoterapia 175:105884
3. Manning SR, La Claire II JW (2013) Isolation of polyketides from *Prymnesium parvum* (Haptophyta) and their detection by liquid chromatography/mass spectrometry metabolic fingerprint analysis. Anal Biochem 442(2):189–195
4. Maheshwaran L et al (2024) Phytochemical testing methodologies and principles for preliminary screening/qualitative testing. Asian Plant Res J 12(5):11–38
5. Rodríguez-Berríos RR et al (2023) Extraction, isolation, characterization, and bioactivity of polypropionates and related polyketide metabolites from the Caribbean Region. Antibiotics 12(7):1087
6. Thomas TA Tirumale S (2022) Production of a polyketide pigment by *Fusarium chlamydosporum*. J Pure Appl Microbiol 16(2):1318–1329
7. Wills RH, Tosin M, O'Connor PB (2012) Structural characterization of polyketides using high mass accuracy tandem mass spectrometry. Anal Chem 84(20):8863–8870
8. Zhang B et al (2020) Amphotericin B biosynthesis in *Streptomyces nodosus*: quantitative analysis of metabolism via LC–MS/MS based metabolomics for rational design. Microb Cell Fact 19(1):18

Check for updates

Chapter 6

Identification and Measurement of Non-Ribosomal Peptides

Sarika Kumari and Natarajan Amaresan

Abstract

Non-ribosomal peptides (NRPs) are structurally diverse and biologically active compounds produced by microorganisms via non-ribosomal peptide synthetases (NRPSs). This chapter presents an integrated approach for the measurement and identification of NRPs that combines both laboratory and computational methods. It covers analytical techniques such as agar diffusion bioassay, radioimmunoassay (RIA), high-performance liquid chromatography (HPLC), and liquid chromatography-mass spectrometry (LC–MS/MS) for accurate detection and quantification. For genome mining and structural prediction, bioinformatics tools including antiSMASH, NRPSPredictor2, and NaPDoS are described along with the NORINE database for NRP identification. Collectively, these protocols support the discovery and characterization of novel NRPs for pharmaceutical and environmental applications.

Keywords Non-ribosomal peptides, NRPSs, HPLC, antiSMASH, NORINE, LC–MS/MS

1 Introduction

Non-ribosomal peptides (NRPs) are a diverse class of secondary metabolites produced by various species of bacteria, fungi, plants, and higher organisms [1]. Unlike ribosomally synthesized peptides, NRPs are assembled by large multimodular enzymatic complexes known as non-ribosomal peptide synthetases (NRPSs). This process occurs independently of messenger RNA templates or ribosomal machinery [1, 2]. This ribosome-independent biosynthetic mechanism enables the incorporation of a broad range of monomeric units, including D-amino acids, β-amino acids, *N*-methylated residues, and various non-proteinogenic building blocks, which extend beyond the standard 20 proteinogenic amino acids [3]. To date, more than 500 distinct monomers have been identified in the characterized NRPs.

The ability to produce diverse chemicals via a process called biosynthesis is a key feature of a class of enzymes called NRPSs. This biosynthetic versatility is responsible for the remarkable chemical diversity observed in the NRPs. Modular domains within NRPS complexes sequentially activate, modify, and condense specific substrates.

Dhruti Amin et al. (eds.), *Phytoconstituent Analysis*, Springer Protocols Handbooks,
https://doi.org/10.1007/978-1-0716-5400-2_6,

Furthermore, additional tailoring steps such as methylation, glycosylation, acylation, and cyclization help in structural and functional diversity of the initial molecule [4, 5]. Cyclisation, in particular, introduces structural rigidity and bioactivity, and can occur via amide or ester bonds using nucleophilic groups such as the N-terminal amine or hydroxyl-containing side chains. NRPs range in size from simple dipeptides to complex molecules comprising up to 30 amino acid residues. This structural complexity confers a wide range of biological activity. NRPs include clinically significant compounds such as penicillin, vancomycin, daptomycin, the anticancer agent bleomycin, and the immunosuppressant cyclosporin [1, 6]. The high density of functional groups and diverse backbone structures enhances the pharmacological potential of these metabolites, making them a valuable resource for therapeutic development.

Owing to their complexity and therapeutic potential, accurate identification and measurement of NRPs are crucial for their application in natural product discovery, drug development, and microbial ecology. This chapter highlights a range of analytical and molecular techniques used for the detection and quantification of NRPs, including high-performance liquid chromatography (HPLC), RIA, and LC–MS/MS. It also covers the application of web-based bioinformatics tools that allow structural comparisons with known NRP databases.

2 Materials

2.1 Analytical Techniques

2.1.1 Bioassay (Agar Diffusion Assay)

KH_2PO_4 (0.32%), K_2HPO_4 (0.42%), NaCl (0.2%), $(NH_4)_2SO_4$ (0.2%), agar (0.5%), $MgSO_4.7H_2O$ (5%), $FeCl_3$(1%), $CaCl_2$(0.5%), $MnSO_4$(0.1%), NRP sample (e.g., vancomycin), heat inactivated human serum (for standard dilutions), *Bacillus subtilis* (indicator strain), and pH meter.

2.1.2 Radioimmunoassay (RIA)

Vancomycin hydrochloride (e.g., from Eli Lilly), Bovine serum albumin (BSA), Carbodiimide reagent, Freund's complete adjuvant, Chloramine T, Iodine-125 radiolabel, 0.5 M sodium phosphate buffer, Sephadex G-10 Column, Rabbit anti-vancomycin antiserum, and Gamma counter.

2.1.3 High-Performance Liquid Chromatography (HPLC)

Vancomycin standard, Ristocetin sulfate (internal standard), 0.02 M acetic acid, 0.16 M Na_2SO_4, 0.05 M$Na_2B_4O_7$, 8.5% H_3PO_4, 0.2 M Na_2SO_4, 0.01 M Potassium phosphate buffer, acetonitrile, Carboxymethylethyle-Sephadex, polypropylene columns, Waters 6000 A

HPLC pump, Waters U6K injector, Waters 450 UV detector, and μBondapak C18 column.

2.1.4 Liquid Chromatography-Tandem Mass Spectrometry (LC–MS/MS)

Bacillus natto NT-6, BPY medium, Landy medium, NRPs, Acetonitrile (HPLC grade), formic acid, ammonium acetate, anhydrous sodium sulfate, distilled water, and acetic acid.

2.2 Software for the Analysis of NRPSs and Their Products

2.2.1 antiSMASH

http://antismash.secondarymetabolites.org/

2.2.2 NRPSpredictor2

http://nrps.informatik.uni-tuebingen.de/

2.2.3 NaPDoS

http://napdos.ucsd.edu/

2.3 Software for the Analysis of NRPs

2.3.1 Norine

http://bioinfo.lifl.fr/norine/

3 Methods

3.1 Analytical Techniques

3.1.1 Bioassay Method (Agar Diffusion Assay) [7–9]

1. Prepare the minimal salts agar (MSA) medium with 0.32% KH_2PO_4, 0.42% K_2HPO_4, 0.2% NaCl, and 0.2% $(NH_4)_2SO_4$, and then adjust the pH to 7.3.
2. Add 1.5% agar to the buffer and autoclave it. Once cooled to 50°C, aseptically add 0.5% glucose (from a 10% stock solution) and 0.1 mL of a sterile mineral salt mixture (5% $MgSO_4 \cdot 7H_2O$, 0.1% $MnSO_4$, 1% $FeCl_3$, and 0.5% $CaCl_2$) in 100 mL of medium.
3. Grow *Bacillus subtilis* in Luria broth at 37°C for 4–5 h until mid-log phase (~5×10^7 CFU/mL), and standardize to 0.5 McFarland turbidity.
4. Inoculate the agar plates by swabbing evenly with the bacterial suspension using sterile cotton swabs.
5. Prepare NRP sample (e.g., vancomycin) standards by serially diluting the 250 μg/mL stock in heat-inactivated human serum to final concentrations of 50, 25, 12.5, 6.3, 3.1, 1.6, and 0.8 μg/mL.

6. Saturate 6-mm sterile discs with the standard solutions and unknown test samples. Place duplicate discs onto inoculated plates, alternating high and low concentrations with a maximum of 10 discs per plate.
7. Incubate the plates at 35–37°C for 10–12 h until inhibition zones appear.
8. Measure the diameter of inhibition zones (in mm) around each disc using a ruler or caliper.
9. Plot the zone diameters of the standards on semi-log graph paper against their concentrations to generate a standard curve.
10. Determine the concentration of unknown samples by comparing their inhibition zones with the standard curve.

3.1.2 Radioimmunoassay [10, 11]

1. Prepare immunogen: conjugate NRP (e.g., vancomycin) to BSA using carbodiimide. Dialyze the conjugate against 0.01 M PBS (pH 7.8) for 72 h.
2. Emulsify the conjugate in Freund's complete adjuvant and inject into rabbits intramuscularly every month to raise anti-vancomycin antibodies.
3. Label vancomycin: Iodinate vancomycin with iodine-125 using a modified chloramine-T method. Purify the iodinated product using a Sephadex G-10 column eluted with PBS.
4. Prepare reaction mix: In RIA tubes, add:
   - 100 μL of diluted standard or sample (1:101 in PBS)
   - 100 μL of rabbit anti-vancomycin antiserum (1:50 dilution in PBS with 1% normal rabbit serum)
   - 100 μL of ^{125}I-labeled vancomycin (~50,000 cpm)
5. First incubation: vortex tubes and incubate at room temperature for 10 min.
6. Add secondary antibody: add 100 μL of sheep anti-rabbit IgG to each tube. Vortex and incubate again for 20 min.
7. Centrifuge tubes at 3400 rpm for 10 min to pellet antibody-bound complexes.
8. Aspirate the supernatant carefully and count the radioactivity in the pellet using a gamma counter.
9. Generate a standard curve using known vancomycin concentrations. Compare CPM values from unknown samples to determine concentration.

3.1.3 High-Performance Liquid Chromatography [12]

1. Column preparation: Suspend carboxymethyl-ethyl Sephadex overnight in 0.2 M Na_2SO_4 (pH 6.8). Pack 0.8 mL of the swelled resin into a polypropylene column.

2. Sample preparation: Mix 0.5 mL of test sample with 0.5 mL internal standard solution (16 μg/mL ristocetin in 0.02 M acetic acid + 0.16 M Na_2SO_4). Apply the mixture to the prepared column.
3. Column washing: wash the column with two 1 mL volumes of 0.2 M Na_2SO_4 (pH 6.8). Wash with 0.5 mL of 0.5 M sodium tetraborate (pH 9.45).
4. Elution: elute vancomycin (or other NRP) and ristocetin from the column with an additional 0.5 mL of sodium tetraborate buffer. Add 40 μL of 8.5% phosphoric acid to the elute.
5. HPLC analysis: inject 50 μL of the treated elute into HPLC system. Use a C18 column and detect at 120 nm wavelength by using mobile phase (0.01 M phosphate buffer: acetonitrile, 90:10) and a flow rate of 1.5 mL/min.
6. Quantification: Construct a standard curve by plotting peak height ratio (e.g., vancomycin/ristocetin) vs known vancomycin concentrations. Determine unknown sample concentration by comparing peak height ratios to the standard curve.

3.1.4 Liquid Chromatography-Tandem Mass Spectrometry [13, 14]

1. Prepare stock solutions of NRPs (e.g., surfactin or iturin) at a concentration of 1 mg/mL in an acetonitrile and water mixture (8:2, v/v).
2. Dilute these solutions to create working standards of 10 μg/mL with acetonitrile and water mixture (8:2, v/v) containing 0.1% formic acid. Store these solutions at 4°C.
3. Inoculate *Bacillus natto* NT-6 into BPY medium containing beef extract (5.0 g), peptone (10 g), yeast extract (5.0 g), NaCl (5.0 g), and distilled water (1000 mL). Adjust the pH to 7.2 and sterilize the medium at 115°C for 20 min.
4. Incubate the seed culture at 37°C at 150 rpm for 24 h.
5. Transfer 5% (v/v) of the seed culture to a modified Landy medium containing glucose (20 g), sodium L-glutamate (5 g), $MgSO_4.7H_2O$ (0.5 g), KCl (0.5 g), $KH_2PO_4.3H_2O$ (1.0 g), $FeSO_4.7H_2O$ (0.15 mg), $MnSO_4.H_2O$ (5.0 mg), and $CuSO_4.5H_2O$ (0.16 mg). Adjust the pH to 7.2 and add distilled water (1.0 L).
6. Incubate the production culture at 30°C at 160 rpm for 36 h.
7. Centrifuge the culture broth at 5000 rpm for 15 min at 4°C to remove bacterial cells and then collect the supernatant.
8. Mix the supernatant with an equal volume of acetonitrile and water (7:3, v/v) containing 0.1% formic acid.
9. Filter the mixture through a 0.45 μm membrane filter into LC–MS/MS autosampler vials.

10. Inject 10 μL of the prepared sample into an LC-MS/MS system equipped with a Venusil XBP CN column (150 × 2.1 mm, 5 μm, 100 Å), which is maintained at 35°C.
11. Set the mobile phase to solvent A (5 mM ammonium acetate with 0.1% formic acid in water) and solvent B (acetonitrile).
12. Perform gradient elution as follows: begin running the gradient at 60% A at 0 min, decrease to 10% A from 0 to 4 min, hold at 10% A from 4 to 7 min, increase to 60% A from 7 to 7.1 min and then maintain 60% A until 12 min.
13. Maintain a flow rate of 250 μL/min throughout the process.
14. Operate the mass spectrometer in positive ion mode with electrospray ionization (ESI) using a spray voltage of 4500 V, an ion source temperature of 350°C, and a capillary temperature of 270°C.
15. Use selected ion monitoring (SIM) to detect NRPs for their respective $[M+H]^+$ ions (*see* **Note 1**).
16. Prepare a calibration curve by injecting standard solutions ranging from 0.2 to 10.0 μg/mL and plotting peak area versus concentration.
17. Calculate the concentrations of NRPs in the samples using the calibration curve.

3.2 Bioinformatics Tools to Predict NRPSs

3.2.1 antiSMASH

antiSMASH is compatible with several types of genomic input formats including GenBank (mostly), EMBL, and plain FASTA files. If annotated GenBank or EMBL files are unavailable, antiSMASH can process unannotated FASTA sequences and predict the putative coding sequences. Prodigal is employed for gene prediction in bacterial genomes, whereas Glimmer HMM is used for fungal genomes. The tool also allows automated retrieval of genomic data from public databases. By entering a valid NCBI accession number, antiSMASH can fetch the corresponding sequences from GenBank or RefSeq [15].

1. Open the antiSMASH homepage https://antismash.secondarymetabolites.org for bacterial sequences, https://fungismash.secondarymetabolites.org for fungal sequences and http://plantismash.secondarymetabolites.org/ for plant sequences.
2. Enter the email address in the provided field (optional but recommended) (*see* **Note 2**).
3. Choose one of the input methods.

- Enter a GenBank or RefSeq accession number to allow antiSMASH to retrieve the sequence data directly from the NCBI database.
- Upload a local sequence file by clicking the "Upload file" button and selecting a file in either FASTA or GenBank format (*see* **Note 3**).

4. Scroll to the bottom of the page and click the "Submit" button to begin the analysis.
5. Wait for the processing (*see* **Note 4**).
6. Once the antiSMASH job is submitted, identification and analysis of the biosynthetic gene clusters will be performed on the server. If an email address is provided during submission, a notification will be sent upon completion of the analysis to inform the user that the results are available.
7. Each identified biosynthetic gene cluster (BGC) is represented by a colored circle at the top of the results page.
8. Click on a colored circle or the associated cluster number to open the detailed cluster view. This view includes the genomic location of the cluster, the predicted domain architecture, the substrate specificity, and the gene map to facilitate rapid interpretation.

3.2.2 NRPSpredictor2 (http://nrps.informatik.uni-tuebingen.de)

NRPSpredictor2 is a web tool that identifies the substrates or compounds that are likely to be altered by specific A domains in NRPSs. It uses a support vector machine (SVM) algorithm to organize A-domains into four groups based on the different characteristics of the substrates and amino acid details. The accuracy and stability of the predictions are measured using an applicability domain model that performs statistical analyses [16]. Checking A-domain sequences against carefully chosen enriched databases allows for improved substrate prediction [17, 18].

1. Access the NRPSpredictor2 online platform using a web browser and the following URL: http://nrps.informatik.uni-tuebingen.de.
2. Provide the input sequence for analysis using one of the following methods:
   - Paste the FASTA-formatted amino acid sequence of the NRPs into the designated text input box.
   - Upload a FASTA file using the "Choose File" button.
   - Alternatively, manually extract 34-amino acid signature sequences and paste or upload them by selecting "Signature "as the "Filetype" (*see* **Note 5**).

3. Select the predictor model based on the origin of the sequence.
   - The bacterial model is used by default.
   - For sequences of fungal origin, select the "Fungal" option under the "Type of Predictor" setting.
4. Initiate the prediction by clicking the "Submit" button.
5. After submission, a table displaying the prediction results for each identified adenylation (A) domain in the input sequence will be generated.
6. To download the results, click the "Report file" link at the top of the NRPSpredictor2 results page. The file contains a tabular summary of substrate specificity predictions for all detected A domains (*see* **Note 6**).

3.2.3 NaPDoS (http://napdos.ucsd.edu/)

The Natural Product Domain Seeker (NaPDoS) is an online application that uses a phylogenetic approach to identify and categorize condensation (C) domains in non-ribosomal peptide synthetases (NRPSs) By analyzing both genomic and metagenomic DNA sequences, it is possible to identify C domains and categorize protein sequences according to their catalytic roles. These include the starting C-domains, dual condensation-epimerization domains, LCL (L–L condensation) domains, DCL (D–L condensation) domains, and heterocyclization domains [19].

1. Access the NaPDoS web server by visiting http://napdos.ucsd.edu/.
2. Navigate to the "Run Analysis" tab.
3. Select the desired domain type for analysis, either "KS-domain" (ketosynthase) or "C-domain" (condensation domain).
4. Choose the appropriate query type based on the available sequence data:
   - "Predicted protein sequences (amino acids)" for amino acid sequence input.
   - "Predicted coding sequences or PCR products (DNA)" for DNA sequences from PKS or NRPS genes.
   - "Genome or metagenome contigs (DNA)" for DNA sequences derived from genomes or metagenomes.
5. Provide the input data by pasting the query sequence into the text box in FASTA format or by uploading a sequence file.
6. Initiate the analysis by clicking the "SEEK" button, then selecting "Submit Job" on the subsequent page.
7. Optional parameters for domain identification such as e-value thresholds, minimum alignment length, and the number of

displayed hits can be configured under "Advanced Settings" before submission (*see* **Note 7**).

8. Upon submitting the job, the initial results are presented in a table format. For the "Genome or Metagenome contigs (DNA)" mode, a summary table with domain coordinates is displayed. Additional information can be accessed by selecting "Get more info". For protein or coding sequence queries, a detailed table is shown directly, which contains domain classification and pathway similarity data.
9. Information on pathways for hits that are similar to known biosynthetic clusters can be accessed via hyperlinks within the results table.
10. For an extended analysis, select individual or multiple hits.
11. The results table can be downloaded in tab-delimited text format by selecting the "DOWNLOAD" option.
12. Additional downloadable content is available at the bottom of the results page including:
    - Identified NRPS C-domain in FASTA format.
    - NaPDoS alignments or the query sequence with the best hits from the database.
    - Phylogenetic trees in SVG graphics or Newick format for use with external tools such as Dendroscope (*see* **Note 8**).

3.3 Bioinformatic Tools to Predict NRPs

3.3.1 Norine (http://bioinfo.lifl.fr/norine/)

The NORINE database is a detailed and manually curated online resource that focuses solely on non-ribosomal peptides (NRPs) It allows users to study, identify, and compare NRPs based on various criteria such as composition, structure, biological activity, and generating organisms [20, 21].

Annotation-Based

1. Open the NORINE website in a web browser.
2. Click on the "Annotation Search" tab.
3. In the search form, enter information in one or more of the following fields:
    - Peptide name
    - Peptide category
    - Monomer name
    - Structure type
4. Use the autocomplete suggestions to select valid terms.

5. Use operators like AND, OR, or NOT to combine multiple search fields.
6. Submit the query to retrieve a list of matching peptides.
7. Click on any peptide name to access its detailed page, which contains information about its structure, the organisms it is associated with, its taxonomy, references, and external links.
8. Export the search results using the HTML, XML or JSON download options.

Structure-Based

1. Navigate to the "Structure Search" tab on the NORINE homepage.
2. Use the graphical editor to build the peptide structure:
   - Select the monomer from the clustered list on the left.
   - Drag and place each monomer in the empty box on the right.
   - Connect the monomers to represent the full structure including linear, cyclic, or branched configurations.
3. Submit the query to retrieve peptides from the database that match the full or partial structure. Review the list of structurally similar NRPs and access the detailed page for further comparisons or information.

Monomer-Based

1. Navigate to the "Monomer" tab on the NORINE homepage.
2. Enter the code, formula, and weight of the monomers present in the predicted peptide.
3. Execute the search to find NRPs that contain the specified monomers.
4. Analyze whether the peptides share a similar monomeric composition with known compounds to determine relatedness or novelty.

4 Notes

1. The presence of multiple $[M+H]^+$ ions for each lipopeptide is due to the existence of homologous with different fatty acid chain lengths, which confirm the peak identity using MS/MS fragmentation patterns if needed.
2. Providing an email address ensures that users are promptly notified when the analysis is complete, which is particularly useful for long or queued jobs.

3. GenBank files contain annotated gene information that improves the quality and specificity of the antiSMASH domain prediction and visualization.
4. Processing times depend on the genome size, input format, and server queue. Therefore, it may be preferable to install antiSMASH locally for very large metagenomes or draft genomes.
5. The signature sequence format should follow the structure: "vntsfdgsvfdgfilfgeihvygptestvyaty<TAB>domain 1".
6. NRPSpredictor2 uses multiple algorithms, including nearest-neighbor analysis and support vector machines, trained on curated signature databases.
7. Parameter customization is recommended for highly divergent or partial sequences to improve accuracy.
8. The Newick format enables flexible downstream phylogenetic analysis and integration using various evolutionary biology software platforms.

References

1. Süssmuth RD, Mainz A (2017) Nonribosomal peptide synthesis—principles and prospects. Angew Chem Int Ed 56:3770–3821
2. Izoré T, Cryle MJ (2018) The many faces and important roles of protein–protein interactions during non-ribosomal peptide synthesis. Nat Prod Rep 35:1120–1139
3. Walsh CT, O'Brien RV, Khosla C (2013) Nonproteinogenic amino acid building blocks for nonribosomal peptide and hybrid polyketide scaffolds. Angew Chem Int Ed 52:7098–7124
4. Al Toma RS, Brieke C, Cryle MJ et al (2015) Structural aspects of phenylglycines, their biosynthesis and occurrence in peptide natural products. Nat Prod Rep 32:1207–1235
5. Horsman ME, Hari TP, Boddy CN (2016) Polyketide synthase and non-ribosomal peptide synthetase thioesterase selectivity: logic gate or a victim of fate?. Nat Prod Rep 33:183–202
6. Yim G, Thaker MN, Koteva K et al (2014) Glycopeptide antibiotic biosynthesis. J Antibiot (Tokyo) 67:31–41
7. Hubbard BK, Walsh CT (2003) Vancomycin assembly: nature's way. Angew Chem Int Ed 42:730–765
8. Pupin M, Esmaeel Q, Flissi A et al (2016) Norine: a powerful resource for novel nonribosomal peptide discovery. Synth Syst Biotechnol 1:89–94
9. Walker CA, Kopp B (1978) Sensitive bioassay for vancomycin. Antimicrob Agents Chemother 13:30–33
10. Crossley KB, Rotschafer JC, Chern MM et al (1980) Comparison of a radioimmunoassay and a microbiological assay for measurement of serum vancomycin concentrations. Antimicrob Agents Chemother 17:654–657
11. Fong KL, Ho DH, Bogerd L et al (1981) Sensitive radioimmunoassay for vancomycin. Antimicrob Agents Chemother 19:139–143
12. Ghasemiyeh P, Vazin A, Zand F et al (2020) A simple and validated HPLC method for vancomycin assay in plasma samples: the necessity of TDM center development in Southern Iran. Res Pharm Sci 15:529–540
13. Brozmanová H, Kacířová I, Uřinovská R et al (2017) New liquid chromatography-tandem mass spectrometry method for routine TDM of vancomycin in patients with both normal and impaired renal functions and comparison with results of polarization fluoroimmunoassay in light of varying creatinine concentrations. Clin Chim Acta 469:136–143
14. Deng Q, Wang W, Sun L et al (2017) A sensitive method for simultaneous quantitative determination of surfactin and iturin by LC-MS/MS. Anal Bioanal Chem 409:179–191
15. Blin K, Shaw S, Medema MH et al. (2024) The antiSMASH database version 4: additional genomes and BGCs, new sequence-based searches and more. Nucleic Acids Res 52:D586–D589
16. Röttig M, Medema MH, Blin K et al (2011) NRPSpredictor2—a web server for predicting

NRPS adenylation domain specificity. Nucleic Acids Res 39:W362–W367
17. Stachelhaus T, Mootz HD, Marahiel MA (1999) The specificity-conferring code of adenylation domains in nonribosomal peptide synthetases. Chem Biol 6:493–505
18. Challis GL, Ravel J, Townsend CA (2000) Predictive, structure-based model of amino acid recognition by nonribosomal peptide synthetase adenylation domains. Chem Biol 7:211–224
19. Ziemert N, Podell S, Penn K et al (2012) The natural product domain seeker NaPDoS: a phylogeny based bioinformatic tool to classify secondary metabolite gene diversity. PLoS One 7:e34064
20. Caboche S, Pupin M, Leclere V et al (2007) NORINE: a database of nonribosomal peptides. Nucleic Acids Res 36:D326–D331
21. Flissi A, Duban M, Jacques P et al (2023) Norine: bioinformatics methods and tools for the characterization of newly discovered nonribosomal peptides. In: Burkart M, Ishikawa F (Eds.)Eds. Non-ribosomal peptide biosynthesis and engineering. Springer, US, New York, NY, 303–318

Chapter 7

Qualitative and Quantitative Analysis of Steroids

Sarika Kumari and Natarajan Amaresan

Abstract

Plant steroids, including phytosterols, brassinosteroids, phytoecdysteroids, and steroidal saponins, play essential roles in plant growth, development, and stress regulation. Their accurate characterization is often challenging because of their low concentrations, conjugated forms, and complex plant tissues. This chapter provides standardized methods for both qualitative and quantitative analyses of plant steroids, covering sample preparation, extraction, hydrolysis, and instrumental detection. This chapter compiles protocols that include LC-MS, spectrophotometric assays, GC-FID/GC-MS, UPLC-qTOF, and LC-MS/MS to enable accurate detection and quantification in various plant tissues.

Keywords Plant steroids, Phytosterol, LC-MS, GC-MS/GC-FID, LC-MS/MS, UPLC-MS/MS

1 Introduction

Plant steroids are a diverse group of isoprenoid-derived metabolites that naturally occur in most plant tissues [1, 2]. They contribute to membrane structure, growth regulation, and stress response [3]. The major classes are phytosterols, brassinosteroids, phytoecdysteroids, and steroidal saponins. These classes share a common sterane backbone but differ in their functional groups and conjugation patterns. Owing to their roles in plant physiology, nutritional importance, and potential pharmacological value, these compounds have been increasingly studied [4–6].

Analytical investigations of plant steroids present several challenges due to their low natural abundance and complex mixtures [7]. Structurally similar analogs often occur together within the same sample matrix. Many steroids exist in conjugated forms, such as esterified, glycosylated, or acylated, which makes them difficult to extract without prior hydrolysis or specialized cleanup steps. Their strong associations with lipids, pigments, polysaccharides, and other compounds make chromatographic separation and mass spectrometric detection challenging. Therefore, it is essential to perform robust

Dhruti Amin et al. (eds.), *Phytoconstituent Analysis*, Springer Protocols Handbooks,
https://doi.org/10.1007/978-1-0716-5400-2_7,

sample preparation, including optimized extraction, saponification or hydrolysis, and solid-phase purification, before instrumental analysis.

Plant steroids can be analyzed using qualitative and quantitative methods [8, 9]. Qualitative screening methods such as LC-MS and high-resolution UPLC-qTOF are used. Quantitative techniques, including the Liebermann–Burchard assay, GC-FID/GC-MS, LC-MS/MS, and UPLC-MS/MS, enable accurate measurement of individual compounds [10]. This chapter presents standardized, broadly applicable protocols for both qualitative and quantitative analyses of plant steroids in a wide range of plant tissues and matrices.

2 Materials

2.1 Extraction

1. Plant material (fresh or dried) from selected species and tissues.
2. Ethanol.
3. Methanol (HPLC grade).
4. Acetonitrile.
5. Ethyl acetate.
6. 1-butanol.
7. Diethyl ether.
8. n-hexane.
9. Chloroform.
10. Formic acid.
11. Potassium hydroxide.
12. Oasis HLB SPE cartridges internal standards (ß-cholesterol, botulin, or deuterated steroids).
13. Mortar and pestle.
14. Syringe filters (0.45 μm PTFE).

2.2 Qualitative Analysis

2.2.1 LC-MS

1. LC-MS system.
2. Symmetry C18 analytical column (150 × 2.1 mm, 3.5 μm).
3. HPLC-grade solvents.
4. Water with 0.1% formic acid (mobile phase A).
5. Acetonitrile with 0.1% formic acid (mobile phase B).
6. Sample vials.
7. Nitrogen gas.

8. Calibration and tuning standards for the ESI source.
9. Autosampler syringes.

2.2.2 UPLC-qTOF

1. UPLC system equipped with qTOF mass spectrometer with an ESI source.
2. UPLC C18 BEH column, 100 × 2.1 mm, 1.7 μm particle size.
3. Mobile phase solvents:
   - LC-MS grade water with 0.1% formic acid.
   - LC-MS grade acetonitrile with 0.1% formic acid.
4. Autosampler vials with PTFE/silicon septa.
5. 0.22 μm PTFE or PVDF syringe filters.
6. Milli-Q water.
7. Vortex mixer and microcentrifuge.
8. Sonication bath.
9. Standards of plant steroids.
10. Temperature controller autosampler.
11. Nitrogen gas supply.

2.3 Quantitative Analysis

2.3.1 Spectrophotometric (Liebermann–Burchard)

1. ß-sitosterol or selected plant steroid standard (1 mg/mL stock in chloroform).
2. Chloroform.
3. Liebermann–Burchard reagent (acetic anhydride + concentrated sulfuric acid).
4. Amber glass vials.
5. UV-Vis spectrophotometer.
6. Quartz or glass cuvettes.

2.3.2 UPLC-MS/MS

1. UPLC system.
2. Triple quadrupole mass detector with ESI source.
3. UPLC column (HSS C18, 100 × 1 mm, 1.8 μm).
4. Solvents (LC-MS grade).
5. Water + 0.1% formic acid (mobile phase A).
6. Acetonitrile + 0.1% formic acid (mobile phase B).
7. Internal standards, calibration standards.
8. Sample extract solvent (e.g., methanol).
9. 0.22 μm membrane filters.

10. Glass vials.
11. Centrifuge.
12. Analytical balance.

2.3.3 GC-MS/GC-FID

1. Fresh plant tissues.
2. Hot air oven.
3. Analytical balance.
4. Ethanol (95%).
5. Whatman No.1 filter paper.
6. Aqueous ethanol (50% v/v).
7. Sodium hydroxide solution.
8. Water bath.
9. n-hexane (HPLC grade).
10. GC autosampler.
11. GC-FID system equipped with:
    - Nonpolar capillary column (e.g., HP-1, 30 m × 0.32 mm × 0.25 μm).
    - Injector capable of 280–300°C.
    - Detector (FID).
12. Helium carrier gas.
13. Stock standard solutions prepared in ethanol.
14. Glass volumetric flasks.
15. Vortex mixer.
16. Syringe filters.

2.3.4 LC-MS/MS

1. Plant samples.
2. Internal standards.
3. 2 M ethanolic potassium hydroxide solution.
4. Deionized water.
5. n-hexane (HPLC grade).
6. Isopropanol (HPLC grade).
7. Reference sterol standards.
8. Amber autosampler vials.
9. 0.22 μm PTFE or PVDF syringe filters.
10. Glass tubes with Teflon-lined caps.
11. Vortex mixer.
12. Heating block or water bath.
13. Nitrogen evaporator.

14. Analytical balance.
15. LC-MS/MS system equipped with a reversed-phase C18 column, APCI+ion source, triple quadrupole mass analyzer.
16. Mobile phases.

3 Methods

3.1 Extraction

3.1.1 Liquid–Liquid Extraction (LLE) [11, 12] (see Note 1)

1. Add three volumes of ethyl acetate to the sample; vortex vigorously.
2. Allow layer separation and collect the organic phase.
3. Repeat the extraction three times.
4. Extract residue with three volumes of water-saturated 1-butanol.
5. Combine organic layers and evaporate solvent under reduced pressure.
6. Reconstitute dried residue in 500 μL of 5% methanol.
7. Filter through a 0.45 μm syringe filter.

3.1.2 Solid Phase Extraction (SPE) [13, 14]

1. Condition the Oasis HLB cartridge with distilled water.
2. Load diluted sample containing internal standard.
3. Wash column with 2 mL 0.1% formic acid to remove sugars and polar impurities.
4. Elute steroid fraction with:
   - 2 mL 80% methanol + 0.1% formic acid.
   - 2 mL 100% methanol (to remove remaining bound compounds).
5. Evaporate solvent and re-dissolve residue in 500 μL 5% methanol.
6. Filter with a 0.45 μm filter.

3.1.3 Saponification-Based Extraction [15, 16] (see Note 2)

1. Weigh 1 g of powdered plant sample into a reflux tube.
2. Add 40 mL ethanol + 10 mL of 500 g/L KOH.
3. Add 50 μg internal standard (e.g., β-cholestanol per sample).
4. Reflux at 80°C for 2 h with constant stirring (*see* **Note 3**).
5. Transfer mixture to a separatory funnel.
6. Wash the reflux tube with 5 × 2 mL water and pool into the funnel.

7. Extract unsaponifiable fraction:
   - First extraction: 20 × 2 mL diethyl ether.
   - Second extraction: 20 × 2 mL n-hexane.
8. Combine organic layers.
9. Wash the organic phase three times with 20 mL of 5% NaCl solution (*see* **Note 4**).
10. Continue washing with Milli-Q water until the phenolphthalein indicator remains colorless.
11. Dry collected organic layers using anhydrous sodium sulfate.
12. Evaporate to dryness using a rotary evaporator at 45°C.
13. Reconstitute dried extract in 10 mL chloroform and filter (0.48 μm syringe filter).

3.2 Qualitative Analysis

3.2.1 LC-MS [17]

1. Prepare the LC-MS system by connecting the Surveyor HPLC to the Thermo LCQ Advantage Max ion-trap mass spectrometer and ensuring the ESI source is calibrated.
2. Install the Symmetry C18 column (150 × 2.1 mm, 3.5 μm) and equilibrate it with initial mobile phase conditions.
3. Prepare mobile phase A using water with 0.1% formic acid and mobile phase B using acetonitrile with 0.1% formic acid.
4. Set the HPLC flow rate to 0.4 mL/min and program the linear gradient from 2.5% B to 60% B over 25 min.
5. Inject 1 μL of each sample into the LC system and allow chromatographic separation to occur under the programmed gradient.
6. Operate the mass spectrometer in negative ESI mode and set the spray voltage to 3.9 kV, the capillary voltage to −47 V, and the tube lens to −60 V.
7. Adjust the sheath gas flow to 65 arbitrary units and the auxiliary gas flow to 10 arbitrary units.
8. Acquire full-scan MS data over an m/z range of 150–2000 throughout the chromatographic run.
9. Enable data-dependent MS/MS so that ions exceeding 4×10^5intensity automatically undergo CID fragmentation for structural analysis.
10. Record all chromatograms and mass spectra, process the data, and identify the steroid compounds present in the samples.

3.2.2 UPLC-qTOF [18]

1. Equilibrate BEH C18 UPLC column (100 × 2.1 mm, 1.7 μ m) as per manufacturer instructions.
2. Set mobile phases: A = 5% ACN in water + 0.1% FA; B = ACN + 0.1% FA.
3. Use gradient: start 100% A → to 72% A at 22 min → to 0% A by 23 min → re-equilibrate.
4. Set flow rate to 0.3 mL/min, column temperature to 35°C, autosampler to 12°C; inject 4–10 μL.
5. Run qTOF in ESI (+) and/or ESI (−) mode, scan m/z 50–1500, source at 125°C, desolvation at 275°C, capillary at 3.4 kV.
6. Acquire full-scan MS and MS/MS (ramped collision energies) to obtain molecular ions and fragmentation patterns.
7. Use generated accurate masses and MS/MS spectra to identify sterols, BRs, and saponins via databases and standards.

3.3 Quantitative Analysis

3.3.1 Spectrophotometric (Liebermann–Burchard) [19]

1. Prepare β-sitosterol (or chosen steroid) stock at 1 mg/mL in chloroform.
2. Prepare working standards (e.g., 0.1, 0.25, 0.5, 0.75, 1.0 mg/mL) by serial dilution.
3. Keep standards in amber vials at −20°C; prepare fresh working dilutions before assay.
4. Pipette aliquots of standards and sample SS into 10 mL volumetric flasks or glass tubes. (Aliquot size chosen so final concentration falls within calibration range— e.g., 1 mL of standard + reagent).
5. Add 2 mL LB reagent to each flask/tube (use identical reagent volume for standards and samples).
6. Mix gently (*see* **Note 5**).
7. Allow reaction to develop at room temperature (20–25°C) for 5 min.
8. Make up to the final volume (if using a volumetric flask) with chloroform and mix.
9. Measure absorbance at 625 nm against a chloroform blank. Record absorbance at 5 min and optionally monitor up to 20 min to ensure stability.
10. Run each sample and standard in triplicate for precision.
11. Construct calibration curve: Absorbance (A) vs. Concentration (mg/mL) for standards. Fit linear regression: $A = mC + b$.
12. Determine sample concentration from regression:

$$C_{sample} = \frac{A_{sample} - b}{m}$$

13. Convert to content per mass of original plant:

$$\text{Steroid}\left(\text{mg/g dry weight}\right) = \frac{C_{sample} \times V_{SS}}{m_{sample}}$$

3.3.2 UPLC-MS/MS [11, 20]

1. Prepare the UPLC system by connecting it to the TQD triple quadrupole mass detector and equilibrate the Acquity HSS C18 column (100 × 1 mm, 1.8 μm) with the initial mobile phase conditions.
2. Prepare mobile phases: solvent A (water containing 0.1% formic acid) and solvent B (acetonitrile containing 0.1% formic acid).
3. Set the chromatographic gradient to run from 15% to 65% solvent B over 17 min at a flow rate of 0.1 mL/min.
4. Set the injection volume to 1 μL and operate the autosampler in partial-loop overfill mode.
5. Turn on the mass spectrometer and set it to negative electrospray ionization (ESI−) mode.
6. Configure MS/MS parameters: capillary voltage 3.0 kV, cone voltage 35 V, source temperature 140°C, desolvation temperature 350°C, desolvation gas flow 600 L/h, and cone gas flow 50 L/h.
7. Prepare calibration standards of the target plant steroid(s) at multiple concentrations and add a fixed concentration of internal standard to each tube.
8. Prepare sample extracts by dissolving the dried sample in an appropriate solvent, filtering through a 0.22 μm membrane, and adding the same internal standard concentration used for standards.
9. Inject standards and samples into the UPLC system under identical conditions.
10. Acquire chromatograms and measure the peak areas of the analyte and internal standard in MRM mode.
11. Plot a calibration curve using peak area ratio (analyte/internal standard) against the known concentrations of the standards.
12. Determine the slope of the calibration curve and use it for quantification of the analytes in unknown samples.

13. Calculate the concentration of the steroid in the sample using the formula:

$$C_{\text{sample}} = \frac{A_{\text{sample}}/A_{\text{IS}}}{\text{Slope}} \times \text{Dilution factor}$$

14. Convert the calculated value into desired units, such as μg/mg dry weight of plant material or μg/mL extract.
15. Repeat all injections in triplicate to ensure accuracy and report results as mean ± standard deviation.

3.3.3 GC-MS/GC-FID [17, 21, 22]

1. Collect fresh plant tissues such as leaves, callus, or suspension-cultured cells and record their weight before processing.
2. Dry the collected samples in a hot air oven at 50°C for 12–16 h until constant weight is obtained.
3. Transfer the dried biomass into extraction tubes and add 95% ethanol at a ratio of 10 mL solvent per gram of sample.
4. Heat the mixture at 70°C for 1 h to extract total steroids, including phytosterols, diterpenes, and other neutral sterols.
5. Cool the extract and filter it through Whatman No.1 filter paper to remove particulate matter, collecting the clear filtrate.
6. Evaporate the solvent under reduced pressure or a gentle nitrogen stream until dryness and re-dissolve the residue in aqueous ethanol (50% v/v) to a final volume of 2 mL (*see* **Note 6**).
7. Add 10% NaOH solution to the extract to obtain a final concentration of 1% NaOH (v/v) to saponify esterified steroids and facilitate the release of free sterols.
8. Heat the alkaline mixture at 70°C for 30 min to complete saponification of bound phytosterols and related plant sterols.
9. Cool the saponified mixture and extract lipophilic steroid components using n-hexane fractions and evaporate them to dryness under nitrogen or low-temperature vaccum evaporation.
10. Reconstitute the dried extract in a small volume of n-hexane or isooctane suitable for GC analysis and transfer to GC vials.
11. Inject 1.0 μL of the prepared sample into a GC-FID system equipped with a nonpolar capillary such as HP-1 (30 m × 0.32 mm × 0.25 μm), ensuring the injector temperature is set to 280°C, detector temperature to 280–300°C, carrier gas helium at 1.0 mL/min, and split ratio 10:1.

12. Perform chromatographic separation using a temperature program optimized for plant steroids, typically starting at 220°C (hold 3 min) and ramping at 10°C/min to 280–300°C, holding at the final temperature for 10–15 min to elute high molecular weight sterols.
13. Quantify plant sterols using calibration curves constructed from standards prepared as stock solutions in ethanol or isopropanol.
14. Prepare calibration standards using serial dilutions across a wide concentration range suitable for plant steroid quantification.
15. Inject calibration standards under identical GC conditions and record peak areas to generate individual calibration curves for each steroid using linear regression.
16. Quantify sample steroids by comparing their GC-FID peak areas to calibration curves and express results as μg or mg/g dry weight of plant material.

3.3.4 LC-MS/MS [23] (see Note 7)

1. Weigh the sample (e.g., 20 mg of plant tissue) into a clean glass tube and record the mass (*see* **Note 8**).
2. Add internal standard solution (e.g., 10 μl of [2,2,3,4,4,6-d6]-cholesterol at 250 μg/mL in isopropanol).
3. Add 2 mL of 2 M ethanolic KOH to the tube.
4. Cap the tube and incubate at 80°C for 60 min to saponify sterol esters.
5. Remove from heat and allow to cool to room temperature.
6. Add 2 mL deionized water to the cooled mixture.
7. Add 3 mL n-hexane, vortex 1 min to partition unsaponifiable into the organic phase.
8. Centrifuge at 3000 × *g* for 5 min at 4°C and collect the upper hexane layer into a clean vial.
9. Repeat the hexane extraction two more times with fresh 3 mL portions and combine the hexane extracts.
10. Evaporate solvent to dryness under reduced pressure or a gentle nitrogen stream at ≤40–45°C.
11. Reconstitute the dried residue in isopropanol.
12. Filter the reconstituted solution through a 0.22 μm PTFE or PVDF syringe filter into an amber autosampler vial (*see* **Note 9**).
13. Install a reversed-phase C18 analytical column on the LC system.
14. Set mobile phase to acetonitrile:methanol (99:1, v/v) and equilibrate the system; set flow rate to 600 μL/min.

15. Set column temperature (30–40°C) and autosampler temperature as appropriate.
16. Configure the mass spectrometer for positive-ion atmospheric pressure chemical ionization (APCI+). Set source temperature to ~ 75 V. (*see* **Note 10**).
17. Program the MS to acquire selective reaction monitoring (SRM/MRM) transitions for each target analyte and for the internal standard (*see* **Note 11**).
18. Inject 3 μL of each sample into the LC-MS/MS system and acquire data in SRM mode.
19. Prepare 1.000 mg/mL stock solutions of each reference sterol standard in isopropanol and store at −20°C.
20. Prepare serial working standards by diluting stock solutions to obtain the range: 0.001, 0.010, 0.025, 0.250, 0.500, 1.000, 5.000, 10.000, 25.000, and 50.000 μg/mL.
21. Prepare calibration standards by mixing 200 μL aliquots of each working standard with 10 mL internal standard working solution (i.e., 250 g/mL), add 2 mL 2 M ethanolic KOH, and process these calibration aliquots through the same saponification, extraction, evaporation, and reconstitution steps as the samples (*see* **Note 12**).
22. Inject processed calibration standards and construct calibration curves plotting analyte/internal standard peak area ratios versus nominal concentrations; apply appropriate regression and weighting.
23. Determine limits of detection (LOD) and lower limit of quantification (LLOQ) from signal-to-noise ratios and from calibration data.
24. Quantify analyte concentrations in samples using the calibration curve and express results in mg steroid per kg sample.

4 Notes

1. Useful for medium-to-non-polar steroids such as phytosterols and sapogenins.
2. Saponification is essential to break esterified lipid complexes and release total free steroids, e.g., sterols, stanols, brassinosteroids, and sapogenins.
3. Do not exceed 80°C, as higher temperatures cause brassinosteroid oxidation and breakdown of thermolabile sterols.
4. Incomplete neutralization or washing causes alkali carryover, which destroys GC columns and suppresses MS signals.

5. Avoid vigorous vortexing after adding LB reagent, as high mixing intensity can cause rapid oxidation and unstable color development.
6. GC-MS requires complete removal of the solvent before derivatization, because ethanol, methanol, and water interfere with silylation and broaden peaks.
7. This method is optimized for heat-labile or polar plant steroids that cannot undergo high-temperature or derivatization-based analysis.
8. Use glass tubes, as plastic tubes may leach contaminants when treated with ethanolic KOH, leading to suppression in GC-MS and LC-MS signals.
9. Use PTFE/PVDF filters only, as nylon filters may retain non-polar sterols and saponins due to hydrophobic interactions.
10. Brassinosteroids ionize better in negative ESI, while phytosterols ionize better in APCI. Select the correct mode.
11. MRM transitions must be optimized for each class of steroids because collision energy varies significantly between phytosterols, brassinosteroids, and saponins.
12. Always prepare calibration standards in the same solvent and processed through the same extraction steps to correct matrix effects.

References

1. Tarkowská D, Strnad M (2018) Isoprenoid-derived plant signaling molecules: biosynthesis and biological importance. Planta 247:1051–1066
2. Kumari S, Priya P, Misra G et al (2013) Structural and biochemical perspectives in plant isoprenoid biosynthesis. Phytochem Rev 12:255–291
3. Du Y, Fu X, Chu Y, et al (2022) Biosynthesis and the roles of plant sterols in development and stress responses. Int J Mol Sci 23:2332.
4. Heftmann E (1975) Functions of steroids in plants. Phytochemistry 14:891–901.
5. Kreis W, Mller-Uri F (2010) Biochemistry of sterols, cardiac glycosides, brassinosteroids, phytoecdysteroids and steroid saponins. In: Wink M (ed) Biochemistry of plant secondary metabolism. Wiley-Blackwell, Oxford, pp 304–363
6. Tarkowská D, Strnad M (2016) Plant ecdysteroids: plant sterols with intriguing distributions, biological effects and relations to plant hormones. Planta 244:545–555
7. Tiwari P, Pandey R, Singh R, et al (2020) Role of natural products as alternative of synthetic steroidal drugs. In: Patra JK, Shukla AC, Das G (eds) Advances in pharmaceutical biotechnology. Springer Singapore, pp 77–89
8. Alamgir ANM (2018) Methods of qualitative and quantitative analysis of plant constituents. In: Therapeutic use of medicinal plants and their extracts: volume 2. Springer International Publishing, Cham, pp 721–804
9. Gracelin DHS, Britto A, Kumar B (2013) Qualitative and quantitative analysis of phytochemicals in five Pteris species. Int J Pharm Pharm Sci 5:105–107.
10. Abidi SL (2001) Chromatographic analysis of plant sterols in foods and vegetable oils. J Chromatogr A 935:173–201.
11. Kowalczyk M, Pecio Ł, Stochmal A et al (2011) Qualitative and quantitative analysis of steroidal saponins in crude extract and bark powder of *Yucca schidigera* Roezl. J Agric Food Chem 59:8058–8064
12. Bligh EG, Dyer WJ (1959) A rapid method of total lipid extraction and purification. Can J Biochem Physiol 37:911–917
13. Quitério E, Grosso C, Ferraz R, et al (2022) A critical comparison of the advanced extraction

techniques applied to obtain health-promoting compounds from seaweeds. Mar Drugs 20:677.
14. Jenner AM and Brown SHJ (2017) Sterol analysis by quantitative mass spectrometry. In Gelissen IC, Brown AJ (eds) Cholesterol homeostasis. Springer New York, NY, pp 221–239
15. Lagarda MJ, García-Llatas G, Farré R (2006) Analysis of phytosterols in foods. J Pharm Biomed Anal 41:1486–1496.
16. Ms U, Ferdosh S, Haque Akanda MdJ et al (2018) Techniques for the extraction of phytosterols and their benefits in human health: a review. Sep Sci Technol 53:2206–2223
17. Makin HLJ, Honour JW, Shackleton CHL et al (2010) General methods for the extraction, purification, and measurement of steroids by chromatography and mass spectrometry. In: Makin HLJ, Gower DB (eds) Steroid analysis. Springer Netherlands, Dordrecht, pp 163–282
18. Heinig U, Aharoni A (2014) Analysis of steroidal alkaloids and saponins in Solanaceae plant extracts using UPLC-qTOF mass spectrometry. In Rodríguez-Concepción M (ed) Plant isoprenoids. Springer, New York, NY, pp 171–185
19. Kanwar MK, Bhardwaj R, Arora P, et al (2012) Plant steroid hormones produced under Ni stress are involved in the regulation of metal uptake and oxidative stress in Brassica juncea L. Chemosphere 86:41–49.
20. Kanwar MK, Bajguz A, Zhou J et al (2017) Analysis of brassinosteroids in plants. J Plant Growth Regul 36:1002–1030
21. Nes WD (2012) Analysis of sterols and other biologically significant steroids. Elsevier
22. Clement LM, Hansen SL, Costin CD et al (2010) Quantitation of sterols and steryl esters in fortified foods and beverages by GC/FID. J Am Oil Chem Soc 87:973–980
23. Mo S, Dong L, Hurst WJ et al (2013) Quantitative analysis of phytosterols in edible oils using APCI liquid chromatography–tandem mass spectrometry. Lipids 48:949–956

Chapter 8

Qualitative and Quantitative Protocols for Flavonoids

Diana Pinto, Filipa Teixeira, Mariana Marques and Francisca Rodrigues

Abstract

Flavonoids, a diverse class of phenolic compounds, play crucial roles in plant physiology and offer numerous health benefits that have been highlighted since ancient times. Recognized for their powerful antioxidant, anti-inflammatory, antimicrobial, and anticancer properties, flavonoids have gained attention in nutraceutical, pharmaceutical, and functional food industries. Therefore, reliable analytical methods are essential for their identification and quantification in biological and plant matrices. This chapter intends to overview the analytical methodologies available to identify and quantify this class of phenolic compounds, addressing challenges and providing a practical resource for researchers harnessing flavonoids' therapeutic and industrial potential. Based on that, extraction is a crucial step, with advanced extraction techniques, including solid-phase and ultrasonic-assisted extraction, providing enhanced yields and stability. Regarding qualitative and quantitative assessments, chromatographic and spectroscopic methods, such as TLC, LC-MS, UV–vis, NMR, and FTIR, are essential tools to be employed. Nevertheless, despite their potential, challenges remain due to structural complexity and matrix variability, being of huge importance the standardization of protocols to ensure reproducibility and regulatory compliance.

Keywords Extraction methods, Chromatography, Spectroscopy, LC-MS, UV–vis

1 Introduction

Flavonoids, a broad class of phenolic compounds, are widely distributed in plants, supporting plant physiology and providing human health benefits [1]. These natural compounds are well-recognized for their bioactive properties, including antioxidant, anti-inflammatory, antimicrobial, and anticancer effects [1, 2]. Given the pro-healthy potential, researchers have driven innovation in flavonoids for nutraceuticals, functional foods, and pharmaceutical applications [2]. Therefore, reliable qualitative and quantitative methods are pivotal in the identification and quantification of flavonoids in plants and biological matrices, encouraging their valorization for different industrial purposes [3, 4]. These secondary metabolites are categorized into various subclasses, including flavanols, flavanones, flavones, flavonols, isoflavones, and anthocyanins, exhibiting unique

Dhruti Amin et al. (eds.), *Phytoconstituent Analysis*, Springer Protocols Handbooks,
https://doi.org/10.1007/978-1-0716-5400-2_8,

structure and biological properties [1]. Beyond defense mechanisms in plants, flavonoids have been linked to several human health benefits, modulating cellular signaling pathways and combating oxidative stress, therefore reducing the risk of cardiovascular diseases, cancer, and diabetes [1, 5].

The growing recognition of flavonoids as therapeutic agents increased the demand for robust analytical techniques to identify and quantify these bioactive compounds. Nevertheless, their structural complexity and concentration variability within natural matrices pose challenges that must be addressed by the implementation of advanced analytical techniques [6]. The qualitative and quantitative analysis of flavonoids encompasses several steps, initiating by the sample preparation and compounds' extraction, followed by the application of specific detection and quantification methods [6, 7]. Sample preparation is a critical step in flavonoids' detection, involving extraction processes and influencing the accuracy of the methods [7]. The selection of solvent and the optimization of pH and temperature are crucial to prevent degradation, maximize yield, and enhance flavonoids solubility and stability [6, 7]. Advanced techniques, such as solid-phase extraction and ultrasonic-assisted extraction, are increasingly used to improve efficiency and reproducibility [8, 9]. Noteworthy, identification and quantification protocols need to guarantee accuracy, reproducibility, and sensitivity, while being versatile and easily applicable to a broad variety of samples [6]. The selection of a suitable method relies on the flavonoids' nature, matrix composition, and study objectives [9].

The qualitative detection of flavonoids is performed by preliminary assays to identify their presence in plant-derived extracts and biological samples [7, 10]. Colorimetric chemical tests, such as Shinoda, ferric chloride, lead acetate, and sodium hydroxide assays, are simple and cost-effective methods to identify flavonoids [11]. Although these methods provide rapid screening, they are complemented by chromatographic techniques, such as thin-layer chromatography (TLC), for separating and tentatively identify flavonoids based on their elution patterns and interaction with specific reagents [12]. Regarding quantification methods, advanced spectrophotometric and chromatographic methods are employed [3, 4, 6]. Aluminum chloride assay is a widely used spectrophotometric technique based on the formation of complexes between aluminum ions and hydroxyl groups from flavonoids, allowing the determination of total flavonoid content [3, 13]. Liquid chromatography (LC) combined with ultraviolet (UV), diode array (DAD), and mass spectrometry (MS) detectors provides high sensitivity and specificity for the identification and quantification of individual flavonoid compounds in complex matrices [4, 6]. Additionally, spectroscopic techniques, including UV–visible spectroscopy, nuclear magnetic resonance (NMR), and Fourier-transform infrared (FTIR) spectroscopy are powerful tools to explore flavonoids structural and functional properties, complementing chromatographic methods [6, 14]. Capillary electrophoresis, a less common technique, also represents a valuable alternative for the separation and quantification of flavonoids, especially in highly polar matrices [15].

The validation of standardized protocols for flavonoid detection is essential to guarantee reproducibility and comparability across studies, enabling the identification of biomarkers and the assessment of bioactivity and bioavailability [1]. Considering their application in the development of nutraceuticals and functional foods, these methods should comply with regulatory and quality control guidelines [6]. This chapter aims to explore the different available qualitative and quantitative protocols for flavonoid analysis, providing a practical guide for different methodologies. Special emphasis will be placed on the analytical challenges, providing a valuable resource for researchers seeking to harness the potential of flavonoids for health and industrial applications.

Sample preparation is a critical step in the accurate identification and quantification of flavonoids. These phytochemical compounds are often present at low concentrations in complex matrices, including plant-based materials, foods, and biological samples [16, 17]. Proper sample preparation procedures are essential for the effective extraction, preservation, and isolation of flavonoids, minimizing degradation and matrix interferences. These steps are crucial for obtaining reliable analytical results [18].

The primary goal of sample preparation is the successful detection and analysis of flavonoids [6]. This procedure concentrates the target compounds, eliminates undesirable matrix components, and improves sensitivity and specificity in further analyses, producing consistent results [6, 7]. Therefore, the accurate analysis of flavonoids requires careful sample preparation, considering their chemical complexity and challenges posed by natural matrices [7].

Flavonoids exhibit significant chemical diversity, being identified as aglycones, glycosides, and polymeric forms, with distinct solubility and stability [18]. Considering the challenges of sample preparation, the complexity of sample matrices, which may contain interfering compounds (e.g., carbohydrates, lipids, and proteins), hinders the extraction process [8]. Furthermore, flavonoids are highly sensitive to environmental factors, being susceptible to hydrolysis, oxidation, and photodegradation [4, 8]. Another significant challenge focuses on ensuring high recovery rates without affecting their chemical structure [1, 7]. Some key considerations in sample preparation include:

a) **Sample collection and storage:** adequate collection and storage practices are crucial to retain the integrity of flavonoids [8]. For instance, samples should be processed quickly to prevent from degradation, and further stored under suitable conditions (e.g., freezing or in inert atmospheres) to minimize the exposure to oxygen, light, and heat [4, 6]. Freeze-drying or storage at a low temperature (−80°C) is usually recommended to preserve flavonoids' stability [19].

b) **Extraction methods:** flavonoids) [20]. Different extraction techniques may be employed depending on the sample type and compounds' structure, such as:

- **Conventional solid-liquid extraction**: commonly used solvents include water and organic solvents (e.g., acetone, ethanol, and methanol), which can be acidified to improve the recovery of flavonoid glycosides [13, 21].
- **Ultrasound-assisted extraction (UAE)** improves extraction efficiency and reduces the solvent use and extraction time [19, 22].
- **Microwave-assisted extraction (MAE)** uses microwaves to interact with polar molecules, producing heat that disrupts the cell walls and enhances the solubility and diffusion of the target compounds, providing rapid and high-yield extractions, while avoiding the thermal degradation through optimization [21, 23].
- **Supercritical fluid extraction (SFE)** fluids, primarily carbon dioxide (CO_2), which simultaneously exhibit properties of both liquids and gases, such as high diffusivity and low viscosity. These characteristics enable efficient solvent penetration into the sample matrix, while selectively dissolve the target compounds [24].
- **Subcritical Water Extraction (SWE)**: this eco-friendly technique utilizes water at elevated temperatures (100–374°C) and pressures below the supercritical conditions (normally 10–100 bar) to maintain its liquid state. Under these conditions, the water dielectric constant (usually 80 ε) significantly decreases to approximately 30 ε, exhibiting similar behavior to organic solvents (e.g., methanol), arising as efficient and economically attractive for industrial processes [20, 25].

c) **Purification and fractionation**: After extraction, flavonoids may require purification to remove non-target compounds using additional procedures (e.g., liquid-liquid extraction, solid-phase extraction, and column chromatography) and recover flavonoids based on their molecular weight, polarity, or possible interactions [6, 7, 10].

d) **Preventing degradation**: flavonoids) are prone to degradation during extraction, antioxidants (e.g., ascorbic acid) can be added to extraction solvents to stabilize these compounds. Moreover, reduced light and low temperature can be applied to minimize the risks of oxidation and thermal degradation [8, 23].

e) **Concentration**: as flavonoids are often present in low amounts, a concentration step is required to guarantee the successful detection using analytical methods. Some procedures, such as rotary evaporation, lyophilization, or drying under nitrogen gas, can be conducted to prevent excessive heat that may compromise flavonoids integrity [3, 8].

f) **Analytical compatibility**) requires filtration and degassing; spectrophotometric assays entail clear and particle-free solutions; and MS involves the reduction of salt content in the sample to avoid ion suppression [3, 4].

In general, effective sample preparation is vital for the identification and quantification of flavonoids, ensuring accuracy and reproducibility of the results and addressing various challenges, such as chemical diversity, matrix complexity, and compound degradation [6, 7, 9].

2 Materials

2.1 Qualitative Analysis of Flavonoids

2.1.1 Shinoda Test

1. Sample extract diluted in isopropyl alcohol.
2. Preparation of test sample: To 0.5 mL of the sample add 1 mL of anhydrous ethanol.
3. HCl addition and observation: Add three drops of concentrated HCl to the sample solution and observe any color changes.
4. Magnesium addition: If no color change is observed in the previous step, introduce small pieces of metallic magnesium into the tube and analyze the mixture for color changes once more (*see* **Note 15**).
5. Preparation of negative control: In a separate test tube, combine 1 mL of anhydrous ethanol, three drops of concentrated HCl, and 0.5 mL of isopropyl alcohol. Note that no color change should be registered.
6. Isopropyl alcohol (used as a negative control).
7. Test tubes

2.1.2 Ferric Chloride Test

1. Aqueous flavonoid extract solution;
2. 2% Ferric chloride ($FeCl_3$) solution: Weigh 2 g of $FeCl_3$ and dissolve in a graduated cylinder containing 100 mL of distilled water. Stir the solution using a magnetic stirrer until fully dissolved (*see* **Note 6**).

3. Preparation of negative control: In a separate test tube, combine 1 mL of distilled water with 1 mL of 2% FeCl3 solution.
4. Test tubes
5. Magnetic stirrer

2.1.3 Lead Acetate Test

1. Aqueous flavonoid extract solution;
2. 1% Lead acetate solution: Weigh 1 g of lead acetate ($Pb(C_2H_3O_2)_2$) and dissolve in 100 mL of distilled water. Stir gently until fully dissolved.
3. Preparation of test sample: Add 1 mL of 1% lead acetate solution to 1 mL of aqueous extract in a test tube.
4. Preparation of negative control: In a separate test tube, combine 1 mL of distilled water with 1 mL of 1% lead acetate solution.
5. Test tubes
6. Magnetic stirrer

2.1.4 Sodium Hydroxide Test

1. Sample extract diluted in isopropyl alcohol.
2. 10% Sodium hydroxide solution: Dissolve 10 g of NaOH in distilled water and then complete the volume to 100 mL. Mix thoroughly until fully dissolved.
3. Preparation of test sample: Add three drops of 10% sodium hydroxide solution to 1 mL of the aqueous extract solution in a test tube.
4. Preparation of negative control: In a separate test tube, mix 1 mL of isopropyl alcohol with three drops of 10% sodium hydroxide solution.
5. Test tubes.
6. Magnetic stirrer.

2.2 Thin-Layer Chromatography (TLC)

1. Sample extract (dissolved in an appropriate solvent, e.g., water, ethanol, and methanol);
2. Mobile phase: to select the adequate mobile phase for TLC consult the scientific literature for the specific flanovoid to be detected. Common mobile phases for flavonoid detection include ethyl acetate, formic acid, acetic acid, and water in a 100:11:11:27 ratio [12]. To prepare this mixture, add 100 mL of ethyl acetate, 11 mL of formic acid, 11 mL of acetic acid, and 27 mL of distilled water to a TLC chamber.

   - Standard flavonoid compounds (e.g., quercetin, rutin, kaempferol): Weigh a small amount (e.g., 1-2 mg) of the flavonoid standard and dissolve it in an appropriate solvent (e.g., ethanol), using a small amount to reach a concentration of 1–2 mg/ mL.

3. Aluminum chloride ($AlCl_3$) (for flavonoid detection): Weigh 1 g of aluminum chloride ($AlCl_3$) and dissolve it in approximately 50 mL of ethanol. Stir gently until fully dissolved and complete the volume to 100 mL with ethanol.
4. Silica gel 60 F_{254}, aluminum-backed plates.
5. Glass capillary tubes (for sample spotting).
6. Thin-layer Chromatography (TLC) chamber.
7. UV lamp.
8. Magnetic stirrer.

2.3 Spectrophotometric Assays (Aluminum Chloride Assay)

1. UV–vis spectrophotometer.
2. 15 mL glass tubes.
3. Quartz or glass cuvettes (1 cm path length).
4. Vortex mixer.
5. Analytical balance.
6. Flavonoid standards (pure flavonoid standards for calibration, such as quercetin, rutin, catechin, or kaempferol).
7. Sample extract.
8. Solvents (selected based on flavonoid solubility, e.g., distilled water, ethanol, and methanol) (*see* **Note 4**).
9. Sodium acetate (CH_3COONa).
10. Aluminum chloride ($AlCl_3$).

2.4 Liquid Chromatography (LC) Coupled with UV, DAD, and MS Detectors

Flavonoid standard stock solution: select an adequate flavonoid standard (e.g., quercetin, rutin, kaempferol, catechin) and dissolve 1 mg in 1 mL of methanol (*see* **Note 11**). Flavonoid working standard solutions: prepare serial dilutions of the standard stock solution (0.1–1000 μg/mL) for the calibration curve.

1. Sample solution preparation: extract flavonoids from the sample using an adequate technique and solvent, and filter or centrifuge to remove solid residues.
   - if the sample is liquid, filter the solution through a 0.22 or 0.45 μm PTFE/Nylon filter and dilute, if necessary (e.g., 1:10 dilution), in the mobile phase.
   - if the sample is dried, for example, by lyophilization, dilute it in the mobile phase, centrifuge at 10,000–14,000 rpm for 10 min, filter the supernatant through a 0.22 or 0.45 μ m PTFE/Nylon filter, and dilute, if necessary (e.g., 1:10 dilution), in the mobile phase.
2. High-Performance Liquid Chromatography (HPLC) or Ultra-High-Performance Liquid Chromatography (UHPLC) system.

3. Detector systems: Solvent A (aqueous phase): water and 0.1% (*v/v*) formic acid (*see* **Note 12**): add 1 mL of formic acid to 1 L of water, filter using 0.22 μm PTFE or nylon filter, and store at 4°C (use within 1 week).
   Solvent B (organic phase): acetonitrile and 0.1% formic acid (*see* **Note 13**): add 1 mL of formic acid to 1 L of acetonitrile, filter using 0.22 μm PTFE filter, and store at room temperature.
4. Internal standard solution: select a stable isotope-labeled flavonoid (e.g., quercetin-d9, kaempferol-d3, caffeine-d9) and dissolve 10 μg in 1 mL of methanol (*see* **Note 14**).
5. UV: for absorbance-based quantification, typically at 280 or 370 nm.
6. DAD: for scanning across multiple wavelengths simultaneously (e.g., 280, 360, 370 nm for flavonoids), providing more detailed spectral information.
7. MS detector: provides structural identification and quantification based on mass-to-charge ratio (*m/z*).
8. HPLC or UHPLC column (*see* **Note 1**):
   - C18 reverse-phase column 150 × 4.6 mm, 5 μm particle size.
   - C18 reverse-phase column 100 × 2.1 mm, 1.7 μm particle size.
   - Phenyl-hexyl column.
   - C8 column.
9. MS analyzer (*see* **Note 2**):
   - Quadrupole.
   - TOF (Time-of-Flight).
   - Orbitrap.
10. Binary or quaternary pump.
11. Electrospray Ionization (ESI) source: positive (+) or negative (−) mode (*see* **Note 3**).
12. Nitrogen generator (for drying gas in ESI source).
13. Vacuum pump.
14. Autosampler.
15. Column oven (25–40°C).
16. Degasser (to remove dissolved gases and prevent bubble formation).
17. Vacuum filtration system (for mobile phase filtration).
18. Ultrasonic bath (for degassing solvents and dissolving samples).

19. Analytical balance.
20. pH meter.
21. Water purification system (Milli-Q, 18.2 MΩ).
22. HPLC vials (glass, 1.5 mL, with PTFE/Silicone caps).
23. 0.22 or 0.45 μm PTFE/Nylon filters.
24. Data analysis software:
    - UV: ChemStation, Thermo Xcalibur, or Waters Empower.
    - DAD: ChemStation, Thermo Xcalibur, Waters Empower, OpenLab, or Clarity.
    - MS: MassLynx (Waters), Xcalibur (ThermoFisher), Compass (Bruker), or MetaboAnalyst.

2.5 Nuclear Magnetic Resonance (NMR) Spectroscopy

1. NMR spectrometer (high-field spectrometer operating at ≥ 400 MHz for optimal resolution and sensitivity).
2. Sample tubes (5 mm NMR tubes with high precision and chemical resistance).
3. Centrifuge (for isolating or purifying solutions, if needed).
4. Deuterated solvents (e.g., $CDCl_3$, DMSO-d_6, or D_2O, depending on flavonoid solubility).
5. Internal standard (tetramethylsilane (TMS) or 4,4-dimethyl-4-silapentane-1-sulfonic acid (DSS) used for chemical shift reference): Prepare a solution of the internal standard in the same deuterated solvent at a defined concentration. Ensure that the sample solution is clear and prepare dilutions using the same solvent employed for standards [26, 27].
6. Flavonoid standard (use pure flavonoid standards for calibration, such as quercetin or kaempferol);
7. Flavonoid stock solution: Dissolve a precisely weighed amount of the flavonoid standard selected in an adequate solvent (e.g., $CDCl_3$ or DMSO-d_6) and dilute to prepare a known concentration within NMR limit of detection.
8. Buffer solution (required for flavonoid stabilization or pH adjustments, such as phosphate buffer in D_2O).

2.6 Fourier-Transform Infrared (FTIR) Spectroscopy

1. Flavonoid standard solution (e.g., quercetin, rutin, kaempferol): Dissolve a known amount of the flavonoid standard in a suitable solvent and prepare dilutions for the calibration curve.
2. Sample solution: Extract flavonoids from the sample using an adequate technique, filter or centrifuge to remove solid residues, and dilute it to obtain a concentration within the limit of detection.

3. KBr pellet preparation: Mix a small amount of flavonoid sample with KBr powder and compress the mixture into a thin pellet using a hydraulic press [27, 28].
4. Solvents (to be selected based on flavonoid solubility, e.g., water, ethanol, and methanol).
5. Buffer solution (used to stabilize flavonoids in solution, e.g., phosphate buffer at pH 6–7).
6. FTIR spectrometer (equipped with a suitable detector, such as pyroelectric (e.g., Deuterated Triglycine Sulfate, DTGS) or semiconductor (e.g., Mercury Cadmium Telluride, MCT) detectors, for mid-infrared detection).
7. Attenuated Total Reflectance (ATR) accessory (enables direct measurement without additional sample preparation).
8. Desiccator (prevent moisture contamination).

2.7 Capillary Electrophoresis (CE)

1. Standard flavonoid solution: Dissolve a precise amount (Quercetin/rutin/kaempferol 0.1 g) of each flavonoid standard in 100 mL of an adequate solvent (e.g., methanol) and sonicate.
2. Electrolyte solution: Prepare 10 mM of borate buffer by dissolving 6.18 mg of sodium borate and 6.18 mg of boric acid in deionized water. Adjust the pH of the solution to 8.5 using 0.1 M NaOH or 0.1 M HCl solutions. Once the pH is adjusted, transfer the mixture to a volumetric flask and complete with deionized water to reach a final volume of 1 L.
3. Electrolyte solution/Running buffer (e.g., 10 mM borate buffer, pH 8.5).
4. NaOH (0.1 M) and HCl (0.1 M) (for pH adjustment).
5. Sodium borate.
6. Boric acid.
7. Deionized water.
8. Methanol or acetonitrile.
9. Capillary electrophoresis system with a detector (e.g., photodiode array (PDA) detector).
10. Uncoated fused-silica capillary (ideal dimensions: 50–75 μm internal diameter, 50–75 cm length).
11. pH meter.
12. Ultrasonic bath.
13. Injection system.

3 Methods

3.1 Qualitative Analysis of Flavonoids

3.1.1 Shinoda Test

1. Shinoda test involves the addition of concentrated hydrochloric acid and magnesium to the sample. Such an environment leads to the reduction of the keto group and forms flavonoid-metal complexes that produce an orange or red coloration [29].
2. A red coloration after adding HCl confirms the presence of chalcones and aurones [30]. If no red color is observed initially, the addiction of magnesium and subsequent orange or red coloration confirms the presence of flavones or flavonols, respectively [30].
3. Control validation: The negative control must show no color change, validating the specificity of the assay.

3.1.2 Ferric Chloride Test

1. The ferric chloride test detects tannins, phenolic compounds that chelate metal ions. The reaction between ferric ions (Fe^{3+}) and hydroxyl groups in tannins forms colored complexes, which coloration is determined by the specific tannin present in the sample [31]. Tannins are categorized into gallic and catechol tannins. Gallic tannins, also known as gallotannins, are strong antioxidants derived from gallic acid and contain three hydroxyl groups linked to a benzene ring [32]. On the other hand, catechol tannins contain two hydroxyl groups in specific positions on a benzene ring [33]. Despite the differences in their structures, gallic and catechol tannins form colored complexes in contact with metals, resulting in distinct color changes.
2. The formation of a dark-green coloration indicates the presence of catechol tannins, while a deep-blue or black coloration suggests the presence of gallic tannins [11].
3. Control validation: The negative control must show no color change, validating the specificity of the assay.

3.1.3 Lead Acetate Test

1. Lead acetate test is mainly used to test tannins. The formation of a white precipitate in the test sample confirms the presence of tannins [11].
2. Control validation: The negative control must show no precipitation, validating the specificity of the assay.

3.1.4 Sodium Hydroxide Test

1. The Sodium Hydroxide test indicates the presence of coumarins, flavones, and anthocyanins. The change of color to yellow suggests the presence of coumarins and flavones, whilst the

development of a blue or green coloration indicates the presence of anthocyanins (*see* **Note 23**) [11].

2. Control validation: The negative control must show no precipitation, validating the specificity of the assay.

3.2 Thin-Layer Chromatography (TLC)

1. Preparation of test sample: Dissolve in methanol or ethanol to get a 1–2 mg/mL solution.
2. Preparation of the TLC plate: Heat the TLC plate at 100°C for 10 min and let it cool down to room temperature (*see* **Note 16**). Mark a line with a pencil 1 cm from the edge of the plate (baseline).
3. Application of samples to the TLC plate: With a glass capillary tube, spot the samples into the baseline (*see* **Note 17**).
4. Chromatogram: To develop the chromatogram, insert the TLCTLC plate vertically into the chamber containing the selected mobile phase (*see* **Note 18**). Allow the solvent to migrate until it reaches approximately 1 cm from the top edge. With a pencil, mark the solvent front.
5. Application of 1% ethanolic $AlCl_3$ solution: Let the TLC plates air dry and apply 1% ethanolic $AlCl_3$ solution. Visualize the plate using a UV light, at 365 nm [34].
6. Chromatogram visualization: Mark the spots that appear under the UV light.
7. Measure the distance traveled by each of the spotted compounds. The observation of the plate at 365 nm ensures that the detected compounds are flavonoids [34].
8. Calculate the Retention Factor (R_f): R_f is the ratio of the distance migrated to the total distance traveled by the solvent. The Rf value for each sample is calculated using the following Eq. 1:

$$R_f = \frac{\text{distance traveled by the sample}}{\text{distance traveled by the solvent}} \quad (1)$$

9. R_f value interpretation: Compare the R_f values of the compounds in the sample to those of the standards used. The closer the R_f values are, the higher the likelihood that the compounds in the sample are either identical to or chemically similar to the flavonoid standard used.

3.3 Spectrophotometric Assays (Aluminum Chloride Assay)

1. In a 15 mL glass tube, mix [35]:
   a. 0.50 mL of the flavonoid standard or sample or blank solution;
   b. 1.5 mL of the selected suitable solvent;
   c. 0.10 mL CH_3COONa 1 M;

   d. 0.10 mL $AlCl_3$ (10% *w/v*);
   e. 1.8 mL of the selected suitable solvent for a final volume of 5.0 mL.

2. Homogenize the mixture and incubated at room temperature for 30 min;
3. Add the mixture to the cuvette according to the volume stipulated for it (2.5–4 mL);
4. Analyze each mixture using the spectrophotometer UV–vis at 415–440 nm to obtain the absorbance values (*see* **Note 19**).
5. Calibration curve: the absorbance obtained for each flavonoid working standard solution concentration was plotted to create a calibration curve (absorbance *vs* concentration (μg/mL)) corresponding to the following Eq. 2:

$$y = mx + c \tag{2}$$

   where y is the absorbance, x is the flavonoid working standard solution concentration (μg/mL), m is the slope of the calibration curve, and c is the intercept (ideally near zero); Ensure good linearity ($R^2 > 0.99$) for accurate quantification.
6. Quantification: the total flavonoid content (TFC) of each sample can be calculated by inputting the absorbance value obtained into the calibration Eq. 3:

$$x = (y - c)/m \tag{3}$$

   a. where x is the sample flavonoid concentration (μg/mL);
   b. Results are expressed as micrograms of standard flavonoid equivalent per gram of dried extract, e.g., if using quercetin as standard, the results should be expressed as micrograms of quercetin equivalents (QE) per gram of dried extract.

7. Calculating error: Include replicates to evaluate the reproducibility and calculate the standard deviations.

3.4 Liquid Chromatography (LC) Coupled with UV, DAD, and MS Detectors

LC system setup [36]:

1. Set up the HPLCHPLC or UHPLC system with the UV, DAD, and MS detector and the appropriate chromatography column.
2. Column oven: 30–40°C.
3. Flow rate: 0.3–0.5 mL/min (*see* **Note 20**).
4. Injection volume: 10–20 μL.
5. Ionization mode: ESI (−) (*see* **Note 3**).

6. Mass scan range: *m/z* 100–1000.
7. MS detection mode: full scan (for mass determination) or MS/MS fragmentation (for structural confirmation).
8. UV and DAD detection wavelengths: 280, 360, 370 nm (select based on flavonoid absorption maxima).
9. Mobile phase preparation:
   Solvent A: water and 0.1% (*v/v*) formic acid.
   Solvent B: acetonitrile or methanol and 0.1% (*v/v*) formic acid.
10. Gradient elution program:
    - 0–5 min: 90% Solvent A and 10% Solvent B.
    - 5–15 min: 70% Solvent A and 30% Solvent B.
    - 15–25 min: 50% Solvent A and 50% Solvent B.
    - 25–30 min: 10% Solvent A and 90% Solvent B.
    - 30–35 min: 90% Solvent A and 10% Solvent B.
11. The instrument generates raw data files containing retention time (RT), mass-to-charge ratio (*m/z*), absorbance intensity, and peak height or areas under the curve (AUC).

Peak Detection and Identification with UV detector

12. Peak detection: identify peaks in chromatograms corresponding to flavonoids based on RT and UV absorbance using LC-UV software (e.g., Chromeleon, Empower, OpenLab, Clarity). Extract total chromatogram (UV chromatogram): the full chromatogram shows all compounds detected at the selected wavelength;
13. Extract specific wavelength chromatograms (EWC): extract chromatograms at relevant flavonoid wavelengths (e.g., 280, 360 nm);
14. Adjust baseline and set peak detection thresholds to avoid noise and false positives.
15. Peak identification: compare peaks with known standards based on RT and UV absorbance spectra to confirm which flavonoids are present in the sample: Retention time matching: compare sample RT with pure flavonoid standards;
16. UV absorption spectra: compare with literature values or spectral libraries (e.g., quercetin max at 370 nm, catechin at 280 nm);
17. Co-elution tests: spiking the sample with a known standard should increase the peak area at the same RT.

Peak Detection and Identification with DAD detector

18. Peak detection: identify peaks in chromatograms corresponding to flavonoids based on RT and UV absorbance using LC-DAD software (e.g., Chromeleon, Empower, OpenLab, Clarity):

19. Extract total chromatogram (DAD chromatogram): displays all detected compounds across the full UV–vis spectrum;
    a. Extract specific wavelength chromatograms: 280 nm (general flavonoids, catechins), 360 nm (flavonols, flavones), and 370 nm (quercetin derivatives);
    b. Define baseline and peak integration to minimize noise and false peaks.
20. Peak identification: compare peaks with known standards based on RT and UV absorbance spectra to confirm which flavonoids are present in the sample:
    a. Retention time matching: compare sample RT with pure flavonoid standards;
    b. UV absorption spectra matching: compare with literature values or spectral libraries (e.g., quercetin max at 370 nm, catechin at 280 nm);
    c. Co-elution tests: spiking the sample with a known standard should increase the peak area at the same RT.

Peak Detection and Identification with MS detector

21. Peak detection: identify peaks in chromatograms corresponding to flavonoids based on RT and *m/z*)-MS software (e.g., Xcalibur, MassLynx, MZmine, or Skyline):
22. Extract total ion chromatogram (TIC): shows all detected ions;
23. Extract extracted ion chromatograms (EIC/XIC): filters data for specific *m/z* values of flavonoids;
24. Define baseline, noise levels, and peak thresholds to ensure correct peak detection.
25. Peak identification: compare detected *m/z* values with theoretical or reference databases to confirm which flavonoids are present in the sample:
26. Exact mass matching: match measured *m/z* to expected molecular weights;
27. Isotope pattern analysis: confirms molecular composition;
28. MS/MS fragmentation patterns: compare with spectral libraries (e.g., METLIN, MassBank, mzCloud);
    a. Retention time matching: if available, compare with authentic flavonoid standards.

Samples Quantification

29. Internal standard correction: to improve accuracy (reduce variability and improve quantification precision), use an internal standard (e.g., an isotope-labeled flavonoid or structurally similar compound) to normalize peak intensities by calculating the response ratio (Eq. 4):

$$\text{Normalized AUC} = \frac{\text{Flavonoid Peak AUC}}{\text{Internal Standard AUC}} \tag{4}$$

30. Calibration curve: record peak AUC of flavonoid working standard solutions and plot peak AUC *versus* concentration (μ g/mL) to generate a linear regression Eq. 5:

$$y = mx + c \tag{5}$$

where y is the peak AUC, x is the flavonoid working standard solution concentration (μg/mL), m is the slope of the calibration curve, and c is the intercept (ideally near zero); Ensure good linearity ($R^2 > 0.99$) for accurate quantification.

31. Calculate Limit of Detection (LOD) and Limit of Quantification (LOQ): use signal-to-noise ratio for this calculation [37].
32. Samples quantification: determine sample concentrations by inputting the sample's peak AUC into the linear regression Eq. 6:

$$x = (y - c)/m \tag{6}$$

a. where x is the sample flavonoid concentration (μg/mL);
b. Convert concentration into mg of flavonoid per gram of sample (mg/g) using dilution factors.
c. Calculating error: evaluate precision and reproducibility of the assay by calculating standard deviations.

3.5 Nuclear Magnetic Resonance (NMR) Spectroscopy

1. Preparation of flavonoid standard solution: Mix the flavonoid stock solution with the internal standard in a 5 mm NMR tube and ensure that the volume is enough to cover the active region of the NMR coil (usually 500–700 μL).
2. Sample solution: Pipette the sample solution into the NMR tube and seal it to prevent solvent evaporation.
3. Instrument calibration: Calibrate the spectrometer using an internal standard diluted in a deuterated solvent (e.g., TMS in $CDCl_3$) and stabilize the temperature, if needed.
4. Data acquisition: Use ^{1}H-NMR or ^{13}C-NMR spectroscopy for flavonoids quantification by adjusting acquisition parameters, namely number of scans (achieving adequate signal-to-noise ratio), spectral width (reaching a wide range of resonances), and relaxation delay (usually 1–5 s for flavonoids), and record the NMR spectrum.

5. Quality check: Verify the baseline stability and peak resolution and identify impurities and solvent peaks [26, 27].
6. Spectral analysis: Assign peaks to specific protons or carbons in the flavonoid structure based on chemical shifts, splitting patterns, and coupling constants, followed by integrating the area under the peaks, which corresponds to flavonoid resonances.
7. Quantification: Compare the integrated peak area of the flavonoid to that of the internal standard and use the following Eq. 7 for calculations:

$$C_{\text{flavonoid}} = C_{\text{standard}} \times \left(A_{\textit{flavonoid}} / A_{\text{standard}}\right) \quad (7)$$

where *C* refers to the concentration, and *A* refers to the integrated area. Normalize the results obtained considering sample volume and equipment's sensitivity, if needed.
8. Calculating error: Include replicates to evaluate reproducibility and calculate standard deviations [26, 27].

3.6 Fourier-Transform Infrared (FTIR) Spectroscopy

1. Sample measurement using ATR: Clean the ATR crystal with an adequate solvent to prevent contamination and place in it a small amount of sample, guaranteeing the contact between sample and crystal. The spectrum will be recorded above mid-infrared range (e.g., 4000–400 cm^{-1}).
2. Sample measurement using KBr pellets: Insert the prepared pellet into the FTIR sample holder and record the spectrum above mid-infrared range.
3. Background measurement: A background spectrum prepared from air or clean ATR crystal should be collected before sample measurement.
4. Spectral acquisition: For acquiring spectra, use resolution settings of at least 4 cm^{-1} and perform average multiple scans (e.g., 16–64) to enhance signal-to-noise ratio. The absorbance values should be within the linear range of the detector [27, 28].
5. Baseline correction: Apply a baseline correction to remove spectral interferences.
6. Peak identification: Identify characteristic peaks of flavonoids based on functional groups. For example, C=O stretches near 1650 cm^{-1}, while OH extends around 3200–3600 cm^{-1}.
7. Calibration curve: utilize spectra from flavonoid standard solutions to plot the calibration curve of peak area or intensity *versus* concentration.
8. Quantification: Analyze the sample spectrum to determine the intensity or area of the characteristic peaks and compare it with

the calibration curve to determine the flavonoid concentration in the sample [27, 28].

3.7 Capillary Electrophoresis (CE)

1. Capillary conditioning: flush the capillary sequentially with deionized water for 5 min, 0.1 M NaOH for 10 min, deionized water for 5 min, and finally running buffer for 5 min. This procedure should be repeated before each analysis (*see* **Note 21**).
2. Define the electrophoretic conditions: buffer 10 mM borate buffer, pH 8.5; voltage 25 kV; capillary temperature: 25°C; detection at 214 nm with PDA detector [38].
3. Sample injection: Inject each sample hydrodynamically at 0.5 psi, with an injection time of 10 s (*see* **Note 22**).
4. Obtaining the electropherogram: plot of detector response (e.g., absorbance or fluorescence) *versus* migration time (with a PDA detector, the absorbance spectrum is registered at multiple wavelengths).
5. Peak identification: Identify characteristic peaks of flavonoids based on their documented absorption spectrum and specific migration time.
6. Quantification: Measure the integrated peak area (or height) for each analyte of interest using software for automated peak integration.
7. Calibration curve: Construct a calibration curve with the data from flavonoid standards (peak area vs concentration). Normalize the results obtained considering sample volume and system sensitivity. Determine flavonoid concentrations using the equation obtained from the calibration curve.
8. Calculate Limit of Detection (LOD) and Limit of Quantification (LOQ): Use signal-to-noise ratio for this calculation [38].
9. Calculating error: Evaluate precision and reproducibility of the assay by calculating standard deviations.

4 Notes

1. The best column for LC-UV-DAD-MS depends on the specific flavonoids being quantified, their polarity, and the resolution required. Overall, the best choice for LC-UV-DAD-MS would be C18 (150 × 4.6 mm, 5 μm) for general use or C18 (100 × 2.1 mm, 1.7 μm) for high resolution with UHPLC system [36, 39]. C18 Reverse-Phase Column (150 × 4.6 mm, 5 μ m) is the most common choice for LC-UV-DAD-MS since it works well for most polar and moderately nonpolar flavonoids, being compatible with UV, DAD, and MS detection,

and even though it runs longer times, it gives good peak. Phenyl-Hexyl Column is better for complex flavonoid mixtures, as it enhances π-π interactions, improving separation of polyphenols and flavonoids (ideally used when structural isomers (e.g., quercetin *vs.* kaempferol) co-elute on C18). C18 Reverse-Phase Column (100×2.1 mm, 1.7 μm) is best when using a UHPLC system, since it requires a UHPLC system with high-pressure capabilities, showing fast analysis times with sharper peaks, high efficiency and great resolution, and a low solvent consumption. C8 Column is mostly used for nonpolar flavonoids, showing shorter retention times compared to C18 for hydrophobic flavonoids. However, it is not the best choice for most flavonoids, which tend to be more polar.

2. Triple Quadrupole is the most sensitive for targeted quantification, being more valuable for routine flavonoid quantification. Quadrupole-TOF provides mass accuracy and detects unknown flavonoids, both valuable for quantification and structural identification. As for Orbitrap, it is the best option for metabolomics and flavonoid characterization, providing ultra-precise mass measurements useful for research studies.
3. Positive ion mode (ESI (+)) detects protonated ($[M+H]^+$) or adducted ions (e.g., $[M+Na]^+$), while negative ion mode (ESI (−)) detects deprotonated ($[M-H]^-$) or anionic species. For the quantification of flavonoids, ESI (−) is preferred due to ability of flavonoids to form stable $[M-H]^-$ ions with higher sensitivity.
4. Methanol and ethanol are commonly used solvents, but some studies use water or dimethyl sulfoxide (DMSO), mostly depending on the chosen extraction technique.
5. Formic acid is better for LC-UV-DAD-MS (especially MS detection) as it improves ionization in ESI mode in LC-MS analysis (better peak shape and ionization efficiency) and shows less baseline noise using UV or DAD detectors [40] .
6. 2% $FeCl_3$ solution should be freshly prepared since it degrades over time due to the presence of Fe^{3+} ions.
7. Avoid prolonged exposure to light, as flavonoids are sensitive to degradation [5].
8. It is advisable to dry the sample to know the correct mass of compounds present in it and, thus, be able to prepare a correct concentration of 1 mg/mL [13, 20] .
9. If absorbance values exceed 1.5, dilute the sample to ensure accurate readings [13].
10. Both blank solutions aim to correct background absorbance.

11. If the selected standard flavonoid has poor solubility in methanol or acetonitrile, a MeOH:ACN (1:1, *v/v*) mixture can be used.
12. Use water and 5 mM ammonium formate as Solvent A (aqueous phase) when using ESI (−) mode.
13. Use methanol and 0.1% (*v/v*) formic acid as Solvent B (organic phase) when attempting to quantify more polar flavonoids.
14. The internal standard is a compound chemically like the analytes (flavonoids), but not naturally present in the sample. It is added in a known amount to correct for instrument variations, sample losses, and matrix effects, improving quantification accuracy.
15. Adding a small piece of metallic magnesium improves the efficiency of the reaction. Gently stirring the test tube helps to accelerate the color change.
16. By warming up the plate the absence of moisture should be ensured, since it interfere with the migration of compounds.
17. Ensure that the mobile phase does not touch the baseline to prevent the dissolution of samples into the solvent. Additionally, cover the chamber after the TLC plate is positioned to avoid the evaporation of the mixture.
18. Space the sample spots properly to ensure the tested compounds do not overlap on the TLC plate.
19. The wavelength at which the assay is carried out must be optimized if the objective is to quantify a specific flavonoid. The complexes formed between aluminum chloride and flavonoids, including kaempferol, rutin, quercetin, quercitrin, etc., have a maximum absorbance at wavelengths between 415 and 440 nm. Different flavonoids form different complexes with different maximum absorption spectra in the region of interest.
20. When using MS detectors, the flow rate needs to be lower than when using UV or DAD detectors. Lower flow rates improve ionization efficiency, prevent vacuum system overload, enhance sensitivity by reducing ion suppression, and ensure proper solvent evaporation.
21. Conditioning the capillary with water and NaOH is essential for cleaning and activating the capillary surface, effectively removing contaminants or residues. The subsequent flush with the running buffer stabilizes the capillary environment, ensuring consistent results by creating optimal conditions for flavonoid separation.
22. To ensure consistency between the results, maintain identical pressure and injection time for each analysis.

23. To confirm the presence of flavonoids, a few drops of HCl can be added to the yellow precipitate. The disappearance of the yellow color indicates the presence of flavonoids, once they suffer reversible color changes in acidic conditions. Ensure the flavonoid standards or sample is completely dissolved to avoid inaccurate quantification. The sample should be free from impurities or degradation products as these components may interfere with quantification. The solvents used should have minimal infrared absorbance in the regions of interest. Regarding solvent effects, choose a solvent that does not overlap with flavonoid peaks and measurement regions of interest.
24. In NMR, incomplete relaxation can lead to inaccurate integrations, emphasizing the importance of using previous relaxation experiments to determine the optimal delay [26].
25. Different flavonoid concentrations should be tested for the calibration curves to validate linearity [26, 27].
26. During storage, protect flavonoid-rich samples from light and air to prevent degradation [5, 27].
27. Due to the sensitivity of flavonoids to moisture, perform all procedures in a dry environment or use a desiccator [13, 26] .
28. Perform replicates for ensuring reproducibility and calculate average values [26, 27].
29. Avoid overlapping peaks from other compounds within complex mixtures [26, 27].
30. Ensures adequate ATR maintenance procedures by periodically cleaning and checking the ATR crystal for damage [27].
31. For all detection and quantification techniques, guarantee the flavonoids concentration in the sample is within the limits of detection or quantification [26, 27].

Acknowledgments

Filipa Teixeira is thankful for her Ph.D. grant (2024.01202.BD) financed by POPH-QREN and subsidized by the European Science Foundation and Ministério da Ciência, Tecnologia e Ensino Superior. Social Europeu. Francisca Rodrigues was supported by FCT - Fundação para a Ciência e a Tecnologia, I.P under the TENURE program (2023.14718.TENURE.016). This work received financial support from project 2024.16143.PEX - Revolutionizing Glioblastoma Treatment with Kiwiberry Byproducts: From Molecular Pathways to Advanced Spheroid Models (https://doi.org/10.54499/2024.16143.PEX) supported by national funds by FCT / MCTES. The authors acknowledge the support from the Aliança para a Investigação e Ensino Superior, and the funding received by FCT - Fundação para a Ciência e a Tecnologia, I.P., in the scope of the project UIDP/04378/2020 (DOI:10.54499/UIDP/04378/2020) and UIDB/04378/2020 (DOI:10.54499/UIDB/04378/2020) of the Research Unit on Applied Molecular Biosciences - UCIBIO and the project LA/P/0140/2020 (DOI:10.54499/LA/P/0140/2020) of the Associate Laboratory Institute for Health and Bioeconomy - i4HB.

References

1. Luo J et al. (2025) Latest research progress on anti-microbial effects, mechanisms of action, and product developments of dietary flavonoids: a systematic literature review. Trends Food Sci Technol 156:104839
2. Cádiz Gurrea MdlL et al (2019) Chapter 4—Effects of nutritional supplements on human health. In: Galanakis CM (ed) Nutraceuticals and natural product pharmaceuticals. Academic Press, pp 105–140
3. Yu Y et al. (2025) Quantitative predictions of protein and total flavonoids content in tartary and common buckwheat using near-infrared spectroscopy and chemometrics. Food Chem 462:141033
4. Zhu X et al. (2024) Qualitative and quantitative analysis of chemical components in Qianggan capsule by UHPLC-Q-TOF-MS/MS and LC-sMRM. J Chromatogr A 1728:465020
5. Rodrigues F et al (2018) 12—Cosmetics. In: Galanakis CM (ed) Polyphenols: properties, recovery, and applications. Woodhead Publishing, pp 393–427
6. Sammani MS, Clavijo S, Cerdà V (2021) Recent, advanced sample pretreatments and analytical methods for flavonoids determination in different samples. TrAC - Trends Anal Chem 138:116220
7. de Rijke E et al. (2006) Analytical separation and detection methods for flavonoids. J Chromatogr A 1112(1):31–63
8. Pradhan S et al. (2023) A report on green extraction procedures for separation of flavonoids and its bio activities. J Herbal Med 41:100716
9. Pinto D et al. (2021) Castanea sativa shells: a review on phytochemical composition, bioactivity and waste management approaches for industrial valorization. Food Re Int 144:110364
10. Li S et al (2021) Qualitative and quantitative analysis on flavonoid distribution in different floral parts of 42 hemerocallis accessions. Front Plant Sci 12:670506
11. Godlewska K et al (2023) Investigation of chemical constituents and antioxidant activity of biologically active plant-derived natural products. Molecules 28(14):5572
12. Kowalska T, Sajewicz M (2022) Thin-layer chromatography (TLC) in the screening of botanicals-its versatile potential and selected applications. Molecules 27(19):6607
13. Rodrigues F et al. (2013) *Medicago* spp. extracts as promising ingredients for skin care products. Ind Crops Prods 49:634–644
14. Krekora M, Nawrocka A (2024) Interactions of gliadins with flavonoids and their glycosides studied with application of FT-Raman spectroscopy. J Cereal Sci 117:103915
15. Zhen X-T et al. (2022) Analysis of flavonoids in citrus fruits by capillary zone electrophoresis coupled with quadrupole time-of-flight mass spectrometry using chemometrics. J Food Compo Anal 106:104275
16. Pinto D et al. (2023) Appraisal of a new potential antioxidants-rich nutraceutical ingredient from chestnut shells through in-vivo assays—a targeted metabolomic approach in phenolic compounds. Food Chem 404:134546
17. Pinto D et al. (2024) Predicting the effects of in-vitro digestion in the bioactivity and bioaccessibility of antioxidant compounds extracted from chestnut shells by supercritical fluid extraction—a metabolomic approach. Food Chem 435:137581
18. Pinto D et al. (2023) Metabolomic insights into phenolics-rich chestnut shells extract as a nutraceutical ingredient—a comprehensive evaluation of its impacts on oxidative stress biomarkers by an in-vivo study. Food Re Int 170:112963
19. Silva AM et al. (2022) Valorization of Kiwiberry leaves recovered by ultrasound-assisted extraction for skin application: a response surface methodology approach. Antioxidants 11(4):763
20. Pinto D et al. (2021) Optimizing the extraction of phenolic antioxidants from chestnut shells by

subcritical water extraction using response surface methodology. Food Chem 334:127521
21. Silva AM et al. (2021) Salicornia ramosissima bioactive composition and safety: eco-friendly extractions approach (microwave-assisted extraction vs. conventional maceration). Appl Sci 11(11):4744
22. Teixeira F et al. (2024) Ultrasound-assisted extraction of bioactive compounds from goji berries: optimization, bioactivity, and intestinal permeability assessment. Food Res Int 188:114502
23. Pinto D et al. (2021) Microwave-assisted extraction as a green technology approach to recover polyphenols from Castanea sativa shells. ACS Food Sci Technol 1(2):229–241
24. Pinto D et al. (2020) Valorisation of underexploited Castanea sativa shells bioactive compounds recovered by supercritical fluid extraction with CO_2: a response surface methodology approach. J CO2 Util 40:101194
25. Silva AM et al. (2022) Influence of temperature on the subcritical water extraction of Actinidia arguta leaves: a screening of pro-healthy compounds. Sustain Chem Pharm 25:100593
26. Lund JA, Brown PN, Shipley PR (2020) Quantification of North American and European Crataegus flavonoids by nuclear magnetic resonance spectrometry. Fitoterapia 143:104537
27. Kustiati U, Wihadmadyatami H, Kusindarta DL (2022) Dataset of Phytochemical and secondary metabolite profiling of holy basil leaf (*Ocimum sanctum* Linn) ethanolic extract using spectrophotometry, thin layer chromatography, Fourier transform infrared spectroscopy, and nuclear magnetic resonance. Data Brief 40:107774
28. Sun J et al. (2023) Fourier transform infrared spectroscopic markers of the scorching degree of Scutellariae Radix related to the variation patterns of flavonoids and carbohydrates. Spectrochim Acta A Mol Biomol Spectrosc 290:122312
29. Suhad S, Humadi S (2024) Department of Pharmacy Department of Pharmacognosy Pharmacognosy Laboratory, Manual First Semister
30. María R et al. (2018) Preliminary phytochemical screening, total phenolic content and antibacterial activity of thirteen native species from Guayas province Ecuador. J King Saud Univ Sci 30(4):500–505
31. Maheshwaran L et al. (2024) Phytochemical testing methodologies and principles for preliminary screening/qualitative testing. Asian Plant Res J 12(5):11–38
32. Hricovíniová Z et al (2021) New unnatural gallotannins: a way toward green antioxidants, antimicrobials and antibiofilm agents. Antioxidants (Basel) 10(8):1288
33. Hosseini M et al. (2025) The multifaceted role of tannic acid: From its extraction and structure to antibacterial properties and applications. Prog Polym Sci 160:101908
34. Gwatidzo L, Dzomba P, Mangena M (2018) TLC separation and antioxidant activity of flavonoids from Carissa bispinosa, Ficus sycomorus, and Grewia bicolar fruits. Nutrire 43(1):3
35. Shraim AM et al. (2021) Determination of total flavonoid content by aluminum chloride assay: a critical evaluation. LWT 150:111932
36. Chen M et al. (2022) Phytochemical analysis, UPLC-ESI-Orbitrap-MS analysis, biological activity, and toxicity of extracts from Tripleurospermum limosum (Maxim.) pobed. Arab J Chem 15(5):103797
37. Sait S et al. (2015) HPLC-UV/DAD and ESI-MSn analysis of flavonoids and antioxidant activity of an Algerian medicinal plant: Paronychia argentea Lam. J Pharm Biomed Anal 111:231–240
38. Memon AF et al. (2017) Simultaneous determination of quercetin, rutin, naringin, and naringenin in different fruits by capillary zone electrophoresis. Food Anal Methods 10(1):83–91
39. Raal A et al. (2012) Content of essential oil, terpenoids and polyphenols in commercial chamomile (*Chamomilla recutita* L. Rauschert) teas from different countries. Food Chem 131(2):632–638
40. Jadeja S, Kupcik R, Fabrik I, Sklenárová H, Lenčo J (2023) A stationary phase with a positively charged surface allows for minimizing formic acid concentration in the mobile phase enhancing electrospray ionization in LC-MS proteomic experiments. The Analyst 148(23):5980–5990. https://doi.org/10.1039/D3AN01508D

Check for updates

Chapter 9

Chemical Tests and Quantitative Analysis of Saponins

Ahmet Beyatli and Hafiz Suleria

Abstract

This chapter provides a comprehensive overview of the chemical tests and quantitative analysis methods for saponins, a diverse group of phytochemicals with significant biological activities. Different qualitative and quantitative procedures are explained step by step for each method, along with important considerations and potential limitations. This can potentially serve as a valuable resource for researchers and analysts working with saponins in various fields, including pharmacology, food science, and natural product chemistry.

Keywords Saponins, Phytochemicals, Qualitative analysis, Quantitative analysis, Spectrophotometry, Chromatography

1 Introduction

Saponins are a diverse and widespread group of phytochemicals found in numerous plant species [1]. The term "saponin" originates from the Latin word "sapo," meaning soap, due to their ability to form stable foams in aqueous solutions [2]. These complex molecules play various ecological roles in plants, including defense against pathogens and herbivores [3]. Beyond their natural functions, saponins have garnered significant interest in multiple industries due to their diverse biological activities and physicochemical properties. In the pharmaceutical sector, saponins exhibit a range of therapeutic potential, including anti-inflammatory, antifungal, antibacterial, anti-parasitic, anticancer, and antiviral activities [4–6].

The structural diversity of saponins is reflected in their wide array of biological activities. This diversity arises from variations in the aglycone structure, the number and types of sugar moieties attached, and the positions of these attachments [7]. The aglycone, also known as the sapogenin, can be classified into two main groups: triterpenoid and steroidal saponins. Both types are derived from the mevalonic acid pathway, with 2,3-oxidosqualene serving as a common precursor [4]. In recent years, there has been growing interest in saponins for

Dhruti Amin et al. (eds.), *Phytoconstituent Analysis*, Springer Protocols Handbooks,
https://doi.org/10.1007/978-1-0716-5400-2_9,

their potential applications in the food, cosmetic, and pharmaceutical industries [5]. Their surfactant properties make them valuable as natural emulsifiers, foaming agents, and stabilizers in food products [5]. Additionally, some saponins have shown promise in lowering cholesterol levels and exhibiting anticancer effects [1].

The biosynthesis of saponins involves a complex pathway that begins with the formation of isopentenyl pyrophosphate (IPP) and dimethylallyl pyrophosphate (DMAPP). These precursors ultimately lead to the production of 2,3-oxidosqualene, which serves as the branch point between primary metabolism and specialized triterpene metabolism in higher plants [7].

As research continues to uncover the multifaceted roles of saponins in plant physiology and their potential applications in various industries, these compounds remain a subject of significant scientific interest. Their structural complexity and functional diversity make saponins a valuable resource for future drug discovery and development of natural products with industrial applications [8]. Therefore, analysis of saponins is crucial for understanding their biological activities and potential health benefits. This chapter outlines the methods used for the qualitative and quantitative analysis of saponins, focusing on traditional chemical tests and modern chromatographic techniques.

2 Materials

2.1 Plant Samples and Standards

1. Plant samples planned for detection of saponins (e.g., *Glycyrrhiza glabra*, *Quillaja saponaria*, *Sapindus mukorossi*, *Tribulus terrestris*).
2. Purified saponin standards (e.g., ginsenoside Rb1, glycyrrhizin).

2.2 Reagents

Methanol (HPLC grade), ethanol (95%), n-butanol, chloroform, acetic acid (glacial), sulfuric acid (concentrated), vanillin, ethyl acetate, acetonitrile (HPLC grade), and ultrapure water (18.2 MΩ·cm at 25°C).

2.3 Solutions and Buffers

1. Vanillin-sulfuric acid reagent: 5% (w/v) vanillin in ethanol mixed with 72% (v/v) sulfuric acid (1:1).
2. Liebermann–Burchard reagent: Acetic anhydride and concentrated sulfuric acid (5:1).
3. Phosphate-buffered saline (PBS), pH 7.4.
4. Mobile phase for HPLC: acetonitrile and water (gradient elution).

3 Methods

Crude saponins will be extracted using 30 mL of 75% ethanol, and the extract will be re-extracted with 100 mL of ethyl acetate. Then solvent will be evaporated [9].

3.1 Qualitative Analysis

Qualitative tests for saponins typically involve the formation of foam upon shaking with water, the hemolytic activity on red blood cells, color reaction tests, or thin-layer chromatorgraphy (TLC). Common methods include:

Foam Test:

1. Mix a small amount of extract (0.5 mL) with distilled water (5 mL) in a test tube and shake vigorously.
2. When a few drops of olive oil were added to the foaming and rapidly agitated, the foam's appearance indicated the presence of saponins [10].

Hemolysis Test:

1. Prepare a 2% suspension of human or animal red blood cells in phosphate-buffered saline (PBS).
2. Prepare a series of dilutions of the plant extract in PBS.
3. Mix equal volumes (typically 0.5 mL) of the red blood cell suspension and each dilution of the plant extract.
4. Incubate the mixtures at 37°C for 30 min.
5. Centrifuge the samples at 3000 rpm for 5 min.
6. Measure the absorbance of the supernatant at 540 nm.
7. Calculate the percentage of hemolysis, with 100% lysis defined as the absorbance of red blood cells treated with distilled water.

The presence of saponins is indicated by hemolysis of the red blood cells, with the degree of hemolysis correlating to saponin concentration [11].

Color Reaction Tests:

1. Salkowski's Test: Mix extract (5 mL) and 1–2 mL chloroform in a test tube and dissolve. 3 mL of concentrated sulfuric acid will be added along the sides of the test tube. A reddish ring at the interphase of the acid and the chloroform layer indicates the presence of triterpenoids/steroids [12].
2. Liebermann–Burchard's Test: Mix 1–2 mL of chloroform with a small amount of extract in a test tube. After dissolving, the solution is evaporated to dryness in a porcelain dish. Then, add 2 mL of acetic anhydride to the dried residue. Carefully add 2 mL of concentrated sulfuric acid along the sides of the test tube. Observe the color change. A dark bluish-green color

indicates the presence of saponins, specifically steroidal saponins [13].

Lead Acetate Test:

1. Take 2 mL of the extract in a test tube. Add a few drops of 1% lead acetate solution to the extract. Observe the mixture for the formation of a white precipitate. The presence of a white precipitate indicates a positive result for saponins [13].
2. Vanillin-Sulfuric Acid Test: Mix 0.25 mL of the plant extract with 0.25 mL of 8% (w/v) vanillin in ethanol. Add 2.50 mL of 72% (v/v) sulfuric acid in water. Incubate the mixture at 60°C for 15 min in a shaking water bath. Cool the solution to room temperature. Measure the absorbance at 540 nm. A positive test for saponins is indicated by the development of a colored product, typically red to purple [14, 15].

Thin-Layer Chromatography (TLC) Analysis

1. Enrich extract with saturated n-butanol, mixing well. Keep the butanol layer, concentrate it, and use this for chromatographic analysis.
2. For chromatography, use a solvent mixture of chloroform, glacial acetic acid, methanol, and water in a 64:34:12:8 ratio to separate the saponins. After separation, expose the chromatogram to iodine vapors.
3. Record the yellow color and hundred times retention factor (hRf) values of the spots that appear when the chromatogram is exposed to iodine vapors [10].

3.2 Quantitative Analysis

Quantitative methods focus on accurately measuring saponin content using various techniques:
Spectrophotometric Methods [16]

1. Extract saponins from plant material using an appropriate solvent (e.g., water, methanol, or ethanol).
2. Prepare a series of standard solutions using a known saponin (e.g., oleanolic acid or aescin) for calibration.
3. React the extract and standards with a color-developing reagent, such as vanillin-sulfuric acid or Liebermann–Burchard reagent.
4. Measure the absorbance of the colored products at a specific wavelength, typically around 528–540 nm.
5. Create a standard curve using the absorbance values of the known saponin concentrations.
6. Calculate the saponin content in the plant sample using the standard curve and accounting for any dilutions.

Chromatographic Methods

1. High-Performance Liquid Chromatography (HPLC): Prepare samples by extracting saponins using methanol. Then inject samples into HPLC equipped with UV detection at specific wavelengths (e.g., 540 nm). Construct standard curves using known concentrations of saponin standards to quantify unknown samples [17].
2. Liquid Chromatography-Mass Spectrometry (LC-MS): Extract saponins from plant material using an appropriate solvent (e.g., methanol or ethanol). Prepare the LC-MS system: Use a C18 reverse-phase column for separation. Employ a gradient elution with water and organic solvent (e.g., acetonitrile or methanol), often with 0.1% formic acid added. Set up the mass spectrometer: Use electrospray ionization (ESI) in positive or negative mode. Operate in multiple reaction monitoring (MRM) mode for quantification. Develop a calibration curve using standard saponins. Analyze samples and standards under optimized conditions. Process data using appropriate software to quantify saponins based on the calibration curve [18].

Other Methods

1. Gravimetric Method: After extraction of saponins from dried plant material using an appropriate solvent (e.g., ethanol or methanol). Dissolve the concentrated extract in water. Perform liquid–liquid extraction using n-butanol to separate saponins from other compounds. Evaporate the n-butanol fraction to obtain crude saponins. Dry the crude saponins to a constant weight. Calculate the saponin content as a percentage of the initial plant material weight [19].
2. Near-Infrared Spectroscopy (NIRS): Dry and grind plant material to a fine powder (e.g., 250μm). Place in a sample holder or use a fiber optic probe for direct measurement. Record NIR spectra, usually in the range of 800–2500 nm. Analyze a subset of samples using a reference method (e.g., HPLC) to determine actual saponin content. Use chemometric techniques such as partial least squares (PLS) regression to correlate spectral data with reference values. Apply preprocessing techniques (e.g., multiplicative scatter correction, standard normal variate) to improve model performance. Use an independent set of samples to validate the calibration model. Apply the validated model to predict saponin content in unknown samples [17].
3. Capillary Electrophoresis (CE): Extract saponins from plant material using an appropriate solvent (e.g., methanol or ethanol). Filter or centrifuge the extract to remove particulates. Use a fused silica capillary, typically 50–75 μm inner diameter. Select an appropriate buffer system, such as borate buffer

with additives such as cyclodextrins. Adjust buffer pH, concentration, and additives to improve separation. Optimize voltage, temperature, and injection parameters. Prepare standard solutions of known saponins. Create a calibration curve using peak areas or heights. Inject the sample using hydrodynamic or electrokinetic methods. Apply separation voltage (typically 14–25 kV) and maintain temperature (usually 25°C). Detect saponins using UV absorbance, often at 210–254 nm. Identify saponins based on migration times and spectral data. Quantify saponins using the calibration curve [20].

4 Notes

1. Ensure all reagents are prepared fresh and stored properly to maintain their efficacy [21].
2. Calibration curves must be generated for each analytical run to ensure accuracy in quantification [22].
3. Consider potential interferences from other phytochemicals present in plant extracts [23].
4. It is important to note that while the Salkowski's test is commonly used for detecting terpenoids and steroids, it can also indicate the presence of saponins due to their steroidal nature [24].
5. Vanillin-sulfuric acid method is widely used for its simplicity, speed, and cost-effectiveness in quantifying total saponin content in plant materials. However, it is important to note that different solvents can interfere with the reaction, so a solvent evaporation step may be necessary to improve accuracy [14].
6. LC-MS method offers high sensitivity and selectivity for saponin analysis, allowing for accurate quantification even in complex plant matrices [25].
7. While gravimetric method is relatively simple and does not require specialized equipment, it is time-consuming and may be less accurate than other quantitative methods [23].
8. It is important to note that NIRS requires a robust calibration model based on a large number of samples with varying saponin contents to ensure accurate predictions [26].
9. CE method can be adapted for various plant extracts and saponin types, making it a versatile tool for quantitative analysis of these compounds [23].

References

1. Sparg S, Light ME, Van Staden J (2004) Biological activities and distribution of plant saponins. J Ethnopharmacol 94:219–243
2. Kregiel D, Berlowska J, Witonska I, Antolak H, Proestos C, Babic M et al (2017) Saponin-based, biological-active surfactants from plants. In: Najjar R (ed) Application and characterization of surfactants. InTech, pp. 184–205
3. Abdelrahman M, Jogaiah S (2020) Bioactive molecules in plant defense: saponins. Springer International Publishing, Switzerland
4. Mugford ST, Osbourn A (2013) Saponin synthesis and function. In: Bach TJ, Rohmer M (Eds.)Eds. Isoprenoid synthesis in plants and microorganisms. Springer, NY, 405–424
5. Timilsena YP, Phosanam A, Stockmann R (2023) Perspectives on saponins: food functionality and applications. Int J Mol Sci 24:13538. https://doi.org/10.3390/ijms241713538
6. Zhu L, Wang S, Wan F, Zhou Y, Wang Z, Fan G et al. (2023) Quantitative analysis of *Camellia oleifera* seed saponins and aqueous two-phase extraction and separation. Molecules 28:2132. https://doi.org/10.3390/molecules28052132
7. Moses T, Papadopoulou KK, Osbourn A (2014) Metabolic and functional diversity of saponins, biosynthetic intermediates and semi-synthetic derivatives. Crit Rev Biochem Mol Biol 49:439–462. https://doi.org/10.3109/10409238.2014.953628
8. Güçlü-Üstündağ Ö, Mazza G (2007) Saponins: properties, applications and processing. Crit Rev Food Sci 47:231–258. https://doi.org/10.1080/10408390600698197
9. Peng J, Yao X, Kobayashi H, Ma C (2007) Novel furostanol glycosides from *Allium macrostemon*. Planta Med 61:58–61. https://doi.org/10.1055/s-2006-958000
10. Alamgir ANM (2018) Methods of qualitative and quantitative analysis of plant constituents. In: Alamgir ANM (ed) Therapeutic use of medicinal plants and their extracts. Volume 2: phytochemistry and bioactive compounds. Springer International Publishing, Cham, pp 721–804
11. Sotheeswaran S (1988) Screening for saponins using the blood hemolysis test: an undergraduate laboratory experiment. J Chem Educ 65:161. https://doi.org/10.1021/ed065p161
12. Njoku OV, Obi C (2009) Phytochemical constituents of some selected medicinal plants. Afr J Pure Appl Chem 3:228–233
13. Dahanayake JM, Perera PK, Galappatty P, Perera HD, Arawwawala LD (2019) Comparative phytochemical analysis and antioxidant activities of Tamalakyadi decoction with its modified dosage forms. Evid Based Complement Alternat Med 2019:6037137. https://doi.org/10.1155/2019/6037137
14. Le AV, Parks SE, Nguyen MH, Roach PD (2018) Improving the vanillin-sulphuric acid method for quantifying total saponins. Technologies 6:84. https://doi.org/10.3390/technologies6030084
15. Du M, Guo S, Zhang J, Hu L, Li M (2018) Quantitative analysis method of the tea saponin. Open J For 8:61–67. https://doi.org/10.4236/ojf.2018.81005
16. Le Bot M, Thibault J, Pottier Q, Boisard S, Guilet D (2022) An accurate, cost-effective and simple colorimetric method for the quantification of total triterpenoid and steroidal saponins from plant materials. Food Chem 383:132597
17. Wang Y, Ma Y, Tao L, Zhang X, Hao F, Zhao S et al. (2022) Recent advances in separation and analysis of saponins in natural products. Separations 9:163. https://doi.org/10.3390/separations9070163
18. El Sayed AM, Basam SM, El-Naggar EM, Marzouk HS, El-Hawary S (2020) LC–MS/MS and GC–MS profiling as well as the antimicrobial effect of leaves of selected Yucca species introduced to Egypt. Sci Rep 10: 17778. https://doi.org/10.1038/s41598-020-74440-y
19. Budan A, Tessier N, Saunier M, Gillmann L, Hamelin J, Chicoteau P et al. (2013) Effect of several saponin containing plant extracts on rumen fermentation in vitro, Tetrahymena pyriformis and sheep erythrocytes. J Food Agric Environ 11:576–582
20. Gackowski M, Przybylska A, Kruszewski S, Koba M, Mądra-Gackowska K, Bogacz A (2021) Recent applications of capillary electrophoresis in the determination of active compounds in medicinal plants and pharmaceutical formulations. Molecules 26:4141. https://doi.org/10.3390/molecules26144141
21. Bower JF, McClung JB, Watson C, et al. (2014) Recommendations and best practices for reference standards and reagents used in bioanalytical method validation. The AAPS journal 16:352–356
22. Danzer K, Currie LA (1998) Guidelines for calibration in analytical chemistry. Part I. Fundamentals and single component calibration (IUPAC Recommendations 1998). Pure Appl Chem 70:993–1014
23. Cheok CY, Salman HAK, Sulaiman R (2014) Extraction and quantification of saponins: A review. Food Res Int 59:16–40
24. Hammed A, Yeasmen N, Orsat V (2025) Eco-Friendly Extraction and Characterization of

Terpenoids From Plants as Functional Food Ingredients: A Review. J Food Biochem 2025, 9746960. https://doi.org/10.1155/jfbc/9746960
25. Jiménez GG, Durán AG, Macías FA, Simonet AM (2021) Structure, bioactivity and analytical methods for the determination of yucca saponins. Molecules 26:5251
26. Agelet LE, Hurburgh CR (2010) A Tutorial on Near Infrared Spectroscopy and Its Calibration. Crit Rev Anal Chem 40:246–260. https://doi.org/10.1080/10408347.2010.515468

Chapter 10

Detection and Quantification Techniques for Coumarins

Cuong Quoc Nguyen, Quang Le Dang, Quang De Tran and Nhu Y. Nguyen-Thi

Abstract

This chapter comprehensively presents the analytical techniques used for the identifications, detection, and quantification of coumarins in various phytochemical resources, including reaction chemistry, thin-layer chromatography (TLC), ultraviolet–visible spectroscopy (UV-Vis), gas chromatography (GC), and liquid chromatography (LC). The chapter provides a step-by-step guide outlining essential, accessible, and cost-efficient methodologies for the detection of coumarin in plant samples, which are designed for straightforward implementation in resource-limited laboratory settings. Furthermore, the chapter details various coumarin derivatives indentified in plant species, including dihydrofuranocoumarins, furanocoumarins, pyranocoumarins, phenylcoumarins, bicoumarins, and among others. The analysis of the presence of various types of coumarins in plant species contributes valuable information on their diverse biological activities, such as antioxidant, anti-inflammatory, anticancer, antifugal, and hepatoprotective properties.

Keywords Secondary metabolite, Coumarins, Extraction techniques, Chromatographic analysis, Spectroscopic detection, Method validation

1 Introduction

Coumarin, a natural compound with a structure consisting of a pyrone ring fused with a benzene ring is known for many novel biological and medical properties [1]. Coumarin was first isolated from tonka beans in 1820 by Alfred Vogel of Munich and was later synthesized in 1868 by William Henry Perkin [2]. Coumarin is biosynthesized in many plants such as cinnamon, oranges, and peaches; and it also has been proven to have profound antioxidant properties that protect the cells from being damaged by radicals (Fig. 1) [1, 3–5]. In addition, coumarin can reduce inflammatory responses and eliminate pathogenic bacteria by its anti-inflammatory and antibacterial properties [6–8].

It has been known that some coumarin derivatives have been shown to suppress the development of cancer cells, thus showing promising results in cancer treatments [9]. Coumarin also plays an important role in anticoagulant medicines, such as Warfarin that is

Dhruti Amin et al. (eds.), *Phytoconstituent Analysis*, Springer Protocols Handbooks,
https://doi.org/10.1007/978-1-0716-5400-2_10,

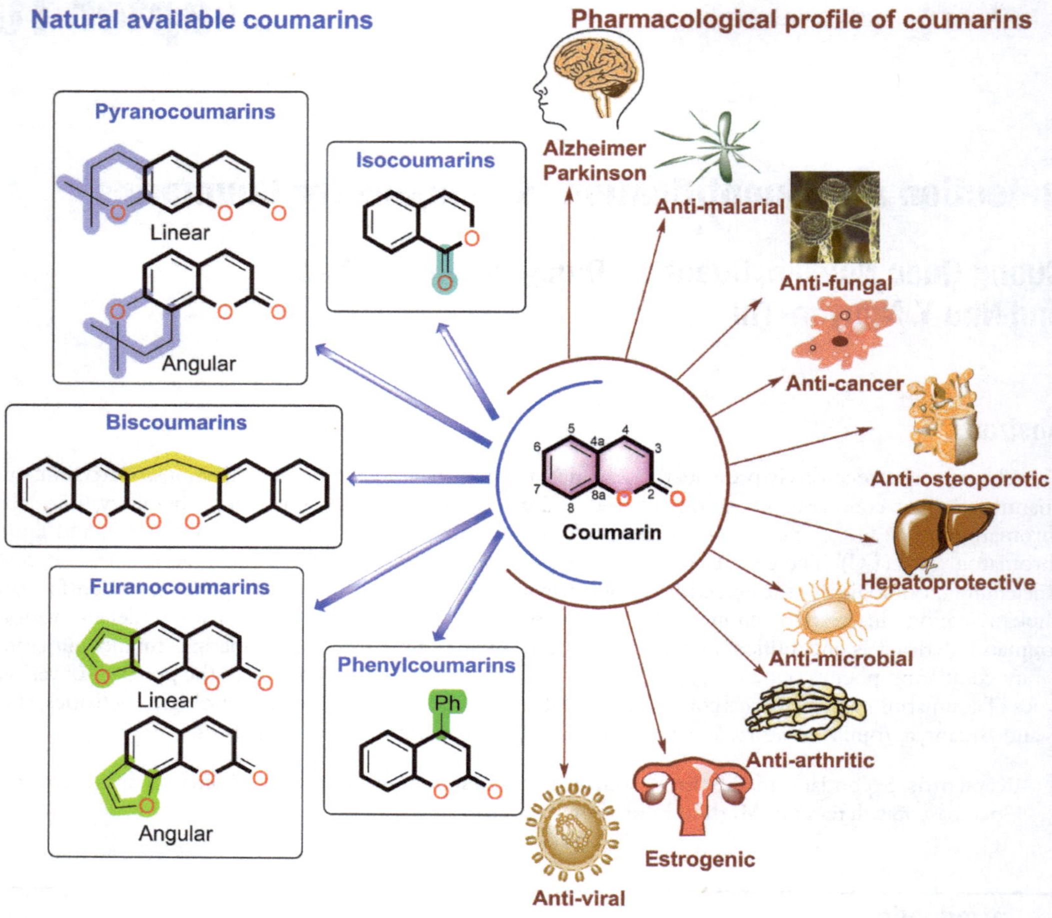

Fig. 1 Main subtypes and therapeutic potentials of coumarin derivatives

used as a rodenticide, and widely in the prevention and treatment of hematology disorders [9]. As such, coumarin has been a major interest in medicine and pharmacology with its extensive applications and potentials yet to be discovered [1, 9].

But it is now well-known that the most suitable analytical technique for coumarins determination is chromatography coupled with detectors including gas chromatography with flame ionization detector (FID) [10, 11] and mass spectrometry (MS) [11, 12]; high-performance liquid chromatography (HPLC) with UV [2], diode-array detector (DAD) [2, 13], and MS [13–15]; high-performance liquid chromatography–diode-array detection–mass spectrometry (HPLC-DAD-MS) [16]; ultra-performance liquid chromatography (UPLC)-Quadrupole-TOF-MS coupled [17], ultra-high-performance liquid chromatography coupled with a mass spectrometry/UV-Vis (UHPLC-MS/UHPLC-UV Vis) [18, 19]. This chapter aims to guide simple, easy-to-perform, and cost-effective methods for the isolation, analysis, and quantification of coumarin compounds from various plant species. This chapter also provides a step-by-step guide for the preparation, analysis, quantification, and isolation of coumarin from plant samples, facilitating effective application in low-budget laboratories. Additionally, the research unveils efficient extraction and analysis techniques for coumarins, offering valuable insights into their diverse biological activities.

2 Materials

Prepare all solutions using ultrapure water, organic solvents, and analytical-grade reagents. Unless otherwise specified, store all reagents at room temperature. Adhere strictly to waste disposal regulations.

2.1 Determination of Coumarins By Opening the Lactone Ring

1. Prepare NaOH 10% (w/v): Precisely weigh 100 g NaOH in a 500 mL beaker. After that, gradually add water into the beaker until all of the NaOH has been dissolved. Add all of the previously made NaOH solution into a 1000 mL volumetric flask and slowly add water until the convex part of the solution surface touches the 1000 mL mark. After that, place the lid on the volumetric flash and start shaking.
2. Prepare HCl 1 M (v/v): Accurately measure 83.33 mL of the HCl stock solution (12 M) using a pipette. Transfer this to a 1000 mL volumetric flask and add distilled or deionized water until the total volume reaches 1000 mL. Cap the flask and mix thoroughly to ensure a homogeneous solution (*see* **Note 1**).

2.2 Preparative Thin Layer Chromatography (TLC)

1. Prepare capillary micropipette: The capillary micropipette is prepared by heating the capillary 0.5 mm on the alcohol burner's flame until the glass has been softened. After that, swiftly pull the capillary from both sides and remove it from the heat source.
2. Prepare the TLC chamber: Create a cost-effective TLC chamber using a glass beaker and a watch glass. Place a piece of filter paper inside the beaker so that it fits snugly, lies flat at the bottom, and does not obstruct visibility.
3. TLC is used by silica gel 60 F254 20×20 cm purchased from Merck, Germany. Using a pencil, lightly mark a baseline approximately 1 cm from the bottom of the TLC plate and an end line 0.5 cm from the top edge.
4. Prepare the mobile phase: Choose and mix the solvents (mobile phase) according to the desired separation.

2.3 Chromatographic and Extraction Techniques for Coumarin

1. Isolation was performed by silica gel column (SiO_2, 60 Å, 15–40 and 40–63 μm) purchased from Merck, Germany (*see* **Note 2**).
2. The nuclear magnetic resonance (NMR) was carried out on a Bruker AM500 FT-NMR spectrometer at 500 MHz using $CDCl_3$ as solvent.

3. Prepare TLC: The TLC process was prepared as described in Sect. 2.2.
4. The elution solvents (hexane, ethyl acetate, and methanol) must meet analytical standards, and are then diluted according to the Hex:EtOAc gradient (1:0–7:3, v/v) and EtOAc:MeOH (1:0–1:1, v/v). Depending on the case, different solvent systems can be used.

2.4 Ultraviolet–Visible Spectroscopy (UV-Vis)

1. Prepare test sample: 0.5 g of powdered herbal material was extracted twice with 10 mL of 50% methanol at 50°C for 40 min. The extract was evaporated, and the residue was dissolved in 3 mL of methanol (*see* **Note 3**).
2. Prepare 0.02% coumarin standard (w/v): The coumarin standard was isolated as described in Sect. 3.2 or purchased commercially (CAS 91-64-5). Coumarin was diluted to a stock concentration of 0.02% by dissolving 2.0 mg of the compound in 10 mL of methanol. Ensure the solution is well-mixed.
3. Prepare 7% (w/v) sodium carbonate solution: Dissolve the 7.0 g of sodium carbonate in a small amount of distilled water. Transfer the dissolved sodium carbonate into a 100 mL volumetric flask. Add distilled water to the flask until the total volume reaches 100 mL, which is well-mixed.
4. Prepare 0.05 N sulfuric acid solution. Carefully measure approximately 1.39 mL of concentrated sulfuric acid using a pipette. Add the measured sulfuric acid to a 500 mL volumetric flask containing about 250 mL of distilled water. Mix the solution gently. Add distilled water to the flask until the total volume reaches 500 mL (*see* **Note 4**).
5. Prepare *p*-nitrobenzenediazonium chloride (1.5 mmol/L) solution. Dissolve the 0.284 g of *p*-nitrobenzenediazonium chloride in a small amount of distilled water. Transfer the dissolved *p*-nitrobenzenediazonium chloride into a 1000 mL volumetric flask. Add distilled water to the flask until the total volume reaches 1000 mL (*see* **Note 5**).

2.5 Preparative Gas Chromatography (GC)

1. Prepare the standard: The coumarin standard was isolated as described in Sect. 3.2 or purchased commercially (CAS 91-64-5). Dissolve coumarin in analytical-grade chloroform to obtain stock solutions. Dilute the stock solutions to achieve concentrations of 0.330, 0.550, 1.370, and 1.72 mg/mL.
2. Prepare test sample: Materials (1.0 g) were macerated with analytical grade hexane (10 mL) (*see* **Note 6**). The filtrate was passed through a glass column (1 cm i.d.) containing a cotton plug, 0.5 g of silica gel, and 100 mg of active charcoal. The column was eluted with 10 mL of hexane and then with 5 mL of chloroform. The fractions were combined and evaporated

to dryness, at room temperature. This procedure was repeated in triplicate. After, the dried extracts were solubilized in a small amount of dried diethyl ether, to which was added an ethereal solution of diazomethane, until the liberation of gases stopped. The samples were evaporated at room temperature and then solubilized in 5.0 mL analytical grade chloroform for GC analysis.

3. The flame ionization detector (FID) temperature at 280°C and column temperature programming from 105°C at 15°C/min to 200°C, and from 200°C at 68°C/min to 240°C (held for 10 min).
4. Hydrogen was used as carrier gas, at an average linear velocity of 65 cm/s.

2.6 Ultra High-Performance Liquid Chromatography (UHPLC)

1. Prepare the standard: The coumarins standard was isolated as described in Sect. 3.2 or purchased commercially (CAS 91-64-5). Dissolve coumarin in analytical grade solvents to obtain stock solutions: 1, 10, 20, 40, 60, 80, and 100 mg/mL.
2. Prepare test sample: The herbal material (1.0 g) is extracted with ethanol:water (1:1; v/v, 10 mL). The extract is then filtered, and the water is evaporated using a lyophilizer (*see* **Notes 7 and 8**).
3. Parameters: A modular Shimadzu LC-10 system comprised of a LC-10AD pump, a CTO-10A column oven, a SPD-10A UV-DAD detector, and a CBM-10A interface. An LC-18 column (250 mm × 4 mm i.d. × 5 mm). UV detection is at 274 nm.
4. Mobile phase: Separations were done in the isocratic mode, using acetonitrile:water (40:60; v/v) at a flow rate of 1 mL/min; with an injection volume ("loop") of 20 mL (*see* **Note 9**).

3 Methods

Carry out all procedures at room temperature unless otherwise specified.

3.1 Determination of Coumarins By Opening the Lactone Ring

1. Add 0.5 mL of sample (Fig. 2) (i.e., plant extracts, *see* **Note 3**) into two test tubes, which are labeled "I" and "II," respectively.
2. 0.25 mL NaOH 10% (w/v) is added into test tube No. I and put two test tubes above the alcohol burner's flame and observed the phenomenon (*see* **Note 10**). Tube no. II as control.
3. Each solution in two test tubes are diluted by adding 1.0 mL water and acidification is done by adding a few drops of HCl 1 M and then observing the phenomenon inside each test tube (Fig. 3) (*see* **Note 11**).

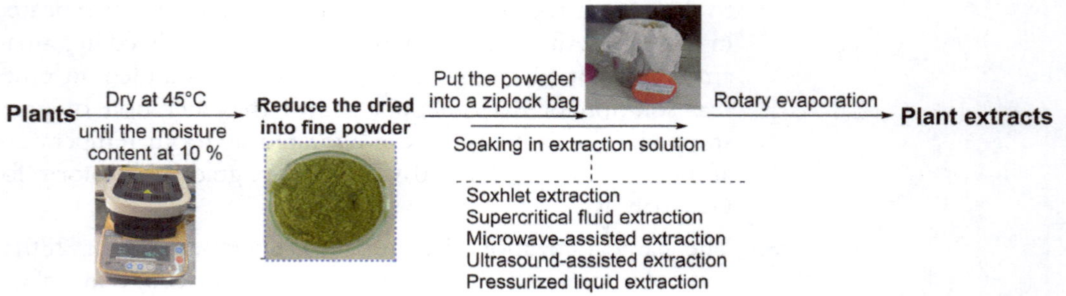

Fig. 2 Describe a general and simple procedure for preparing and assisting the extraction of coumarin-rich extracts from plants

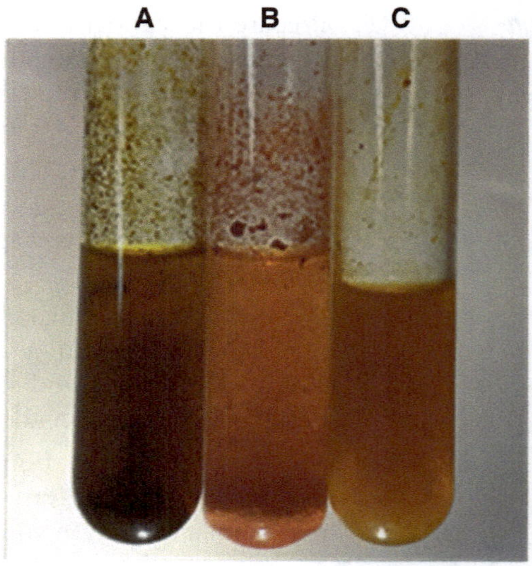

Fig. 3 Determination of coumarin in plant extracts [*Euphorbia thymifolia* (**A**), *Dracaena cambodiana* (**B**), and *Ochna integerrima* (**C**)] by opening the lactone ring reaction

3.2 Determination of Coumarin By TLC

1. Use a capillary tube to apply small spots of the sample solutions onto the marked baseline. Ensure the spots are evenly spaced and small to avoid overlapping (Fig. 4).
2. Pour the mobile phase into a developing chamber to a depth of about 0.5 cm (depending on the case, different solvent systems can be used).
3. Stand the TLC plate upright in the chamber with the baseline above the solvent level. Cover the chamber to prevent evaporation.
4. Allow the solvent to ascend the TLC plate by capillary action until end line.

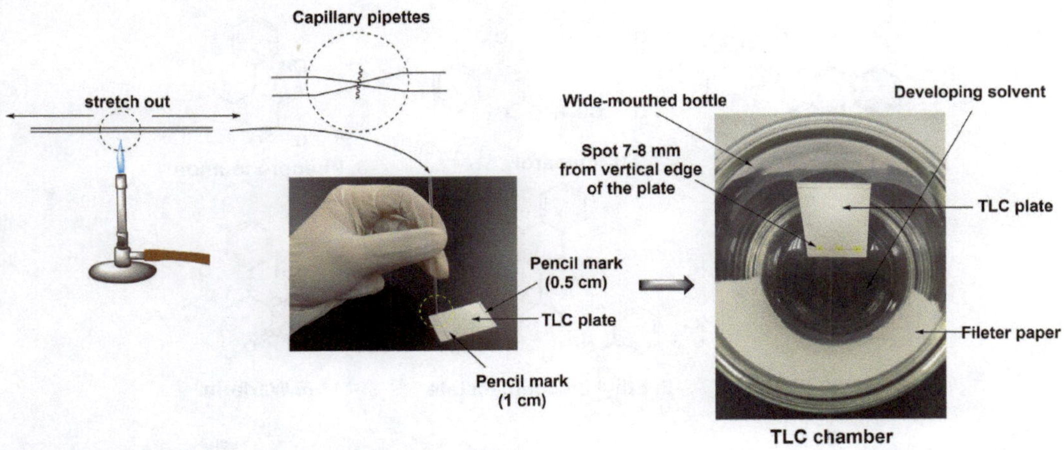

Fig. 4 Visualize the preparation and execution of thin-layer chromatography using a cost-effective TLC chamber made from a glass beaker and a watch glass. Hematocrit capillary tubes from Germany (Length: 120 mm, Width: 84 mm, Height: 50 mm, Nominal capacity: 75 μL) were employed. The TLC plate used was silica gel 60 F254 (20 × 20 cm) sourced from Merck, Germany

5. Remove the plate from the chamber and allow it to dry completely.
6. Using a spray bottle, apply an ethanolic solution (5% v/v) of KOH onto the dried plate.
7. Examine the TLC plate under UV light at 365 nm.
8. Use coumarin as the standard for identification and comparison of the sample spots.
9. Calculate the retention factor (R_f) value for each spot using the formula (Fig. 5) (*see* **Note 12**):

R_f = distance traveled by the spot compound/Distance traveled by the mobile phase.

3.3 Extraction and Chromatographic Approaches for Coumarin Characterization

1. The 4 kg of dried leaves and branches of *Eupatorium fortune* were extracted with MeOH (12 L, three times) at room temperature to obtain 450 g of extract (*see* **Notes 3 and 13**).
2. Suspending 400 g MeOH extract in a 2:8 (v:v) water-methanol mixture, the solution was then partitioned using *n*-hexane and ethyl acetate solvents. This process yielded 40 g of *n*-hexane residue and 26 g of ethyl acetate residue (*see* **Note 14**).
3. The 33.45 g *n*-hexane residue was fractionated on a silica gel column (500 g SiO_2, 40–63 μm, 7 × 32 cm) and eluted using a gradient of Hex: EtOAc (1:0–7:3, v/v) and EtOAc: MeOH (1:0–1:1, v/v) to obtain 20 fractions (labeled from 1 to 20) (*see* **Note 15**) (Fig. 6).
4. Fraction 19 (2.04 g) was chromatographed on a silica gel column (150 g SiO_2, 40–63 μm, 2.5 × 60 cm) with a solvent

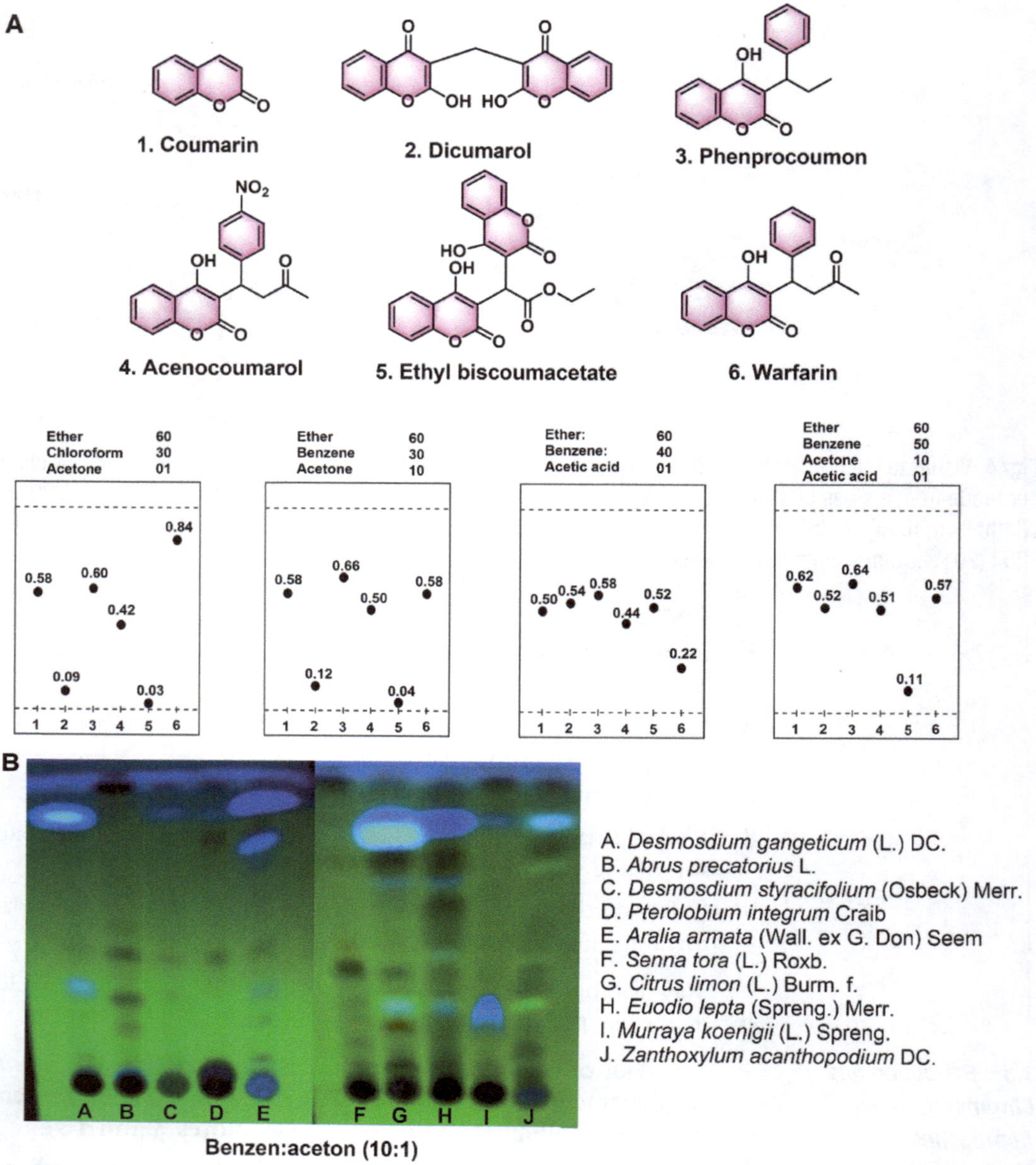

Fig. 5 TLC of coumarin anticoagulants (**A**) and coumarin-rich extracts from plants (**B**) (subfigure B reproduced from [25] under the Creative Commons Attribution-ShareAlike 4.0 International Licence)

mixture of Hex: DCM (10:1, 9:1, 8:2, 7:3, v/v) to produce eight subfractions (labeled from 19.1 to 19.8).

5. White crystals were obtained from subfraction 19.8. The crystals were washed with a Hex: DCM mixture (7:1) to obtain pure coumarin (128 mg) (Fig. 7) (*see* **Note 16**).

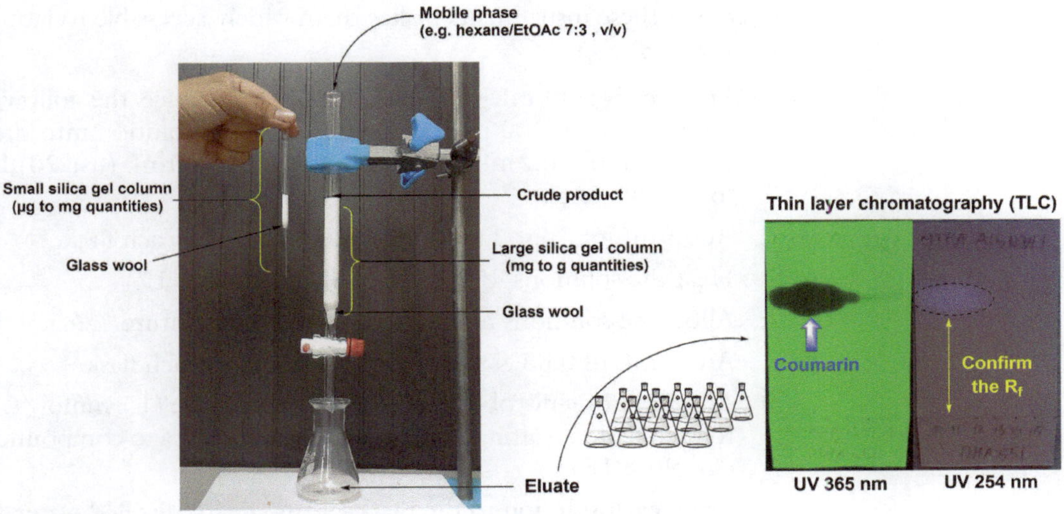

Fig. 6 Silica chromatography columns were prepared for the isolation of coumarins on a small scale (left column) and a large scale (right column) in lab. Glass wool was used to plug the columns, silica gel column (SiO_2, 60 Å, 15–40 and 40–63 μm) and TLC (60 F254 20 × 20 cm) from Merck, Germany. The top of the column is connected to a pressure-driven device, which helps expedite the elution process of compounds from the column. The isolation process was monitored using TLC plates and observed under UV light at 365 nm

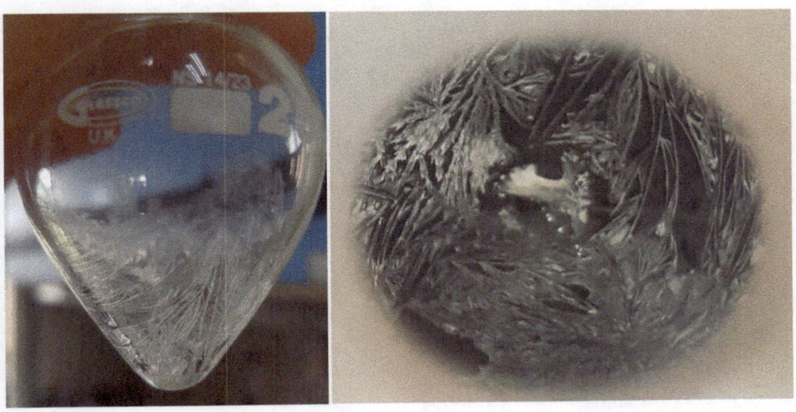

Fig. 7 Coumarin crystals were isolated from the leaves of *Eupatorium fortunei* Turcz

3.4 Quantification of Coumarins By UV-Vis

The analysis of coumarin via UV-Vis spectroscopy has garnered popularity due to its numerous advantages. Firstly, this technique is nondestructive, enabling the reuse of samples for further processing or analysis. Secondly, it allows for rapid measurements, facilitating seamless integration into various experimental protocols. Additionally, the instrumentation is user-friendly, requiring minimal training for operation. Lastly, the cost-effectiveness of both purchasing and operating these instruments makes them widely accessible to laboratories.

1. Prepare five 50-mL-volumetric flasks. Introduce the following volumes of 200 ppm coumarin standard solution into the flasks: 0.1 mL, 0.2 mL, 0.5 mL, 0.8 mL, and 1.0 mL (use 20 μL for sample test).
2. Add 5 mL of 7% sodium carbonate solution to each flask.
3. Heat the solutions for 5 min at 85°C (*see* **Note 17**).
4. Allow the solutions to cool to ambient temperature.
5. Add 5 mL of 0.05 N sulfuric acid solution to each flask.
6. Add 5 mL of *p*-nitrobenzenediazonium chloride (1.5 mmol/L) to each flask, resulting in the formation of a red azo compound (*see* **Note 18**).
7. Bring each solution to the mark (50 mL) with distilled water.
8. Transfer the samples to cuvettes. Measure the absorbance of the resulting solutions at 520 nm by UV-VIS spectrophotometer (Fig. 8).
9. Construct a linear calibration curve y = ax + b from the measured absorbance values (Fig. 9). The coumarin content, expressed in mg/100 g dry weight, is calculated using the formula:

$$C\ (\text{mg Co}/100\ \text{g d.w.}) = \left|\frac{V_{\text{s.flask}}(\text{mL}) \times 10^{-3}}{m_{\text{sample}}(\text{g})} \times F \times C\ (\text{mg/L})\right| \times 100 \frac{\text{mg coumarin}}{100\ \text{g dry weight}}$$

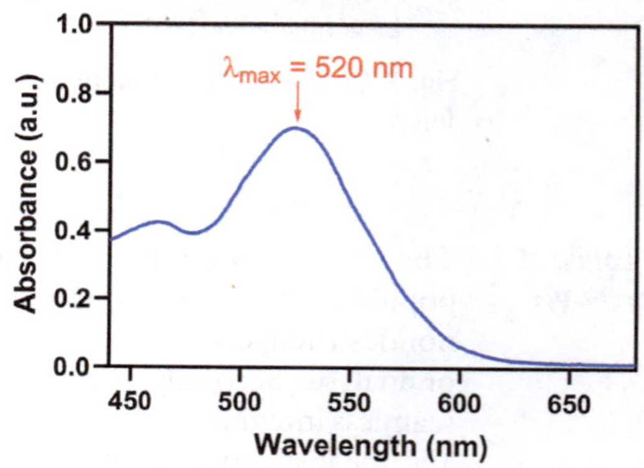

Fig. 8 UV-absorption spectrums of azo coumarin

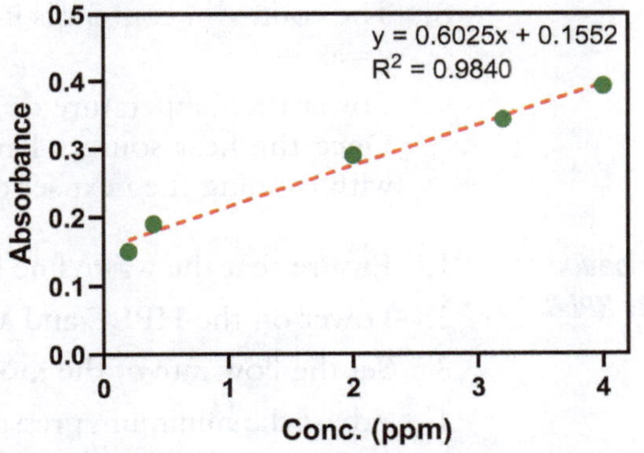

Fig. 9 Standard curves using UV-Vis for quantification of coumarin in plants (For protocols see Methods section)

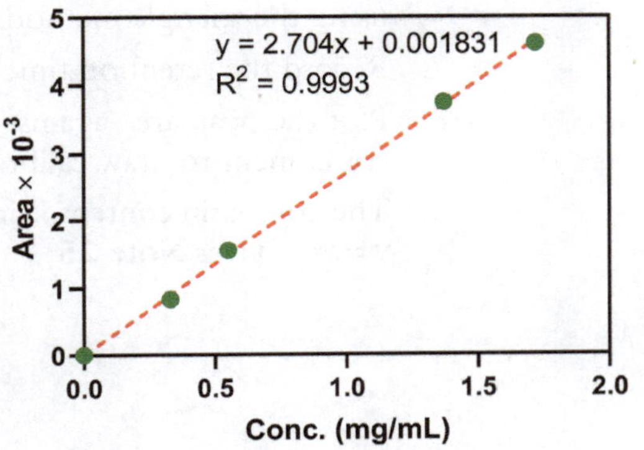

Fig. 10 Standard curves using GC–FID for coumarin (Hydrogen was used as carrier gas, at an average linear velocity of 65 cm/s. The flame ionization detector temperature at 280°C and column temperature programming from 105°C at 15°C/min to 200°C, and from 200°C at 68°C/min to 240°C (held for 10 min)). For protocols see Methods section

3.5 Quantification of Coumarins By GC-FID

1. Ensure the carrier gas cylinder is open, and the chromatography column is installed in the machine.
2. Power on the GC and allow it to warm up. Allow the system to start and run self-checks (no error messages should appear).
3. Set the gas flow rate through the column to 1.2–1.5 mL/min (based on the machine specifications).
4. Use the auto-sampler or manual injection port to introduce the sample into the GC column (25 m × 0.53 mm i.d.) (*see* **Notes 19 and 20**).
5. Record the areas under the peaks for each compound.
6. Determine the peak area on the chromatogram for samples.
7. Plot the mean peak areas obtained from the GC analysis against the injected amount of each standard (Fig. 10).

8. The coumarin content is interpolated from the linear equation y=ax +b
9. Lower the temperature of the injector and other heat sources. Once the heat sources have cooled to below 50°C, proceed with running the next set of samples (*see* **Note 21**).

3.6 Quantification of Coumarins By HPLC

1. Ensure that the waste line is in the waste container.
2. Power on the HPLC and allow it to warm up (*see* **Note 22**).
3. Set the flow rate of the mobile phase to 1 mL/min.
4. Adjust the minimum pressure to 250 psi, and set the maximum pressure to 4000 psi for optimal performance.
5. Use the auto-sampler or manual injection port to introduce the sample into the HPLC column (*see* **Note 23**).
6. Determine the peak area on the chromatogram for samples using the triangle method.
7. Record the retention time of each component (Fig. 11).
8. Plot the peak area against the concentration (mg/L) for each component to draw calibration curves (*see* **Note 24**).
9. The coumarin content is interpolated from the linear equation y=ax+b (*see* **Note 25**).

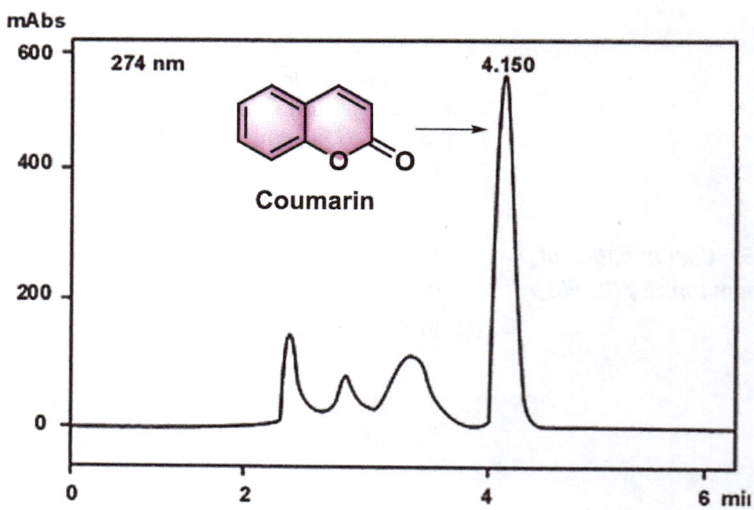

Fig. 11 Chromatogram (HPLC-UV/DAD) of *Mikania glomerata* extract obtained by maceration under sonication. (Reproduced from [21] under the Creative Commons Attribution-ShareAlike 4.0 International Licence)

4 Notes

1. Ensure that appropriate protective equipment is worn and perform the procedure within a fume hood when handling concentrated HCl solution.
2. Depending on the case, different types of normal-phase or reverse-phase silica gel can be used for the purification of various compounds.
3. The test samples used in the manual is the plant extracts. The extract is prepared as per the instruction of the references published paper (Fig. 2) [20].
4. The molarity (N) of concentrated sulfuric acid is usually about 18 N. Always add acid to water, not water to acid, to prevent exothermic reactions.
5. The solution should be prepared in a fume hood and then tightly sealed. The solution can be used effectively within 48 h when stored at a cool temperature of 5°C–10°C. 4-Nitrobenzenediazonium chloride can also be synthesized simply through a diazonium salt formation reaction between 4-nitroaniline and a mixture of $NaNO_2$ and HCl at a temperature of 0°C–2°C.
6. Microwave-assisted extraction and ultrasound-assisted extraction methods can be used to enhance the extraction process.
7. Water extraction will effectively extract polar coumarins compared to using less polar solvents (e.g., hexane), which the concentration process requires specialized equipment. Alternatively, you can heat the extract at 60°C until the water evaporates. However, this process may degrade some heat-sensitive compounds.
8. Extracts obtained by Supercritical Fluid Extraction (SFE)Supercritical Fluid Extraction (SFE) cannot be analyzed directly by HPLC, as they contain high levels of chlorophyll. Cleaning these samples has proven to be unfeasible [21].
9. The solvent must be filtered through a 0.45 μm membrane and degassed by sonication for 30 min. This helps prevent the chromatography column from becoming clogged.
10. After NaOH, 10% (v/v) solution has been added, the test tube No. 1 will become opaque or yellow color or yellow precipitate (Fig. 3).
11. The solution in each test tubes will be returned to its original state upon adding distilled water (transparent). Acidify tube I with a few drops of HCl, and tube I will turn turbid like tube II.

12. Record the observed spots, their R_f values, and any other relevant observations with reagents for future reference, especially during the isolation of compounds from natural sources or synthesis in the laboratory.
13. The extract was filtered using Whatman filter paper to remove fine particles. The solvent was concentrated under reduced pressure using a rotary evaporator at 50°C to remove MeOH.
14. The liquid–liquid extraction was performed in a separatory funnel sequentially with hexane and ethyl acetate. The heavier aqueous layer at the bottom was discarded. The organic solvents were then evaporated at 45°C to obtain the fractionated extracts.
15. The fractions were combined based on the TLC spots with similar R_f values. The solvent was then evaporated to obtain the fractions from 1 to 20.
16. Coumarin was identified through NMR spectroscopy. ^{1}H-NMR [500 MHz, ($CDCl_3$), δ (ppm)): 6.42 (1 H, *d*, J=9.5 Hz, H-3), 7.71 (1 H, *d*, J=9.5 Hz, H-4), 7.5 (1 H, *d*, J=8.0 Hz, H-5), 7.28 (1 H, *t*, J=8.0 Hz, H-6), 7.54 (1 H, *t*, J=8.0 Hz, H-7), and 7.33 (1 H, *d*, J=8.0 Hz, H-8). ^{13}C-NMR (125 MHz, ($CDCl_3$), δ (ppm): 160.7 (C-2), 116,7 (C-3), 143.4 (C-4), 118.9 (C-4a), 127.9 (C-5), 124.4 (C-6), 131.8 (C-7), 116.9 (C8), and 154.1 (C-8a) [22].
17. Place all five volumetric flasks in a thermostatic bath and adjust the temperature at 85°C. This process should be carried out carefully to avoid burns from hot water.
18. The maximum absorption value of the azo coumarin compound is determined to be around 520 nm. The color of the azo coumarin compound changes depending on the environment, ranging from orange-yellow (acidic) to purple (basic). This change can lead to variations in the maximum absorption value.
19. Samples should be dissolved in low boiling solvents.
20. If injecting samples manually, clean the syringe thoroughly with deionized water, then rinse with multiple volumes of the analyte solution, and dry it to avoid any residue buildup that could affect the analysis.
21. GC-FID with a methylation process facilitates the analysis of polar compounds, allowing for more general, economical, and straightforward quantification compared to the GC-MS method [23].
22. Clean and sonicate the solvent filters before running HPLC to prevent residue buildup.
23. If using an automatic sample injector, it is recommended to use the vials provided by the same manufacturer to ensure accuracy and compatibility with the system.

24. The calibration curve shows excellent linearity within the range (1–100 mg/mL). The regression equation, y = 3392.87 + 88593.60x (y = 0.99995), confirms accuracy. With an average standard error under 5% and a detection limit of 5 mg/L, which results indicate low noise, good sensitivity at 274 nm, and efficient analysis within 6 min [21].

25. This analytical technique is not only effective but also versatile, extending its applicability to UPLC systems for the identification of various coumarin derivatives from plants (Fig. 12) [24]. This includes a wide range of coumarins, from highly polar to less polar compounds. The ability to distinguish and quantify these diverse derivatives enhances our understanding of their presence and distribution in plant materials. By employing this advanced technique, researchers can gain deeper insights into the intricate profiles of coumarin compounds, paving the way for further studies on their potential benefits and applications in various fields, such as pharmacology, agriculture, and natural product research.

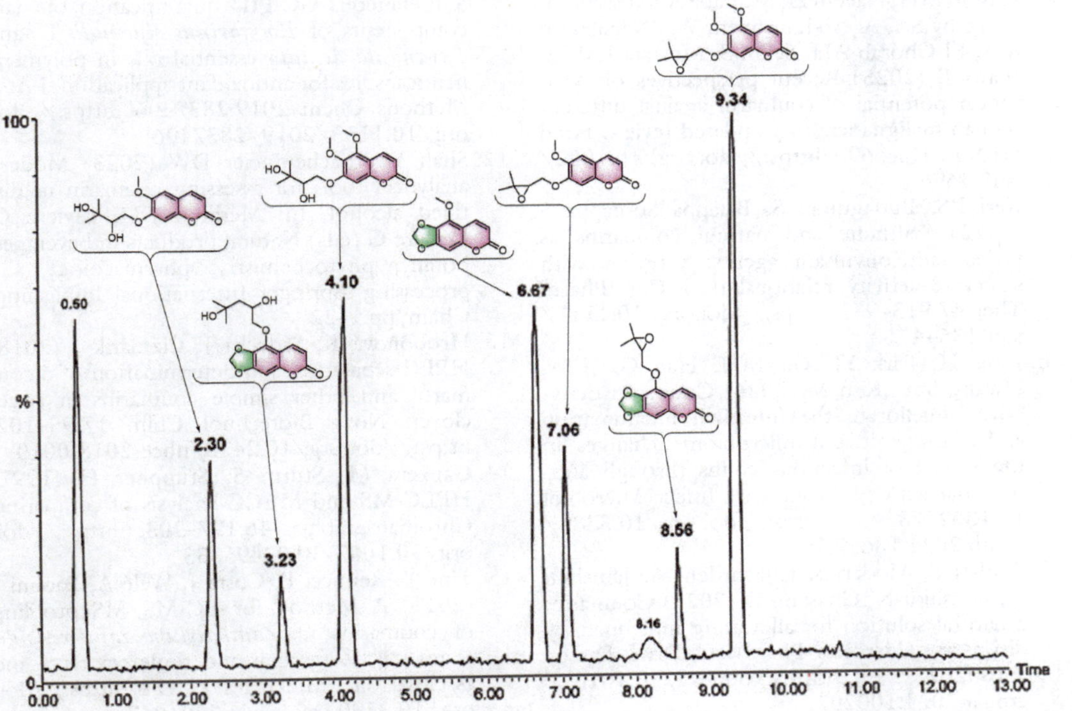

Fig. 12 Chromatographic profile of the aqueous *Pterocaulon balansae* extract by UPLC method. UPLC coupled with ultraviolet (UV) and mass spectrometer (MS/MS) detectors, was performed with the assistance of the ACQUITY UPLC® Columns calculator (Waters). The separation was carried out on Phenomenex Kinetex-C18 UPLC columns at 40°C (C18: 100 mm × 2.10 mm, pore size 1.7 μm, 100Å). (Reproduced from [24] under the Creative Commons Attribution-ShareAlike 4.0 International Licence)

Acknowledgments

This study is funded in part by the Vietnam Ministry of Education and Training (MOET), project number: B2024-TCT-11.

References

1. Matos MJ, Santana L, Uriarte E, Abreu OA, Molina E, Yordi EG (2025) Coumarins—an important class of phytochemicals. In: Phytochemicals—isolation, characterisation and role in human health. IntechOpen
2. Toma AC, Stegmüller S, Richling E (2025) Coumarin contents of tonka (*Dipteryx odorata*) products. Eur Food Res Technol 251:513–517. https://doi.org/10.1007/s00217-024-04648-z
3. Venugopala KN, Rashmi V, Odhav B (2013) Review on natural coumarin lead compounds for their pharmacological activity. Biomed Res Int 2013:963248. https://doi.org/10.1155/2013/963248
4. Shahbaz M, Perween A, Momal U, Imran M, Ul Hassan MH, Naeem H, Mujtaba A, Hussain M, Alsagaby SA, Al Abdulmonem W, Abdelgawad MA, El-Ghorab AH, Selim S, Mostafa EM, Al Jbawi E (2025) Recent perspectives on anticancer potential of coumarin against different human malignancies: an updated review. Food Sci Nutr 13:e4696. https://doi.org/10.1002/fsn3.4696
5. Keri RS, Budagumpi S, Balappa Somappa S (2022) Synthetic and natural coumarins as potent anticonvulsant agents: a review with structure–activity relationship. J Clin Pharm Ther 47:915–931. https://doi.org/10.1111/jcpt.13644
6. Jung H, Park YJ, Gu B-H, Han G, Ji W, Hwang Smi, Kim M (2024) Coumarin derivatives ameliorate the intestinal inflammation and pathogenic gut microbiome changes in the model of infectious colitis through antibacterial activity. Front Cell Infect Microbiol 14:1362773. https://doi.org/10.3389/fcimb.2024.1362773
7. Saadati F, Modarresi Chahardehi A, Jamshidi N, Jamshidi N, Ghasemi D (2024) Coumarin: a natural solution for alleviating inflammatory disorders. Curr Res Pharmacol Drug Discov 7:100202. https://doi.org/10.1016/j.crphar.2024.100202
8. Emam SH, Sonousi A, Osman EO, Hwang D, Kim G-D, Hassan RA (2021) Design and synthesis of methoxyphenyl- and coumarin-based chalcone derivatives as anti-inflammatory agents by inhibition of NO production and down-regulation of NF-κB in LPS-induced RAW264.7 macrophage cells. Bioorg Chem 107:104630. https://doi.org/10.1016/j.bioorg.2021.104630
9. Hussain MK, Khatoon S, Khan MF, Akhtar MS, Ahamad S, Saquib M (2024) Coumarins as versatile therapeutic phytomolecules: a systematic review. Phytomedicine 134:155972. https://doi.org/10.1016/j.phymed.2024.155972
10. Chouchi D, Barth D (1994) Rapid identification of some coumarin derivatives in deterpenated citrus peel oil by gas chromatography. J Chromatogr A 672:177–183. https://doi.org/10.1016/0021-9673(94)80606-3
11. Silva-Flores PG, Pérez-López LA, Rivas-Galindo VM, Paniagua-Vega D, Galindo-Rodríguez SA, Álvarez-Román R (2019) Simultaneous GC-FID quantification of main components of *Rosmarinus officinalis* L. and *Lavandula dentata* essential oils in polymeric nanocapsules for antioxidant application. J Anal Methods Chem 2019:2837406. https://doi.org/10.1155/2019/2837406
12. Shah YA, Lachenmeier DW (2023) Modern analytical tools for assessing coumarin in distilled alcohol. In: Mérillon J-M, Riviere C, Lefèvre G (eds) Natural products in beverages: botany, phytochemistry, pharmacology and processing. Springer International Publishing, Cham, pp 1–22
13. Hroboňová K, Sádecká J, Čižmárik J (2018) HPLC separation and determination of dicoumarol and other simple coumarins in sweet clover. Nova Biotechnol Chim 17:95–102. https://doi.org/10.2478/nbec-2018-0010
14. Ganzera M, Sturm S, Stuppner H (1997) HPLC-MS and MECC analysis of coumarins. Chromatographia 46:197–203. https://doi.org/10.1007/BF02495333
15. Tine Y, Renucci F, Costa J, Wélé A, Paolini J (2017) A Method for LC-MS/MS profiling of coumarins in *Zanthoxylum zanthoxyloides* (Lam.) B. Zepernich and timler extracts and essential oils. Molecules 22:174. https://doi.org/10.3390/molecules22010174
16. Wang B, Liu X, Zhou A, Meng M, Li Q (2014) Simultaneous analysis of coumarin derivatives in extracts of Radix *Angelicae pubescentis* (Duhuo) by HPLC-DAD-ESI-MSn technique. Anal Methods 6:7996–8002. https://doi.org/10.1039/C4AY01468E

17. Lv H, Luo J, Wang X, Kong L (2013) Application of UPLC-quadrupole-TOF-MS coupled with recycling preparative HPLC in isolation and preparation of coumarin isomers with similar polarity from peucedanum praeruptorum. Chromatographia 76:141–148. https://doi.org/10.1007/s10337-012-2368-4
18. Perkowska I, Siwinska J, Olry A, Grosjean J, Hehn A, Bourgaud F, Lojkowska E, Ihnatowicz A (2021) Identification and quantification of coumarins by UHPLC-MS in *Arabidopsis thaliana* natural populations. Molecules 26:1804. https://doi.org/10.3390/molecules26061804
19. Fiorito S, Ianni F, Preziuso F, Epifano F, Scotti L, Bucciarelli T, Genovese S (2019) UHPLC-UV/Vis quantitative analysis of hydroxylated and O-prenylated coumarins in pomegranate seed extracts. Molecules 24:1963. https://doi.org/10.3390/molecules24101963
20. Tram MM, Hoa PTC, Nhon HTN (2022) Nghiên cứu điều kiện chiết coumarin từ lá đơn đỏ (*Excoecaria cochinchinensis*) Effects of extraction conditions on coumarin contents and antioxidant activities of *Excoecaria cochinchinensi*, Vietnamese. T p chí Khoa học Công nghệ và Thực phẩm (HUIT J Sci) 22:47–54
21. Celeghini RMS, Vilegas JHY, Lanças FM (2001) Extraction and quantitative HPLC analysis of coumarin in hydroalcoholic extracts of *Mikania glomerata* spreng: ("guaco") leaves. J Braz Chem Soc 12:706–709. https://doi.org/10.1590/S0103-50532001000600003
22. Trang BTT, Pham T, Trung Hieu P, Huong N, Thoa V, Duyen N, Le Dang Q, Nga B, Trang N (2020) Antifungal activity of coumarin and triterpenoid taraxasteryl acetate from the hexane soluble fraction of the leaves and twigs of eupatorium fortune against sclerotium rolfsii and rhizoctonia solani. Vietnam J Sci Technol 58:17–25. https://doi.org/10.15625/2525-2518/58/6A/15414
23. Young KE, Quinn SM, Trumble S (2012) Comparing gas chromatographic techniques used in fatty acid profiling of Northern Fur Seals (*Callorhinus ursinus*) and Steller Sea Lions (*Eumetopias jubatus*) from Lovushki Island Complex, Russia. Int J Appl Sci Technol 2:1-21
24. Medeiros-Neves B, De Barros FMC, Von Poser GL, Teixeira HF (2015) Quantification of coumarins in aqueous extract of *Pterocaulon balansae* (Asteraceae) and characterization of a new compound. Molecules 20:18083–18094. https://doi.org/10.3390/molecules201018083
25. Huong DTL, Thi PN, Huong LTT, Hoang TN, Thin DB, Thi NN (2012) Định tính coumarin và đánh giá hiệu quả kháng khuẩn của dịch chiết từ một số loài thực vật để chữa bệnh: Kinh nghiệm của người Mường ở Nho Quan, tỉnh Ninh Bình [Determining the coumatin and assensing the anti-bacterial activity of the extractive solution from some medicinal plants by Muong ethnic group based on the lacal people experience in Nho Quan, Ninh Binh province, [Vietnamese]]. Tạp chí Khoa học Trường Đại học Cần Thơ (Can Tho Univ J Sci) 24:140–146

Check for updates

Chapter 11

Qualitative and Quantitative Methods for Tannins

Sapna Chandwani and Natarajan Amaresan

Abstract

Tannins are high molecular weight polyphenolic compounds in the plant kingdom, which are known for their biological activities and ecological importance. These polyphenolic compounds have the ability to bind with proteins and other macromolecules, which play crucial roles in plant defense mechanisms due to their antioxidant, antimicrobial, anticancer, and anti-inflammatory properties. These compounds have also been found to have therapeutic and industrial applications; therefore, the identification and quantification of these compounds in plant extracts are essential to explore their use as food preservatives in food industries to make herbal formulations and to explore their pharmacological potential. Hence, this chapter emphasizes the use of qualitative tests to assess the presence of tannins in plant extracts and a quantitative method to offer reliable and reproducible measurements of total and condensed tannins.

Keywords Tannins, Polyphenols, Phyto-constituents, Folin–Denis method

1 Introduction

Tannins are high molecular weight polyphenolic compounds that are widely distributed across the plant kingdom. Based on their susceptibility to hydrolysis and their chemical structure, these compounds are classified as hydrolyzable tannins (e.g., ellagitannins, gallotannins, etc.) and condensed tannins (proanthocyanidins) [1]. These compounds play an essential role in the management of abiotic and biotic stresses that is they employ a defensive strategy against pathogens and herbivores. These compounds also exhibit antioxidant, antimicrobial, anti-nutritional, and anti-inflammatory effects; therefore, they are of significant interest in several industries such as pharmaceutical, food and beverage, leather tanning, and animal nutrition [1].

To assess the quality, biological activity, and functional properties of plant materials and their derived products, it is essential to accurately determine the amount of tannins. Qualitative analysis is a preliminary screening method to detect the presence of tannins in plant samples by employing simple phytochemical analysis such as the ferric chloride test, gelatin salt precipitation test, lead

Dhruti Amin et al. (eds.), *Phytoconstituent Analysis*, Springer Protocols Handbooks,
https://doi.org/10.1007/978-1-0716-5400-2_11,

acetate test, and goldbeater's skin test. These methods are rapid, cost-effective, and suitable for preliminary detection of tannins [1]. Quantitative methods are used to measure the concentration of tannins in plant extracts or their derived products. These are colorimetric assays that include the Folin–Denis method for the estimation of total tannins, and the vanillin-HCl method for the estimation of condensed tannins. Quantitative methods rely on the principle of chromophoric complex formation, which can be measured using a spectrophotometer at different wavelengths and expressed as standard equivalents such as tannic acid or gallic acid or catechin. Moreover, standardization of the protocol is essential to ensure the reproducibility of the experiment. Appropriate use of the blank, control, and calibration curves are some of the critical parameters to be evaluated. This chapter provides detailed protocols for the qualitative and quantitative analysis of tannins, including the preparation of reagents, procedures, and calculations for accurate analysis of tannins [2].

2 Materials

2.1 Qualitative Analysis of Tannins

1. Plant extract.
2. 1% w/v ferric chloride ($FeCl_3$) solution for the ferric chloride test.
3. 10% lead acetate solution for the lead acetate test.
4. 1% w/v gelatin solution in 10% sodium chloride (NaCl) for gelatin test.
5. 1% ferrous sulfate ($FeSO_4$) solution for Goldbeater's skin test.
6. 2% hydrochloric acid (HCl) for Goldbeater's skin preparation.
7. Goldbeater's skin.
8. Distilled water for the preparation of reagents.

2.2 Quantitative Analysis of Tannins

2.2.1 Requirements for Folin–Denis Method

1. Folin–Denis reagent: For preparation of Folin–Denis reagent, mix 100 g sodium tungstate, 20 g phosphomolybdic acid, 50 mL of 85% phosphoric acid in approximately 750 mL distilled water. Reflux the mixture for 2 h and allow the solution to cool down at room temperature. Make the final volume of the solution to 1 L.
2. 10% sodium carbonate (Na_2CO_3) solution.

3. 100 μg mL^{-1} stock solution of standard gallic acid or tannic acid dissolved in methanol.
4. Distilled water.

2.2.2 Requirements for Vanillin-HCl Method

1. Vanillin-HCl reagent: For preparation of Vanillin-HCl reagent, 1% vanillin is prepared by dissolving it in methanol. The prepared reagent is then mixed with concentrated hydrochloric acid (HCl) in the ratio of 1:1.
2. Standard catechin or tannic acid for analysis of condensed tannins.

2.3 Other Reagents and Equipment's

1. Diluted ammonia solution to precipitate tannins.
2. 70% acetone/methanol/ethanol/petroleum ether for extraction.
3. UV-visible spectrophotometer.
4. Water bath or incubator.
5. Centrifuge.
6. Analytical balance.
7. Mortar and pestle.
8. Whatman No. 1 filter paper.
9. Glassware such as beakers, test tubes, conical flasks, pipettes, and micropipettes.

3 Methods

3.1 Collection of Plant Samples

1. Collect the desired plant parts such as bark, leaves, and fruits. Wash the collected samples thoroughly under running tap water and then with distilled water. Air-dry the samples at room temperature.
2. Once the samples get air-dried, grind the dried plant samples into a fine powder by using a mortar and pestle or a mechanical grinder.
3. Store the dried powder in an air-tight container until further use [3].

3.2 Preparation of Plant Extract

1. Weigh 10 g of grinded plant samples and add 100 mL of 70% methanol or ethanol or 70% aqueous acetone or petroleum ether to it.
2. Macerate the samples with a constant shaking at 150 rpm for 24–48 h.

3. After 24–48 h of maceration, filter the extract through Whatman No. 1 filter paper and store the collected filtrate at 4°C until further use [3].

3.3 Qualitative Screening for Tannins

3.3.1 Ferric Chloride Test

1. Take 2 mL of the plant extract in a test tube and add two to three drops of freshly prepared 1% $FeCl_3$ solution to the extract (*see* **Note 1**).
2. Observe the change in color immediately to differentiate hydrolyzable and condensed tannins.
3. The presence of blue–black color will indicate the presence of hydrolyzable tannins, whereas the green-black or greenish-brown color will indicate the presence of condensed tannins [4].

3.3.2 Gelatin-Salt Test

1. Take 2 mL of the plant extract and inoculate 2 mL of 1% gelatin solution prepared in distilled water containing 10% NaCl (*see* **Note 2**).
2. Stir the mixture thoroughly and gently (*see* **Note 3**).
3. After stirring, allow the mixture to left undisturbed for 10–15 min and then observe for the presence or absence of precipitate or turbidity.
4. Appearance of white or buff-colored precipitate will indicate the presence of tannins in the plant extract [5].

3.3.3 Lead Acetate Test

1. Take 1 mL of the plant extract and add 2 mL of 10% lead acetate solution to it.
2. Allow the solution to stand for 5–10 min and then observe the solution for the formation of a white or bulky precipitate (*see* **Note 4**).
3. White or buff-colored precipitate will indicate the presence of tannins [6].

3.3.4 Goldbeater's Skin Test

1. Prepare thin strips of ox intestine membrane and soak them in 2% HCl for 2 min (*see* **Note 5**).
2. Rinse the prepared strips thoroughly with distilled water and allow them to dry.
3. After drying, dip the prepared strip in the plant extract for about 5 min and then wash it with distilled water.
4. Now, immerse the strip in 1% ferrous sulfate solution for 5 min and observe the change in coloration (*see* **Note 6**).
5. Appearance of brown to black color will indicate the presence of true tannins, i.e., hydrolyzable tannins [7].

3.4 Quantitative Screening for Tannins

3.4.1 Folin–Denis Method for Estimation of Total Tannin Content

1. Weigh 0.5–1.0 g of dried finely grounded plant material and add 10–20 mL of 70% methanol or ethanol or aqueous acetone to it.
2. Constantly shake the mixture in an orbital shaker for 2 h at room temperature and then sonicate the mixture for about 20 min.
3. Filter the mixture through Whatman No. 1 filter paper and keep the sample in refrigerator at 4°C for further analysis.
4. Now, pipette out 1 mL of the extract into a 250 mL volumetric flask and add 5 mL of Folin-Denis reagent to it.
5. Incubate the mixture for about 5 min at room temperature and then add 10 mL of 10% sodium carbonate (Na_2CO_3) solution to it. Make the final volume up to 100 mL with distilled water.
6. Left the solution mixture undisturbed for 30 min at room temperature in order to develop a color.
7. Prepare a standard gallic acid or standard tannins calibration curve (0–100 mg L^{-1} or 0–100 μg L^{-1}) and use water as a blank.
8. Measure the absorbance at 760 nm to determine the concentration of tannins in the given plant extract.
9. Calculate the total tannin content expressed as milligrams or micrograms of tannic acid equivalents (TAE) by the formula:

$$\text{Tannin content (mg or μg TAE g}^{-1}) = \frac{C \times V \times D}{W}$$

where
C = Concentration from calibration (mg or μg mL^{-1})
V = Total volume of the extract (mL)
D = Dilution factor
W = Weight of dried powdered sample (g) [8]

3.4.2 Vanillin Method for Estimation of Condensed Tannin Content (Proanthocyanidins)

1. Prepare the plant sample extract as mentioned above.
2. Prepare standard catechin calibration curve (0–100 μg mL^{-1}) in methanol from stock solution.
3. Pipette out 1 mL of the standard or plant sample extract in a volumetric flask and add 5 mL of vanillin-HCl reagent to it.
4. Shake well and incubate the standard and sample mixture for about 20 min in the dark at room temperature (*see* **Note 7**).
5. To account for the intrinsic sample color, prepare a sample blank by mixing 1 mL of the sample extract with 5 mL of 8% HCl in methanol and omitting the vanillin reagent (*see* **Note 8**).
6. Measure the absorbance of both the test and blank at 520 nm using a spectrophotometer, and the absorbance of blank should

be subtracted from the test (plant extract samples) to obtain corrected values.

7. Plot a calibration curve of absorbance versus standard catechin concentration.
8. To determine the condensed tannin concentration in the sample extract, use the calibration curve, and express the results as mg catechin equivalents (CE) per g of dry plant material by the following formula:

$$\text{Condensed tannin content (mg CE g}^{-1}) = \frac{C \times V}{W}$$

where
C = Concentration from calibration curve ($mg\,mL^{-1}$)
V = Volume of extract used (mL)
W = Weight of dried plant sample used for extraction [9]

4 Notes

1. Ensure that the plant extract is clear; turbidity may mask the occurrence of the reaction.
2. Freshly prepare the gelatin solution and avoid overheating because the high temperature degrades the protein.
3. Do not shake the mixture too vigorously to prevent false turbidity.
4. Wear gloves while preparing a 10% lead acetate solution because it is toxic.
5. Avoid prolonged soaking of the membrane as it can cause non-specific staining.
6. Freshly prepare the ferrous sulfate solution and handle it carefully to avoid oxidation.
7. Incubation time should be consistent for the standards and sample extracts.
8. Use analytical grade methanol, which should be free of impurities as it can hinder the development of color.

References

1. Makkar HP (2013) Quantification of tannins in tree and shrub foliage: a laboratory manual. Springer Science and Business Media, pp 1–102
2. Egbuna C, Ifemeje JC, Maduako MC, Tijjani H, Udedi SC, Nwaka AC, Ifemeje MO (2018) Phytochemical test methods: qualitative, quantitative and proximate analysis. In: Phytochemistry. Apple Academic Press, pp 381–426
3. Nagori M, Rajput D, Choudhary G, Khabiya R (2025) Qualitative and quantitative methods of phytochemical analysis. J Pharmacogn Phytochem 13:143–166
4. Sankhalkar S, Vernekar V (2016) Quantitative and qualitative analysis of phenolic and flavonoid content in *Moringa oleifera* Lam and *Ocimum tenuiflorum* L. Pharmacogn Res 8(1):16

5. Thomas AW, Frieden A (1923) The gelatin-tannin reaction. Ind Eng Chem Res 15(8):839–841
6. Alamgir AN (2018) Methods of qualitative and quantitative analysis of plant constituents. In: Therapeutic use of medicinal plants and their extracts Vol. 2. Progress in drug research, vol 74. Springer, Cham, pp 721–804
7. HonoráPrice P (1924) The gold-beater's skin test for tannins. Analyst 49(574):25–29
8. Pratik DT, Prakash DP, Chandrashekhar KC (2016) Quantitative estimation of tannins from *Pueraria tuberosa* by UV spectrophotometry. Int J Pharm Sci Rev Res 37(1):210–212
9. Bello FA, Henry AA (2015) Storage effects and the postharvest quality of African star apple fruits (*Chrysophyllum africanum*) under ambient conditions. Afr J Food Sci Technol 6(1):35–43

Part II

Advanced Metabolic Profiling Techniques for Phytoconstituent Analysis

Chapter 12

HPLC-MS and SERS Profiling of Opium Alkaloids and DMT: A Nanoparticle-Based Approach

Abhishek Sharma, Roshani Patel, Dhara Patel and Abhishek Sharma

Abstract

Alkaloids are nitrogen contain, naturally occurring compounds present in flowering plants. This chapter focuses on identification and profiling of opium alkaloids and *N, N*-Dimethyltryptamine (DMT) by exploiting advanced chemical techniques involving magnetic solid-phase extraction (MSPE) coupled with High-Performance Liquid Chromatography-Mass Spectrometry (HPLC-MS) and Surface-Enhanced Raman Scattering (SERS). A novel magnetic nanoparticle (Fe_3O_4@SiO_2@mSiO_2@β-CD) modified with β -cyclodextrin was synthesized and employed for the selective extraction of opium alkaloids, which include morphine, codeine, and their intermediates. The extraction efficiency was optimized by varying solvent compositions, pH levels, and adsorption times. The optimized protocol demonstrated high selectivity and recovery rates, making it a promising alternative to conventional sample preparation methods. For DMT detection, gold nanostar (Au NS)-based substrates were fabricated using an inkjet printing technique on functionalized chromatographic paper. Au-NSs allow enhancement of signal intensity during Raman spectrophotometry. The ink stability and enhancement factors of the Au NSs were characterized, and a characteristic Raman peak at 1388 cm^{-1} was identified for DMT detection at low concentrations. The study also evaluated the influence of print cycles and substrate hydrophobicity on SERS (Surface Enhanced Raman Spectroscopy) signal intensity, ensuring reproducibility and sensitivity in real-world applications. The results indicate that Fe_3O_4@SiO_2@mSiO_2@β-CD-based MSPE combined with HPLC-MS provides a reliable method for alkaloid analysis in complex matrices, while Au NS-based SERS enables highly sensitive detection of DMT. Application of MSPE in forensic and pharmaceutical sciences offers robust identification and profiling of psychoactive alkaloids.

Keywords Alkaloids, *β*-cyclodextrin, DMT, Au-nanostars, SERS

1 Introduction

Alkaloids are widespread throughout the plant kingdom, primarily occurring in higher plants like those in the Menispermaceae, Ranunculaceae, Leguminosae, Loganiaceae, and Papaveraceae families. Alkaloids are organic compounds containing one or more nitrogen atoms, often within a heterocyclic ring structure. Typical alkaloids are basic and primarily derived from plants. However, some compounds,

Dhruti Amin et al. (eds.), *Phytoconstituent Analysis*, Springer Protocols Handbooks,
https://doi.org/10.1007/978-1-0716-5400-2_12,

like hordenine, ephedrine, and colchicine, are classified as protoalkaloids because they lack the properties of typical alkaloids [1].

Various alkaloids have been discovered to have diverse biological activities. Emetine is found in plants belonging to the Rubiaceae, Alangiaceae, and Icacinaceae families. This alkaloid inhibits protein synthesis in mitochondria and ribosomes. Emetine also blocks DNA and RNA synthesis leading to its antiviral and anticancer properties. Morphine and Codeine are present naturally in *Papaver somniferum*) extracted from poppy is used as a vasodilator [1].

DMT (*N, N*-Dimethyltryptamine), a pharmacologically active compound found naturally in *Mimosa hostilis*, is an alkaloid of interest in brain research. DMT is metabolized rapidly by Monoamine deoxidase enzyme [2]. Though DMT usage is illicit, its endogenous presence in the mammalian brain with natural role of neurotransmitter and potential role of neuroprotectant and neuroregenerative [2]. With concerns of potential misuse of alkaloids and active neurological research in the pharma industries, the detection and identification of alkaloids becomes necessary.

High-performance Liquid Chromatography (HPLC) coupled with Mass Spectrometry is the most used technique for the identification of analytes. However, purification of the sample with the requirement of pre-concentration and reducing matrix elements in the analyte solution makes HPLC-MS inadequate for analyte (alkaloid) detection. Magnetic solid phase extraction (MSPE) can reduce matrix effect in HPLC-MS. MSPE helps in reducing the time to extract analytes. Sorbents made of Fe (III) Terephthalate have shown high capability to extract alkaloids from bakery samples for detection [3]. Apart from MSPE used in tandem with HPLC-MS, it can also be used with a Raman spectrophotometer to observe surface-enhanced Raman scattering (SERS). SERS involves increasing the Raman signal due to excitation light by the surface of metal generating local surface plasmons [4]. This chapter discusses protocols to detect opium alkaloids and DMT with the help of nanoparticle-based solid phase extraction and subsequent use of HPLC-MS and Raman spectroscopy for respective alkaloids.

2 Materials

2.1 Opium Alkaloids [5]

1) Morphine Standard.
2) Codeine Standard.
3) Thebaine Standard.
4) Oripavine Standard.
5) Noscapine Standard.
6) Papaverine Standard.

7) Ferric chloride 6-hydrate, ≥99%.
8) Ferrous chloride 4-hydrate, ≥99%.
9) Tetraethylorthosilicate (TEOS), ≥98%.
10) Hexadecyl(trimethyl)ammonium bromide.
11) 3-isocyanatopropyltriethoxysilane, 98%.
12) β-cyclodextrin ≥97%.
13) Absolute Ethanol.
14) Formic Acid, 98%.
15) Ammonia, 32% (w/w).
16) Isopropanol, synthesis grade.
17) Toluene, synthesis grade.
18) Pyridine, synthesis grade.
19) Diethyl Ether, synthesis grade.
20) *N, N*-dimethyl formamide (DMF).
21) Acetonitrile, HPLC-MS grade.
22) Methanol, HPLC-MS grade.
23) Ethanol, HPLC-MS grade.
24) Ultra pure water (resistivity: 18.2 MΩ•cm).
25) Nd-Fe-B magnet (5 cm × 5 cm × 2 cm, force: 200 kg).
26) HPLC-MS/MS system.

2.2 N, N -Dimethyltryptamine (DMT) Alkaloid [6]

1) Hydrogen tetrachloroaurate(III) trihydrate ($HAuCl_4.3H_2O$), ≥99.9%.
2) Cetyltrimethylammonium Bromide (CTAB), ≥99.9%.
3) Trisodium Citrate Dihydrate, ≥98%.
4) L-ascorbic Acid, ≥99%.
5) Silver Nitrate ($AgNO_3$), ≥99%.
6) Chloridric Acid, 37%.
7) Whatman 30001-861 Cellulose chromatographic paper.
8) (2-dodecen-1-yl) Succinic Anhydride (DDSA), 95%.
9) 1-Hexanol, 98%.
10) Glycerol, ≥99.5%.
11) Absolute Ethanol, 99.5%.
12) Crystal Violet (CV).
13) Ink Jet Printer.
14) Raman Spectrophotometer.
15) Charged-coupled device detector.
16) Confocal Microscope.

3 Methods

3.1 Opium Alkaloids [5]

1) Stock solution of opioid standards prepared at a concentration of 1 mg/mL in methanol. Working solutions prepared with of concentration 1 mg/L in water/ethanol (3:1, v/v) and 10% formic acid. Solutions stored in photon limiting condition at −20°C.
2) *Preparation of β-cyclodextrin monomer*: Moisture-free β-cyclodextrin (2.5 g) was added to pyridine (50 mL). Add 3- isocyanatopropyltriethoxysilane (1.56 g; stoichiometry 1:3, β-cd:3- isocyanatopropyltriethoxysilane) in β-cyclodextrin solution and place it under magnetic stirring condition (500 rpm) for 24 h at 80°C and inert conditions. Solvent is evaporated at 27°C in vacuum for 24 h and the light-yellow (dried solid) residue is dried (See Note 1).
3) *Magnetite Preparation* [7]: 2.35 g of $FeCl_3.6H_2O$ (2.35 g) and 0.86 g of $FeCl_3.4H_2O$ are dissolved in 40 mL of water (deoxygenated). 30 mL Ammonium hydroxide (5%) added dropwise to ferric and ferrous hydrate solution at 80°C in an inert N_2 gas condition, further stirred vigorously for 30 min. Precipitate, black in color, is separated with the help of a magnet. Precipitate was washed with MilliQ (deionized) water until the pH of water was near neutral.
4) *Preparation of Fe3O4@SiO2* [8]: Fe_3O_4 nanoparticles (1.5 g) were resuspended in isopropanol/MilliQ water (5:1, v/v; 60 mL). Dispersion is ultrasonically dispersed for 20 min. Ammonia (15 mL) solution and TEOS (8 mL) were added sequentially and promptly to Fe_3O_4 suspension and stirred for 24 h at room temperature. Nanoparticles are separated with the help of a magnet and washed with MilliQ water to remove reactants till pH of the washing ultra pure water is close to neutral. NPs are dried at 60°C under vacuum for 24 h.
5) *Preparation of Fe3O4@SiO2@mSiO2* [8]: Fe_3O_4@SiO_2 NPs (3 g) dispersed in MilliQ water/Ethanol (1.33:1, v/v) solution, Ammonia (12 mL), and CTAB (1.4 g). Suspension is stirred for 30 min to homogenize the constituents. TEOS (6.4 mL) is added dropwise to the suspension, placed in stirring conditions for 6 h. Mesostructured silica-coated Fe_3O_4 NPs are separated by magnet and washed with methanol and ethanol multiple times. NPs are separated with magnet and calcined with gradual heating (1°C/min) to 550°C from room temperature and held at 550°C for 6 h.
6) *Functionalization of Fe3O4@SiO2@mSiO2 with β-CD*: Fe_3O_4@SiO_2@mSiO_2 NPs (3 g) mixed with β-CD monomer (25% by

weight, 0.75 g) in 50 mL DMF under magnetic stirring conditions at 500 rpm and 70°C for 24 h in an inert N_2 environment. Material is separated with the help of a magnet and washed with DMF, diethyl ether, and ethanol multiple times before drying in vacuum for 24 h.

7) *Extraction of Opium alkaloids*: Fe3O4@SiO2@mSiO2 (50 mg) dissolved in pH 9 water and ultra sonicated for 1 min with addition of 2 mL alkaloid sample infusion (pH 9) further ultra sonication for 1 min for adsorption equilibrium between NPs and analytes. Separate NPs with help of a magnet and elute analyte from NPs using wash solution (2 mL) of 1% formic acid with ethanol/water (50% v/v) for 1 minute. Recover the analytes in the wash solution and make 5% solution of standards (1 mg/L) with elutes. Analyze the solution by HPLC-MS/MS.
8) *Settings for HPLC-MS/MS: Injection volume set at 10 μl, with 0.25 mL/min flow rate. Column re-equilibration with water and acetonitrile (90-30% water, 0-6 min; 30-90% water, 6-9 min; 90% water; 9-11 min) and 0.1% formic acid. MRM mode selected during ESI+* (positive mode electron spray ionization). Inert nitrogen used as drying and nebulizer gas. Both gas temperatures are set at 350°C with drying at 22 psi and nebulizer at 58 psi. Argon is used as a collision gas at a pressure of 1.90 mTorr and a voltage (detector) of 1480 V. 5000 V held at capillary, with shielding at 600 V and cone voltage at 72 V.

3.2 N, N -Dimethyltryptamine (DMT) Alkaloid [6]

Synthesis of Au Nano Stars :

1) Seeding: Sodium Citrate (1% m/m) 7.5 mL added to 50 mL aqueous boiling solution of $HAuCl_4$ (1 mmol/L). Boil the solution till the color change to red wine (15 min). Store solution at room temperature.
2) Seed growth: Mix 50 μL of $HAuCl_4$ (50 mmol/L), 10 μL of HCl (1 mol/L) with 10 mL of seed solution. Mix 10 μL of $AgNO_3$ (20 mmol/L) and 50 μL Ascorbic Acid simultaneously while stirring the solution for 10 min.
3) Add 100 μL of CTAB buffer (100 mmol/L) under stirring condition for 10 min.
4) Centrifuge the solution, remove excess CTAB buffer, and resuspend in 10 mL ultrapure water.

Fabricating SERS substrate (Inkjet-printed Au Nano Stars) (See Note 2)

1) Activating Whatman chromatographic paper with DDSA: Chromatographic paper dipped in 50 mL DDSA (0.5% m/v) prepared in 1-hexanol. Curing to increase the strength of the material is done at 160°C for 6 min.

2) Activation of base Whatman 30001-861 with DDSA and curing procedure is replicated twice for a single paper.
3) Viscosity of the Gold Nano Star solution is calibrated to mimic the viscous property of commercial ink. Au NSs solution is suspended in a glycerol and ethanol mixture with a ratio of 65:25:10, respectively.
4) Deposit the Au NSs suspension on functionalized Whatman paper using Ink Jet Printing setup.
5) Raman spectroscopy of the analyte. Low concentrations of DMT will give a characteristic peak at 1388 cm^{-1}.

4 Notes

1. Initial study to standardize *β*) monomer formation done in toluene, DMF, and pyridine. Lower yield of monomer was obtained due to the low solubility of *β*-CD in toluene. DMF forms a hydrogen bond with *β*-CD monomer. This makes evaporation of solvent difficult due to the high temperature and the vacuum environment required for its removal. Pyridine, having a lower boiling point and lacking strong interaction with the monomer, is suitable for the preparation of a modified monomer. Characteristic band of 3-isocyanatopropyltriethoxysilane and *β*-CD is observed in ATR-FT-IR at 2270 cm^{-1} [9] and 900–1400 cm^{-1} and 3000 cm^{-1} respectively. Lack of band at 3400 cm^{-1} due to deprotonation of OH^{-1} group in silica funcationalized material indicates bonding between Fe_3O_4 NPs and modified monomer. The amount of *β*-CD monomer added to magnetite was standardized to 25%. Both 100% and 12.5% of *β*-CD monomer resulted in a decrease in pore size availability and functionalized surface, respectively. Adsorption study done with all three forms of substrates, Fe_3O_4@*β*-CD, Fe_3O_4@SiO_2@*β*-CD, and Fe_3O_4@SiO_2@mSiO_2@*β*-CD showed maximum adsorption with mesostructured silica modified NPs with pH at 9. 1 min was optimized for adsorption of opioid standards on magnetized solid substrate. Thebaine, papaverine, and noscapine had 70%, codeine (60%), and morphine (40%) recovery with Fe_3O_4@SiO_2@mSiO_2@*β*-CD. The adsorptions of standards with Fe_3O_4 and with silica modified substrate (Fe_3O_4@SiO_2@mSiO_2) were significantly lower than with Fe_3O_4@SiO_2@mSiO_2@*β*-CD. Desorption of analytes was optimized for 50% methanol, ethanol, and acetonitrile with water with 1% formic acid. Ethanol/water mixture (50% v/v) with 1% formic acid showed the highest elution of standards, 94%, 98%, 97%, 85%, 90%, and 91% for morphine, codeine, thebaine, papaverine,

noscapine, and oripavine, respectively. This procedure was also found reproducible after multiple adsorption experiments done in batches [5].

2. Au NSs preparation should be increased to 500 mL batch due to a large requirement of NSs to fill printer ink cartridge. The stability of Au NSs ink and printed substrate can be assessed by observing a redshift from 690 to 627 nm and a single peak observed at 548 nm, respectively, after 90 days. With increase in print cycle (up to 10) the contact angle (avg.) decreased from 110° to 91°. A single print of substrate shows maximum SERS intensity, while increasing prints cause less surface availability and decreased hydrophobicity of the paper. At high concentration of DMT characteristic spectra of the indole group at 765, 1358, 1581, and 1616 cm^{-1} are observed [6] while at low concentrations ($<10^{-5}$ mol/L) there is an interaction with alkyl amino side chain of DMT molecule [10, 11]. Therefore, for protonation of the alkyl amino group, an analyte solution is prepared at pH 5.8. Characteristic adsorption of DMT is observed at 1388 cm^{-1}.

References

1. Debnath B, Singh WS, Das M, Goswami S, Singh MK, Manna K, Das S (2018) Role of plant alkaloids on human health: a review of biological activities. Mater Today Chem 9:56–72. https://doi.org/10.1016/j.mtchem.2018.05.001
2. Barker SA (2018) N, N-dimethyltryptamine (DMT), an endogenous hallucinogen: past, present, and future research to determine its role and function'. Front Neurosci 12:536. https://doi.org/10.3389/fnins.2018.00536
3. Casado-Hidalgo G, Martínez-García G, Morante-Zarcero S, Pérez-Quintanilla D, Sierra I (2022) New validated method for the determination of six opium alkaloids in poppy seed-containing bakery products by high-performance liquid chromatography-tandem mass spectrometry after magnetic solid-phase extraction. J Agric Food Chem 70(24):7594–7606. https://doi.org/10.1021/acs.jafc.2c01664
4. Langer J et al. (2020) Present and future of surface-enhanced Raman scattering. ACS Nano 14(1):28–117. https://doi.org/10.1021/acsnano.9b04224
5. Casado-Hidalgo G, Morante-Zarcero S, Pérez-Quintanilla D, Sierra I (2023) Magnetic material based on mesostructured silica functionalized with β-cyclodextrin to extract opium alkaloids in poppy seed infusions. Adv Sample Prep 6:100056. https://doi.org/10.1016/j.sampre.2023.100056
6. Huertas-Montoya E, Barros A, Oliveira L, Aponte P, Oliveira A, Shimizu F, Sussulini A, Sigoli F, Mazali I (2024) Detection of psychoactive N,N-dimethyltryptamine alkaloid based on surface-enhanced Raman scattering using gold nanostars in flexible inkjet-printing paper substrates. J Braz Chem Soc 35. https://doi.org/10.21577/0103-5053.20240105
7. Zhang H-F, Shi Y-P (2012) Magnetic retrieval of chitosan: Extraction of bioactive constituents from green tea beverage samples. Analyst 137:910–916. https://doi.org/10.1039/C1AN15873B
8. Casado-HidalgoG,Pérez-QuintanillaD,Morante-Zarcero S, Sierra I (2021) Mesostructured silica-coated magnetic nanoparticles to extract six opium alkaloids in poppy seeds prior to ultra-high-performance liquid chromatography-tandem mass spectrometry analysis. Foods 10(7):1587. https://doi.org/10.3390/foods10071587
9. National Center for Biotechnology Information (2025) PubChem compound summary for CID 90613, triethoxy(3-isocyanatopropyl)silane. https://pubchem.ncbi.nlm.nih.gov/compound/Triethoxy_3-isocyanatopropyl_silane. Accessed 30 Jan 2025

10. Wu X, Cañamares MV, Kakoulli I, Sanchez-Cortes S (2022) Chemical characterization and molecular dynamics simulations of bufotenine by surface-enhanced Raman scattering (SERS) and density functional theory (DFT. J Phys Chem Lett 13(25):5831–5837. https://doi.org/10.1021/acs.jpclett.2c01300
11. da Fonseca BG, Costa LAS, Sant'Ana AC (2018) Insights of adsorption mechanisms of Trp-peptides on plasmonic surfaces by SERS. Spectrochim Acta A Mol Biomol Spectrosc 190:383–391. https://doi.org/10.1016/j.saa.2017.09.039

Chapter 13

Terpenoids: Advanced Chromatographic Profiling

Roshani Patel, Abhishek Sharma and Abhishek Sharma

Abstract

The enormous diversity of terpenes found in nature is generated by enzymes known as terpene synthases or cyclases. Some are also known for their ability to convert a single substrate into multiple products. An enormous range of plant substances is covered by the word "terpenoid," a term which is used to indicate that all such substances have a common biosynthetic origin. A considerable number of quite different functions have been ascribed to plant terpenoids. Although terpenoids are derived biogenetically from the molecule of isoprene, which does occur as a natural product, this substance is not the in vivo precursor. The mainly terpenoid essential oils comprise the volatile, steam-distillable fraction responsible for the characteristic scent, odor, or smell found in many plants. They are commercially important as the basis of natural perfumes and also of spices and flavorings in the food industry. Most natural "terpenoids" have cyclic structures with one or more functional groups (hydroxyl, carbonyl, etc.), so that the final steps in synthesis involve cyclization and oxidation or other structural modifications. Terpenoids are normally extracted from plant tissues with light petroleum, ether, or chloroform and can be separated by chromatography on silica gel or alumina using the same solvents. For their isolation from plant tissues, mono- and sesquiterpenes are nowadays separated by extraction into ether, petroleum, or acetone. The classic procedure for essential oils is separation from fresh tissue by steam distillation. The methods such as metabolic profiling involve the comprehensive analysis of metabolites in a biological sample to understand metabolic pathways, identify bioactive compounds, and explore variations in metabolite levels under different conditions.

Keywords Terpenoids, Metabolic profiling, High-performance liquid chromatography, Ultra processing liquid chromatography, Photodiode array

1 Introduction

Terpenoids are organic, primary bioactive compounds found in essential oils of plant extracts and are derived from isoprene, a 5-carbon atom. These are also known as isoprenoids. Terpenoids are the largest class of plant secondary metabolites. These substances demonstrate remarkable biological activity and are frequently used as medicines, food additives, and scent molecules. The complex nature of terpenoids and different functional groups has a wide range of

Dhruti Amin et al. (eds.), *Phytoconstituent Analysis*, Springer Protocols Handbooks,
https://doi.org/10.1007/978-1-0716-5400-2_13,

pharmacological effects on the human body, including the treatment, prevention, and reduction of discomfort associated with a number of chronic illnesses [1].

Chemically, terpenoids are generally lipid-soluble and are located in the cytoplasm of plant cells. Terpenoids are normally extracted from plant tissues with light petroleum, ether, or chloroform and can be separated by chromatography on silica gel or alumina using the same solvents. Most terpenoids originate from plants and are known as precursors of many essential compounds such as vitamins, hormones, and medicines, apart from providing plants with their distinct fragrance and flavor properties [2]. Essentially all terpenoids are derived by the condensation of multiple units of IDP and its isomer DMADP by prenyl-converting enzymes; further conjugation or cyclization steps significantly enhance the chemodiversity in nature [3].

A major factor behind terpenoid diversity is the ability of these highly promiscuous enzymes to synthesize multiple products from a single substrate through a complex reaction cascade. Besides the main product, almost half of identified mono- and sesquiterpene synthases generate substantial amounts of different products with potential to be classified as multiproduct synthases [4, 5]. It is believed that the ability to form a large number of products from single substrate is mainly due to the electrophilic reaction cascades controlled by these enzymes [4, 5].

The value in identifying and quantifying isoprenoids may be illustrated with two examples. First, for some decades, isoprenoid biosynthesis has been an important target site for bleaching herbicides, and quantitative structure–activity relationships (QSAR) need to be determined [6]. Secondly, the health-promoting properties of certain antioxidant isoprenoids such as carotenoids and vitamin E have resulted in intense interest in determining and manipulating their levels in plants [7]. The amenability of crops to genetic manipulation and the availability of isoprenoid biosynthetic genes [8, 9] enables the isoprenoid content of food crops such as tomato [10] and rice [11] to be altered.

In order to assess the utility of genes and fully evaluate the effects of these manipulations, rapid and comprehensive isoprenoid analyses are required. The extraction, separation, and identifcation of a wide range of compounds in a single measurement (metabolic profiling) will be essential for the progression of such studies [12]. Traditionally, analytical biochemists have focused their efforts on optimizing extraction, separation, and detection methodologies for specific compounds rather than for a range of metabolites in a single sample. Advances in DNA technologies have facilitated the genetic manipulation of metabolic pathways and physiological processes in plants.

Metabolic profiling involves the comprehensive analysis of metabolites in a biological sample to understand metabolic pathways, identify bioactive compounds, and explore variations in metabolite levels under different conditions. Recent advances in chromatographic techniques have significantly enhanced the ability to analyze with high sensitivity and specificity. High-performance liquid chromatography (HPLC), gas chromatography (GC), and liquid chromatography-mass spectrometry (LC-MS) are effective methods for metabolic profiling. These techniques allow the separation of complex mixtures, the identification of individual compounds, and the quantification of terpenoid concentrations in various plant tissues under different environmental conditions. A photodiode array detector (PDA), often coupled with chromatography, is a powerful tool for this purpose because it captures the full UV-visible spectrum of analytes in real time. A PDA detector measures the absorbance of light across a range of wavelengths (typically 190–800 nm) as analytes pass through the detector.

2 Materials

Chemicals and solvents were obtained from Merck (Lutterworth, Leicestershire, UK):

1. Ammonium acetate.
2. HPLC-grade solvents.
3. MS medium (Murashige and Skoog).
4. 0.5% sucrose.
5. Wild-type *Arabidopsis thaliana* (ecotype Columbia).
6. 0.2% phytagel.
7. Distilled water.
8. Chloroform.
9. Mortar and pestle.
10. Microcentrifuge tube or screw-capped Pyrex tube.
11. Tris-HCL (include NaCl).
12. Pasteur pipette.
13. Ultra Turrax T25 (Jankel & Kunkel, Starufen, Germany).
14. 60% KOH.
15. Buffer (50 mM Tris-HCl).
16. Ethyl acetate.
17. Water system (Watford, Hertfordshire, UK) consisting of a no. 616 pump, no. 996 diode array detector, and no. 717 auto-sampler or a Waters Alliance 2600 S system with no. 996 diode array.
18. Water Millennium software.

19. No. 7955 column oven (Jones Chromatography, Hengoed, Mid-Glamorgan, UK).
20. A reverse-phase C18 Nucleosil 5 mm (250 × 4.6 mm or 150 × 4.6 mm) column (Jones Chromatography) coupled to a 5 mm (10 ×4.6 mm) Phase Sep C18 guard column (Phase Separations, Deeside, Clwyd, UK).
21. Acetonitrile-based mobile phase (either methanol, water, isopropanol, or ethyl acetate).
22. Reverse-phase C30, 5 mm column (250 × 4.6 mm) coupled to a 20 × 4.6 mm C30 guard (YMC Inc., Wilmington, NC, USA).
23. Mobile phase: (A) methanol, (B) water/methanol (20/80 by volume) containing 0.2% ammonium acetate, (C) *tert*-methyl butyl ether.
24. Inert Sil 5 mm, 250 × 4.6 mm column with identical guard unit (10 × 4.6 mm) purchased from Chrompak (Walton, Surrey, UK).
25. 0.5% ethanol in n-hexane.
26. 20% ethyl acetate in hexane.

3 Methods

3.1 Plant Material, Growth Conditions, and Application of Inhibitors

1. Wild-type *Arabidopsis thaliana* (ecotype Columbia) seeds are surface-sterilized and germinated on Murashige and Skoog (MS) medium.
2. Seedlings are grown under controlled environmental conditions (*see* **Note 1**).
3. Inhibitors are prepared in methanol as 100× stock solutions and stored at −20°C.
4. Media containing inhibitors are prepared by adding an appropriate aliquot of the stock solution to obtain a final concentration of 100 mM.
5. Methanol is used as the solvent control.
6. Grown plants are harvested, gently washed with distilled water (dH_2O), frozen, and stored at −20°C until further use.

3.2 Extraction of Isoprenoids

1. Freeze-dried *Arabidopsis* tissue (1–10 mg) is ground into a fine powder using a mortar and pestle.
2. Extraction is carried out in microcentrifuge tubes using methanol (100 μL).

3. Tris-HCl buffer (50 mM, pH 7.5) containing 1 M NaCl (100 μL) is added, followed by incubation at 4°C for 10 min.
4. Chloroform (400 μL) is added, and the mixture is incubated on ice for 10 min, followed by centrifugation at 3000 × *g* for 5 min at 4°C.
5. The hypophase is removed using a Pasteur pipette, and the aqueous phase is re-extracted with chloroform.
6. The combined organic phases are dried by centrifugal evaporation and stored at −20°C prior to HPLC analysis.

3.3 Isoprenoid Separation and Detection by HPLC-PDA

1. The dried HPLC sample residue is dissolved in ethyl acetate.
2. Chromatographic separation is performed using a Waters HPLC system (*see* **Note 2**).
3. Data are analyzed using Waters Millennium32 software.
4. Separation is carried out using a reverse-phase C18 Nucleosil column (*see* **Note 3**).
5. In addition, a normal-phase Inertsil column is used (*see* **Note 4**).
6. For the normal-phase column, mobile phases consist of either 0.5% ethanol in n-hexane run isocratically or a gradient elution from 20% ethyl acetate in hexane to 100% ethyl acetate over 25 min.

3.4 Quantification

1. Peak areas of standards are determined at wavelengths corresponding to their maximum absorbance using the Waters Millennium32 software.

4 Notes

1. A 16 h photoperiod is provided using fluorescent lights at an intensity of 250 μmol m^{-2} s^{-1}.
2. The HPLC system consists of a Waters system (Watford, Hertfordshire, UK) equipped with a no. 616 pump, no. 996 photodiode array detector, and no. 717 autosampler, or alternatively a Waters Alliance 2600S system fitted with a no. 996 photodiode array detector.
3. A reverse-phase C18 Nucleosil column (5 μm; 250 × 4.6 mm or 150 × 4.6 mm; Jones Chromatography) coupled with a C18 guard column (5 μm; 10 × 4.6 mm; Phase Separations, Deeside, Clwyd, UK) is used.
4. A normal-phase Inertsil column (5 μm; 250 × 4.6 mm) with an identical guard unit (10 × 4.6 mm), purchased from Chrompak (Walton, Surrey, UK), is used.

References

1. Das DD, Sharma N, Chawla V, Chawla PA (2023) Current trends of analytical techniques for bioactive terpenoids: a review. Crit Rev Anal Chem 54(8):2984–3000
2. Fraser PD, Pinto MES, Holloway DE, Bramley PM (2000) Application of high-performance liquid chromatography with photodiode array detection to the metabolic profiling of plant isoprenoids. Plant J 24(4):551–558
3. Brandt W, Bräuer L, Günnewich N, Kufka J, Rausch F, Schulze D, Schulze E, Weber R, Zakharova S, Wessjohann L (2009) Molecular and structural basis of metabolic diversity mediated by prenyldiphosphate converting enzymes. Phytochemistry 70:1758–1775
4. Gambliel H, Croteau R (1984) Pinene cyclases I and II. Two enzymes from sage (Salvia officinalis) which catalyze stereospecific cyclizations of geranyl pyrophosphate to monoterpene olefins of opposite configuration. J Biol Chem 259:740–748
5. Degenhardt J, Köllner TG, Gershenzon J (2009) Monoterpene and sesquiterpene synthases and the origin of terpene skeletal diversity in plants. Phytochemistry 70:1621–1637
6. Böger P, Sandmann G (1989) Inhibition of carotenoid biosynthesis by herbicides. In: Böger P, Sandmann G (eds) Target sites of herbicide action. CRC Press, Boca Raton, FL, pp 25–32
7. Grusak MA, DellaPenna D (1999) Improving the nutrient composition of plants to enhance human nutrition and health. Annu Rev Plant Physiol Plant Mol Biol 50:133–161
8. Cunningham Jr. FX, Gantt E (1998) Genes and enzymes of carotenoid biosynthesis in plants. Annu Rev Plant Physiol Plant Mol Biol 49:557–583
9. Scolnik PA, Bartley GE (1996) A table of some cloned plant genes involved in isoprenoid biosynthesis. Plant Mol Biol Report 14:305–319
10. Römer S, Fraser PD, Kiano JW, Shipton CA, Misawa N, Schuch W, Bramley PM (2000) Elevation of the provitamin A content of transgenic tomato plants. Nat Biotechnol 18:666–669
11. Ye X, Al-Babili S, Kloti A, Zhang J, Lucca P, Beyer P, Potrykus I (2000) Engineering the pro-vitamin A (β-carotene) biosynthetic pathway into (carotenoid-free) rice endosperm. Science 287:303–305
12. Trethewey RN, Krotzky AJ, Willmitzer L (1999) Metabolic profiling: a Rosetta Stone for genomics?. Curr Opin Plant Biol 2:83–85

Chapter 14

Phenolic Acids: High Resolution Profiling Techniques

Snaa Mistry and Ankit Chaudhary

Abstract

Phenolic acids constitute a crucial class of plant secondary metabolites belonging primarily to hydroxybenzoic and hydroxycinnamic acid derivatives. Owing to their antioxidant, antimicrobial, anti-inflammatory, and chemopreventive properties, they are extensively investigated in plant metabolomics, food chemistry, pharmacognosy, and natural product research. Accurate profiling of phenolic acids, however, requires high-resolution analytical platforms due to their low abundance, structural similarity, extensive conjugation, and strong matrix interference. This chapter presents a comprehensive, laboratory-ready protocol for the high-resolution profiling of major phenolic acids using UHPLC–PDA–MS/MS, HPLC–DAD, LC–QTOF–MS, LC–Orbitrap–MS, and GC–MS after derivatization. The workflow includes sample collection, extraction, clean-up using solid-phase extraction (SPE), chromatographic separation, mass-spectrometric acquisition, and data analysis.

Keywords Hydroxybenzoic acids, Hydroxycinnamic acids, UHPLC–PDA–MS/MS, HPLC–DAD, QTOF, Orbitrap, GC–MS

1 Introduction

Phenolic acids represent an important subgroup of polyphenols widely distributed in fruits, vegetables, cereals, legumes, herbs, and medicinal plants [1]. They occur as free acids, glycosides, esters, and complex conjugates linked to cell-wall polysaccharides and organic acids [2]. Their biosynthesis follows the shikimate and phenylpropanoid pathways and contributes to plant defense, pigmentation, stress responses, and interaction with microorganisms [3].

Structurally, phenolic acids are categorized into:

1. **Hydroxybenzoic acids:** (C6–C1 backbone; e.g., gallic, vanillic, protocatechuic, syringic acids) [4].
2. **Hydroxycinnamic acids:** (C6–C3 backbone; e.g., caffeic, ferulic, p-coumaric, sinapic acids) [4].

These compounds often co-exist in closely related structural forms, including positional isomers, methoxylated derivatives,

Dhruti Amin et al. (eds.), *Phytoconstituent Analysis*, Springer Protocols Handbooks,
https://doi.org/10.1007/978-1-0716-5400-2_14,

geometric isomers (E/Z), and conjugates (e.g., chlorogenic acid) [5] (*see* **Note 1**), and are sensitive to pH and light, so sample handling must minimize degradation (*see* **Note 2**). Such complexity necessitates high-resolution analytical technologies that provide:

- sub-2 μm UHPLC separation.
- mass accuracy <5 ppm.
- high-resolution MS/MS fragmentation.
- UV and PDA spectral confirmation.
- improved sensitivity at trace levels [6].

Traditional analytical approaches, such as TLC, TLC, colorimetric assays, or classical HPLC, lack the selectivity and sensitivity required for detailed molecular analysis [7]. High-resolution liquid chromatography coupled with tandem mass spectrometry (UHPLC–HR–MS/MS) has become the gold standard for comprehensive phenolic acid characterization [8].

This chapter provides detailed laboratory instructions for extraction, purification, chromatographic separation, mass spectrometric detection, and data interpretation of phenolic acids using modern analytical instrumentation(Tables 1 and 2).

1.1 Chemical Structures and Physicochemical Properties

.

2 Materials

2.1 Chemicals and Reagents

1. HPLC/MS-grade methanol and acetonitrile.
2. Ultrapure water (0.22 μm filtered).
3. Formic acid (≥99%).
4. Phenolic acid analytical standards (gallic, caffeic, ferulic, etc.).
5. Internal standard (e.g., 3,5-dichloro-4-hydroxybenzoic acid).
6. SPE cartridges: C18, polymeric reversed-phase, or hydrophilic-lipophilic balance (HLB) type.
7. Nitrogen gas for MS.
8. Amber vials.
9. Syringe filters (0.22 μm PTFE).

2.2 Equipment

1. UHPLC system with PDA detector and MS/MS.
2. HPLC–DAD.
3. QTOF–MS or Orbitrap-MS.
4. GC–MS with EI source.
5. Analytical balance.

Table 1
Major phenolic acids and their structures [9, 10]

Phenolic acid	Structural description	SMILES
Gallic acid	3,4,5-trihydroxybenzoic acid	O=C(O)c1cc(O)c(O)c(O)c1
Protocatechuic acid	3,4-dihydroxybenzoic acid	O=C(O)c1ccc(O)c(O)c1
Vanillic acid	4-hydroxy-3-methoxybenzoic acid	COc1cc(C(=O)O)ccc1O
Syringic acid	3,5-dimethoxy-4-hydroxybenzoic acid	COc1cc(OC)c(C(=O)O)c(O)c1
p-Coumaric acid	4-hydroxycinnamic acid	O=C(O)/C=C/c1ccc(O)cc1
Caffeic acid	3,4-dihydroxycinnamic acid	O=C(O)/C=C/c1ccc(O)c(O)c1
Ferulic acid	4-hydroxy-3-methoxycinnamic acid	COc1cc(/C=C/C(=O)O)ccc1O
Sinapic acid	3,5-dimethoxy-4-hydroxycinnamic acid	COc1cc(OC)/C=C/C(=O)Occc1O
Chlorogenic acid	Caffeoyl ester of quinic acid	COC(=O)C(O)C1CC(O)(C(O)C1O)

Note: Acidic pH (<3) and light protection are essential to prevent degradation, especially for caffeic and chlorogenic acids (*see* **Note 2**)

Table 2
Physicochemical properties of phenolic acids [11, 9]

Phenolic acid	pKa	logP	λ_{max} (nm)	Notes
Gallic acid	4.4	0.7	260–275	Highly polar, strong antioxidant
Protocatechuic acid	4.5	1.1	260–285	Moderate polarity
Vanillic acid	4.3	1.3	260	Methoxy ↑ lipophilicity
Caffeic acid	4.6	1.2	320–325	Strong UV absorbance
Ferulic acid	4.6	1.5	320–330	Excellent MS response
Sinapic acid	4.5	1.7	325–335	Multiple methoxy groups
Chlorogenic acid	3.4	−1.0	325–330	Requires HR-MS for isomers

6. Sonicator bath.
7. Nitrogen evaporator.
8. Vortex mixer.
9. High-speed centrifuge.
10. Cryogenic grinder or mortar and pestle.
11. SPE vacuum manifold.

3 Methods

3.1 Sample Preparation

3.1.1 Plant Material Collection and Storage

1. Collect fresh plant material (leaves, fruits, roots, seeds, or whole tissues) ensuring minimal environmental stress during harvesting.

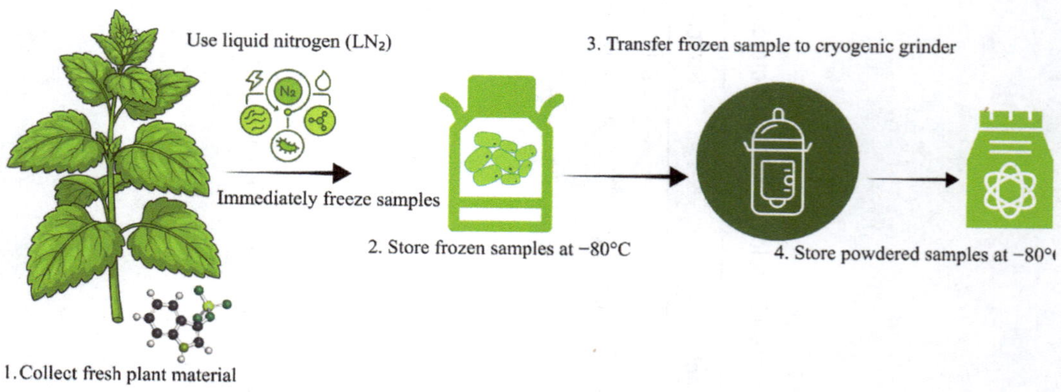

Fig. 1 Cryogenic sample preparation workflow for preserving phenolic acid integrity in plant tissues

2. Immediately freeze samples in liquid nitrogen to prevent enzymatic oxidation of phenolic acids (*see* **Note 2**).
3. Store the frozen material at −80°C in airtight containers until extraction.
4. Avoid repeated freeze–thaw cycles to prevent degradation (Fig. 1).

3.1.2 Tissue Grinding

1 Transfer frozen plant material to a cryogenic grinder.
2 Pulverize to a fine homogeneous powder under liquid nitrogen.
3 Alternatively, use a pre-chilled mortar and pestle.
4 Store powdered samples at −80°C until analysis.

3.2 Extraction of Phenolic Acids [12]

3.2.1 Solvent Extraction

1 Weigh approximately 100 mg of plant powder into a microcentrifuge tube.
2 Add 1.5 mL of 80% methanol acidified with 0.1% formic acid.
3 Vortex vigorously for 1 min to suspend the sample uniformly.
4 Sonicate for 20 min at room temperature to enhance extraction efficiency.
5 Centrifuge at 10,000 rpm for 10 min.
6 Transfer the supernatant to a fresh tube.
7 Repeat the extraction once more and combine both supernatants.
8 Evaporate the solvent under nitrogen at ≤35°C.
9 Reconstitute the dried extract in 500 μL of 50% methanol.
10 Filter the solution through a 0.22 μm PTFE syringe filter into amber vials (*see* **Note 3**).

3.3 Solid-Phase Extraction (SPE) Cleanup [13]

3.3.1 Cartridge Conditioning

1 Condition the SPE cartridge with 5 mL methanol.
2 Equilibrate with 5 mL ultrapure water.

3.3.2 Sample Loading

1 Dilute the extracted residue with water (1:3 v/v) to reduce solvent strength.
2 Load the sample slowly onto the cartridge to ensure proper adsorption.

3.3.3 Washing

1 Wash the cartridge with 5% methanol to remove sugars, pigments, and other hydrophilic matrix components.

3.3.4 Elution

1. Elute phenolic acids with 80% methanol acidified with 0.1% formic acid.
2. Collect the eluate and evaporate under nitrogen.
3. Reconstitute in 300–500 μL of injection solvent (usually 20%–50% methanol) (*see* **Note 4**).

3.4 UHPLC–PDA–MS/MS Method [14]

3.4.1 Chromatographic Separation

1. Install a C18 UHPLC column (sub-2 μm particle size, 100 × 2.1 mm).
2. Maintain the column at 40°C for consistent retention times.
3. Mobile phase A: Water + 0.1% formic acid.
4. Mobile phase B: Acetonitrile + 0.1% formic acid.
5. Start with a low percentage of organic phase (≈5% B).
6. Increase B gradually using a controlled gradient to resolve phenolic acids (12–20 min typical).
7. Use a flow rate of 0.3 mL/min.
8. Inject 2–5 μL of sample depending on expected analyte concentration (*see* **Note 5**).

3.4.2 PDA Detection

1. Set PDA detection wavelengths at 260 nm and 320 nm (universal for phenolic acids).
2. Use full PDA spectral acquisition (200–400 nm) for spectral confirmation.
3. Integrate peaks using extracted PDA chromatograms.

3.4.3 MS/MS Detection

1. Operate the electrospray ionization (ESI) source in **negative ion mode**.
2. Optimize the capillary voltage (≈2.5 kV), sheath gas, and auxiliary gas.
3. Set the capillary temperature to ~320°C.
4. Scan a mass range of m/z 100–1000.
5. Acquire full-scan MS followed by data-dependent MS/MS.
6. Use collision energies between 20 and 40 eV depending on the compound.
7. Enable dynamic exclusion to avoid repetitive fragmentation of abundant ions.
8. Identify phenolic acids using retention time, accurate mass, and characteristic fragment ions.

3.5 HPLC–DAD Method [15]

3.5.1 Separation Parameters

1. Use a conventional C18 column (4.6 × 250 mm, 5 μm).

2. Maintain flow at ~1.0 mL/min.
3. Prepare mobile phases identical to UHPLC (water + FA and acetonitrile + FA).
4. Apply a slower gradient (typically 30–40 min) to achieve adequate resolution.

3.5.2 Detection

1. Monitor 260 nm for hydroxybenzoic acids.
2. Monitor 320 nm for hydroxycinnamic acids.
3. Record full UV spectra for purity assessment.

3.6 LC–QTOF–MS and LC–Orbitrap–MS (High-Resolution Profiling) [16]

3.6.1 Instrument Setup

1. Use ESI negative mode as the primary ionization method.
2. Set resolving power to ≥30,000 FWHM (QTOF) or ≥60,000 FWHM (Orbitrap).
3. Calibrate the instrument with standard calibration solutions before analysis.

3.6.2 Acquisition

1. Acquire accurate-mass full-scan spectra across m/z 100–1000.
2. Trigger auto-MS/MS using the top 5–10 most intense precursor ions.
3. Adjust collision energies (20–40 eV) based on the degree of fragmentation needed.
4. Collect both centroid and profile data for optimal interpretation.

3.6.3 Data Analysis

1. Perform peak detection, deconvolution, alignment, and normalization using software, such as MZmine, XCMS, Compound Discoverer, or MassHunter.
2. Identify phenolic acids using:
   - Exact mass.
   - MS/MS fragmentation.
   - Retention time (RT) matching.
   - Isotopic pattern analysis.
3. Confirm structures using diagnostic fragment ions (*see* **Note 6**).

3.7 GC–MS Analysis After Derivatization [17]

3.7.1 Derivatization

1. Ensure complete dryness of samples (even trace water inhibits derivatization).

2. Add 100 µL of BSTFA + 1% TMCS to the dried residue.
3. Incubate at 70°C for 30 min.

3.7.2 GC–MS Conditions

1. Use a non-polar DB-5MS column.
2. Start oven temperature at ~80°C and increase gradually to 300°C.
3. Use helium as carrier gas.
4. Operate MS in electron ionization (EI) mode.

3.7.3 Identification

1. Identify peaks using retention index matching.
2. Confirm using EI fragmentation patterns (*see* **Note 7**).

4 Notes

1. Use high-resolution UHPLC–MS/MS for isomeric phenolics and complex plant matrices.
2. Freeze samples immediately after harvest; store at −80°C; protect light-sensitive compounds; avoid repeated freeze–thaw cycles to prevent enzymatic degradation.
3. Filter samples with PTFE cautiously; some phenolic acids may adsorb; avoid strong acid loss.
4. SPE cleanup removes matrix interferences, sugars, organic acids, and pigments, improving MS signal.
5. Use internal standards and matrix-matched calibration for accurate quantification; isotopically labeled ferulic acid is recommended for reliability.
6. High-resolution MS (QTOF/Orbitrap) is essential for structural elucidation of chlorogenic acid isomers; methoxylated derivatives may ionize better in positive mode if needed.
7. Dry samples fully before GC derivatization to ensure complete reaction.
8. Maintain acidic conditions (pH <3) during extraction to prevent degradation [18, 19].
9. Narrow-bore columns and controlled gradients enhance separation of isomers and improve sensitivity [20].
10. Include pooled QC samples to monitor reproducibility; avoid gradient fluctuations and high column backpressure to maintain retention times [21].
11. Protect light-sensitive phenolics (e.g., caffeic acid) using amber vials and a refrigerated autosampler [22].

12. Validate linearity over multiple concentrations for method accuracy [23].
13. Avoid metal tubing and plasticware for certain phenolics to prevent chelation or adsorption [24].
14. Monitor column backpressure and gradient stability regularly to maintain reproducible chromatography [25].
15. Use high-resolution UHPLC–MS/MS and combined MS/MS+UV spectra for structural elucidation of complex phenolics (Table 3) [24, 20].

5 Troubleshooting

Table 3
Common problems and solutions

Problem	Cause	Solution
Poor peak shape	Column overloading	Reduce injection volume
Low MS sensitivity	Matrix suppression	Perform SPE cleanup
Retention time shift	Gradient instability	Re-equilibrate column
Peak tailing	Secondary interactions	Increase formic acid %
No derivatization (GC–MS)	Residual moisture	Extend drying time

.

References

1. Prabhu S, Molath A, Choksi H, Kumar S, Mehra R (2021) Classifications of polyphenols and their potential application in human health and diseases. Int J Physiol Nutr Phys Educ 6:293–301
2. De La Rosa, LA, Moreno-Escamilla, JO, Rodrigo-García, J, Alvarez-Parrilla, E, (2019). Phenolic compounds, Postharvest physiology and biochemistry of fruits and vegetables. Elsevier, Amsterdam, Netherlands
3. Tohge T, Fernie AR (2017) An overview of compounds derived from the shikimate and phenylpropanoid pathways and their medicinal importance. Mini Rev Med Chem 17:1013–1027
4. Da Silva APG, Sganzerla WG, John OD, Marchiosi R (2025) A comprehensive review of the classification, sources, biosynthesis, and biological properties of hydroxybenzoic and hydroxycinnamic acids. Phytochem Rev 24:1061–1090
5. Vuolo, MM, Lima, VS, Junior, MRM, (2019). Phenolic compounds: Structure, classification, and antioxidant power, Bioactive compounds. Elsevier, Amsterdam, Natherlands
6. Eugster, PJ, (2013). Strategies for high resolution profiling of natural extracts: UHPLC-MS and physicochemical approaches for early metabolite identification. Université de Genève, Geneva, Switzerland
7. Loescher CM, Morton DW, Razic S, Agatonovic-Kustrin S (2014) High performance thin layer chromatography (HPTLC) and high performance liquid chromatography (HPLC) for the qualitative and quantitative

analysis of Calendula officinalis—Advantages and limitations. J Pharm Biomed Anal 98:52–59
8. Chen Y, Zhou Z, Yang W, Bi N, Xu J et al. (2017) Development of a data-independent targeted metabolomics method for relative quantification using liquid chromatography coupled with tandem mass spectrometry. Anal Chem 89:6954–6962
9. Silva FA, Borges F, Guimarães C, Lima JL, Matos C et al (2000) Phenolic acids and derivatives: studies on the relationship among structure, radical scavenging activity, and physicochemical parameters. J Agric Food Chem 48:2122–2126
10. Zhang B, Lan W, Wang Z, Shao Z, Xie J (2024) Modified chitosan with different phenolic acids: characterization, physicochemical properties, and biological activity. Food Chem 441:138337
11. Luo Y, Li Y-C, Wang M, Zhou L, Meng F-B et al (2023) Effects of grafting methods and raw materials on the physicochemical properties and biological activities of phenolic acids grafted oat β-glucan. Food Res Int 173:113250
12. Stalikas CD (2007) Extraction, separation, and detection methods for phenolic acids and flavonoids. J Separation Sci 30:3268–3295
13. Michalkiewicz A, Biesaga M, Pyrzynska K (2008) Solid-phase extraction procedure for determination of phenolic acids and some flavonols in honey. J Chromatogr A 1187:18–24
14. Gonçalves J, Silva CL, Castilho PC, Câmara JS (2013) An attractive, sensitive and high-throughput strategy based on microextraction by packed sorbent followed by UHPLC-PDA analysis for quantification of hydroxybenzoic and hydroxycinnamic acids in wines. Microchem J 106:129–138
15. Zeb A (2015) A reversed phase HPLC-DAD method for the determination of phenolic compounds in plant leaves. Anal Methods 7:7753–7757
16. Wang C-F, Li L (2022) Instrument-type effects on chemical isotope labeling LC-MS metabolome analysis: quadrupole time-of-flight MS vs. Orbitrap MS. Anal Chim Acta 1226:340255
17. Lin D-L, Wang S-M, Wu C-H, Chen B-G, Liu RH (2008) Chemical derivatization for the analysis of drugs by GC-MS-A conceptual review. J Food Drug Anal 16:1
18. Clifford MN (2000) Review—Chlorogenic acids and other cinnamates—nature, occurrence, dietary burden, absorption and metabolism in man. J Sci Food Agric 80:1033–1043.
19. Crozier A, Jaganath IB, Clifford MN (2009) Dietary phenolics: chemistry, bioavailability and effects on health. Nat Prod Rep 26:1001–1043.
20. Spaggiari D, Fekete S, Eugster PJ, Veuthey JL, Geiser L, et al (2013) Contribution of various types of liquid chromatography–mass spectrometry instruments to band broadening in fast analysis. J Chromatogr A 1310:45–55.
21. Broadhurst D, Goodacre R, Reinke SN, Kuligowski J, Wilson ID, et al (2018) Guidelines and considerations for the use of system suitability and quality control samples in mass spectrometry assays applied in untargeted clinical metabolomic studies. Metabolomics 14:72
22. Naczk M, Shahidi F (2004) Extraction and analysis of phenolics in food. J Chromatogr A 1054:95–111.
23. ICH (2005) Validation of analytical procedures: text and methodology Q2(R1). ICH Harmonised Tripartite Guideline 1:15.
24. Robbins RJ (2003) Phenolic acids in foods: an overview of analytical methodology. J Agric Food Chem 51:2866–2887.
25. Snyder LR, Kirkland JJ, Dolan JW (2009) Introduction to modern liquid chromatography. John Wiley & Sons, Hoboken

Check for updates

Chapter 15

Glycosides: Advanced Glycosides Profiling Methods

Priyang Kanojiya, Anjali Desai, Rukhsar Ansari and Sanket Ray

Abstract

Plant glycosides are naturally occurring, most fascinating, diverse biomolecules known for their broad range of biological and pharmacological activities. Glycosides, linked to sugars, are generously present in medicinal plants, which enhances their stability and bioactivity. They are complex and present in low quantities and that's where advanced profiling has advantages compared to traditional analytical approaches. The present study includes the most advanced, sensitive, and recent methods for profiling of glycosides. Profiling methods like solvent extraction, RP-HPLC, GC–MS, LC–MS, and NMR, each with have been included in this chapter with simplification. RP-HPLC ensures sharp separation of glycosides based on polarity and hydrophobicity. GC–MS allows identification of volatile glycosides by mass splitting patterns. LC–MS facilitates enhanced sensitivity and precise mass detection of glycosides. NMR spectroscopy validates structural elucidation details through different spectral correlations of carbons and protons.

Keywords Plant glycosides, Solvent extraction, RP-HPLC, GC–MS, LC–MS, NMR

1 Introduction

Glycosides are plant-derived, naturally occurring compounds in which a sugar moiety (glycone) is linked with non-sugar moiety (aglycone) with a glycosidic bond. The glycosides help plants defend themselves against various plant pathogens. Glycosides have the ability to show anti-oxidant as well as anti-inflammatory activity against plant pathogens, and so far, they act as bioactive compounds. Extraction of these bioactive compounds is important step that starts with sample processing followed by the extraction process. One of the traditional extraction process is solvent extraction process (Maceration) [1] and other are microwave assisted extraction (MAE), ultrasound assisted extraction (UAE), supercritical fluid extraction (SFE). For quantification of glycosides, HPLC, RP-HPLC, GC–MS, LC–MS/MS, and NMR analysis is mostly used [2–6]. RP-HPLC helps in the identification of flavonoid glycosides like wogonoside with fine and clear separation. GC–MS detects several glycosides such as acteoside,

Dhruti Amin et al. (eds.), *Phytoconstituent Analysis*, Springer Protocols Handbooks,
https://doi.org/10.1007/978-1-0716-5400-2_15,

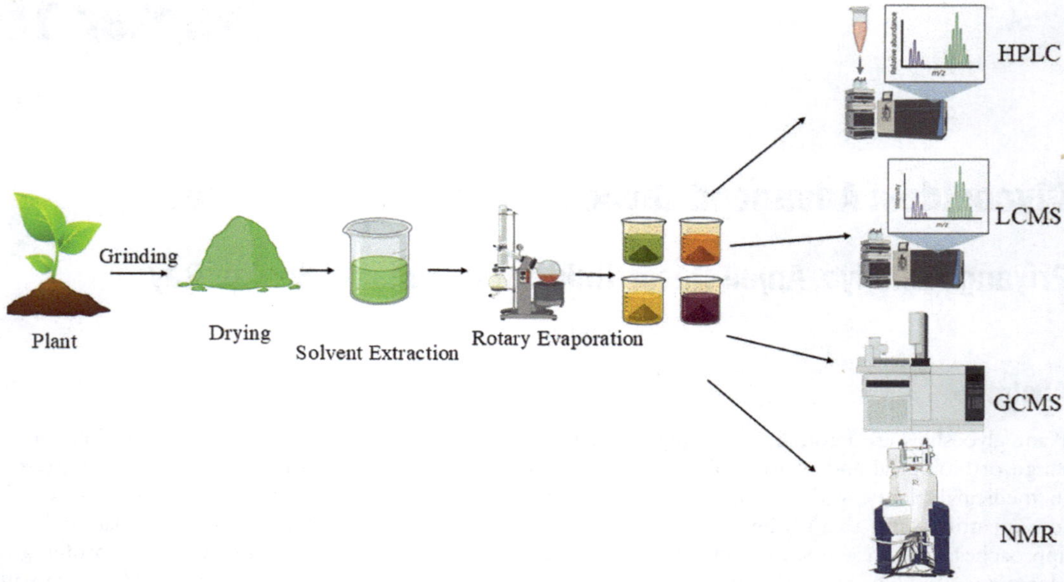

Fig. 1 Schematic representation of plant sample preparation and analytical profiling of plant-derived glycosides using advanced analytical techniques

calycosin-7-O-glucoside. LC-MS permits the detection of glycosides like rebaudioside and saponins. NMR analysis confirms the structures of glycosides such as sarmentosin, chiisanenside, and ranuncoside (Fig. 1).

2 Materials

2.1 Solvent Extraction Method—Maceration [1]

1. Any dried plant material or part (bark, leaves, roots).
2. Solvents: 70% (v/v) ethanol, 75% (v/v) methanol or hydrochloric mixture.
3. Maceration container made up of glass with an airtight lid.
4. Mortar and Pestle for grinding.
5. Muslin cloth/Whatman filter paper no. 1.
6. Funnel and clean bottles (amber-colored).

2.2 Reverse Phase—HPLC Analysis [2]

1. Agilent 1100 series HPLC system (degasser, quaternary pump, autosampler, column compartment, PDA (Photodiode Array Detector), DAD (Diode Array Detector).
2. Merck LaChrom HPLC System.
3. Zorbax SB-C18 column (150×2.1 mm, 3.5 μm) [2].

4. LiChrospher 100 RP-18 column (125 × mm, 5 μm) [2].
5. HPLC-grade water, 8%–80% acetonitrile (HPLC-grade), methanol (HPLC-grade) and 0.1% formic acid (v/v) (Analytical grade).
6. Syringe filters (0.22 μm) and sterile vials.
7. Standard companies: Naringenin, Naringenin-7-glucoside (Sigma Aldrich and Extrasynthese), Rutin, Apigenin (Carl Roth GmbH) [2, 7].

2.3 Gas Chromatography–Mass Spectroscopy (GC–MS)

1. Solvents and chemicals—80% methanol, petroleum ether, helium gas.
2. Instruments—Agilent 7890 GC and MS Detector, Elite 5MS column, (5%-Phenyl)-methyl polysiloxane (HP-5MS column) (30 × 0.25 mm × 0.22 μm), rotary evaporator and sterile syringe filter 0.22 μm.
3. Standards: National Institute of Standards and Technology (NIST, United States), National Institute of Health (NIH) Environmental Protection Agency (EPA) [8–10].
4. Retention indices calculation: n-alkane series C8-20.
5. Standards: Costunolide (NIST library), Clinacoside, Acetoside (NIST/EI-MS library), Sarasapogenin (NIST library) [8, 9, 11].

2.4 Liquid Chromatography–Mass Spectroscopy (LC–MS)

1. Solvents and Chemicals: 99.9 % (v/v) methanol, 99.9 % (v/v) acetonitrile, 0.1% and 10% (v/v) formic acid, deionized water, and acetate.
2. LC Columns: YMC-Triart C18 column (150×2.0 mm, 3 μm), Acquity BEH C18 column (50×2.1 mm, 1.7 μm), Mediterranean Sea C18 column (150×4.6 mm, 5 μm), Kinetex PFP C18 column (50×2.1 mm, 1.7 μm), Xterra MS C18 (150×3.0 mm, 5 μm).
3. Instruments: Triple quadrupole MS, Orbitrap MS, Quadrupole Time-of-flight (QTOF) MS, Single quadrupole MS, Electrospray Ionization (ESI), Ultra- High Performance Liquid Chromatography (UHPLC) system.
4. Standards: Rebaudioside (Sigma, ChromaDeX), flavonoids (mz cloud database) [12, 13].

2.5 Nuclear Magnetic Resonance (NMR)

1. Solvents: CD_3OD (Deuterated Methanol), DMSO-d_6 (Deuterated Dimethyl Sulfoxide), MeOD/D_2O (3:2, v/v) (Deuterated methanol and Deuterium Oxide), Pyridine-d_5.
2. NMR tubes (5 mm), analytical balance, syringes, micropipettes.
3. Instruments: $^1H/^{13}C$/2D NMR (400–600 MHz; Bruker or Varian).
4. Internal standard: Caffeine [14], Tetramethylsilane (TMS).

3 Methods

For the extraction of plant materials, there are several methods, among them, three: Solvent Extraction Method, MAE, UAE, and SFE are commonly used.

3.1 Solvent Extraction Method—Maceration Process [1, 5, 15]

1. Take the desired amount of plant material & grind into tiny particles by using a mortar and pestle or plant tissue, which will increase interaction time with analytes.
2. Transfer the grinded material into a clean, dry and waterproof glass container vessel. Add a sufficient amount of solvent (70% (v/v) ethanol, 75% (v/v) methanol) until the plant material is submerged. Generally, 1:10 or 1:15 (w/v) of plant to solvent is employed.
3. Seal the vessel tightly with the lid & allow the mixture to stand at room temperature (25–30°C) for a maceration period of 4–5 days. Put the container in a dark area to avoid direct sunlight.
4. After the maceration period, filter the liquid extract through Whatman filter paper no.1 into a clean, dry, and concentrate using a rotatory evaporator at 40°C in reduced pressure conditions.
5. Re-dissolved the dried residue in 1 mL solvent and filter through 0.22 μm syringe filter before inject (*see* **Note 1**).
6. Transfer the final filtrate into a sterile, amber-colored bottle and label it.
7. Store the bottle in a dry, cool and dark location [1].

3.2 Reversed Phase–HPLC (RP-HPLC) Method [2–4]

Use previously extracted glycoside fractions for all analytical investigations.

1. **Mobile Phase Preparation**
   - For RP-HPLC, three solvent preparations are required (Solvent A, Solvent B, Solvent C) (Table 1) (*see* **Note 2**).
2. **Instrument Cluster**
   - Connect the system to HPLC column (Zorbox SB-C18 or LiChrospher 100 RP-18), set the temperature of the column at 30°C (*see* **Note 2, 4**).
3. **Gradient Program—Linear gradient method** (Table 2).
4. **Injection and Detection**
   - Take 5 μL of the sample and inject it into the system. The range of the PDA detector should be 220–500 nm.
5. **Data Analysis**
   - As the run reaches the end, look for a chromatogram and identify the peaks by comparing it with standards.

Table 1
RP-HPLC mobile phase preparations and gradient programs (Steps 2 and 4) (*see* Note 3)

RP-HPLC	Solvent A	Solvent B	Solvent C	Gradient program	Flow rate and temperature	Detector	References
Protocol 1	Water	Acetonitrile (8%)	–	Acetonitrile and water final mobile phase. Starting with A – 95%	0.4 mL/min 30°C	PDA	[4]
Protocol 2	Water + 0.1% formic acid	Acetonitrile: water:formic acid	–	20%–55% B for 37.7 min, 55%–95% B for 1.3 min, isocratic - 95% B for 6 min	0.7 mL/min 30°C	DAD/PDA	[3]
Protocol 3	0.1% Formic acid in water	0.1% Formic acid in acetonitrile	0.1% Formic acid in methanol	Start with 95% A, 2% B, 3% C, final with 43% B, 57% C, for 40 min	0.4–1.2 mL/min 30°C	PDA	[2]
Protocol 4	0.2% Formic acid in water	Acetonitrile	–	8%–18% B (5 min), 18%–20% B (5-min), 20%–30% B (10 min), 30%–50% B (14 min), 50%–80% B (17 min)	0.6 mL/min 25°C	DAD-TOF	[7]

Table 2
GC-MS methods for plant glycosides

GC–MS	Extraction solvent	GC-column	Flow rate of gas	Injector temperature	Oven program	MS settings	References
Protocol 1	80% methanol	HP-5MS	1 mL/min (He)	250°C	70–300°C rate of 5°C/min, 5 min hold	EI 70 eV, 40–800 m/z	[8]
Protocol 2	80% methanol	HP-5MS	1 mL/min (He)	250°C	70–300°C	EI 70 eV, 50–550 m/z	[9]
Protocol 3	80% methanol	HP-5MS	1 mL/min (He)	250°C	60–300°C	EI 70 eV, 40–600 m/z	[10]
Protocol 4	70% methanol	HP-5MS	1 mL/min (He)	250°C	70–300°C	EI 70 eV, full scan	[16]
Protocol 5	75% methanol	Elite 5MS	1 mL/min (He)	250°C	70–76°C rate of 1°C/min (hold), 76–300°C - rate of 6°C/min, 5 min hold	EI 70 eV, 0–800 m/z	[11]

3.3 Gas Chromatography–Mass Spectroscopy (GC–MS) Method [8–11, 16]

3.3.1 GC–MS Analysis

1. Take 1 μL of filtered sample and inject in split-less mode and maintain the temperature of the injector at 250°C (*see* **Note 5**).
2. For separation, use a HP-5MS column (30×0.25 mm×0.22 μm) and helium gas as carrier with a constant flow rate of 1.0 mL/min.
3. Maintain the temperature of the oven at 60–70°C and ramp at 3–5°C/min up to 280–300°C.
4. For mass spectroscopic analysis of glycosides, set electron ionization at 70 eV and ion source temperature 230–250°C and mass scan range 40–800 m/z [8, 10] (*see* **Note 6**).
5. Identify and compare the peaks with standards (*see* **Note** 7).

3.4 Liquid Chromatography–Mass Spectroscopy (LC–MS) Method [6, 12, 13, 17, 18]

1. Inject 5–10 μL of filtered extract into LC–MS instrument (*see* **Note 9**).
2. Use a reverse-phase C18 column (YNC-Triart, Kinetex PFP, Acquity BEH) and maintain the temperature of the column at 30–40°C [6, 17, 18].
3. Use the appropriate mobile phase mentioned in Table 3 (*see* **Note 8**).
4. Set the gradient program based on your study requirement (Table 3).
5. Maintain the mobile phase at a flow rate between 0.2 and 0.7 mL/min.
6. Begin by optimizing the ion source and then operate the full MS System (Table 3).
7. Identify glycosides and MS fragmentation patterns with standards (Sigma, mz cloud database).

3.5 Nuclear Magnetic Resonance (NMR)—Method [14, 19–22]

1. Weigh 20–30 mg of plant extract and dissolve in 500 μL of MeOD/D_2O (3:2, v/v) or in any suitable deuterated solvent (*see* **note 11**).
2. Sonicate the mixture for 10 min for homogeneity and centrifuge at 1400 g for 2 min.
3. Transfer the clear supernatant to a 5 mm NMR tube.
4. Record the spectra using high-resolution 400–600 MHz NMR spectrometer (Table 4).
5. Process the spectra using software like TopSpin/MestReNova with accurate phase and baseline correction.
6. For quantitation, consider the ratio of the analyte peak to the internal standard.

Table 3
LC–MS methods for the detection of plant glycosides (*see* Note 10)

LC–MS	Columns	Mobile phase	Flow rate and temperature	MS instrument and ion mode	Scan	References
Protocol 1	YMC-Trait C18	0.1% formic acid in water and 99.9% acetonitrile	0.2 mL/min, 35°C	Shimadzu LCMS-8040 (Triple Quad), ESI	Full scan (m/z 100–1600)	[6]
Protocol 2	Acquity BEH C18	0.1% formic acid in water and 99.9% acetonitrile	0.4 mL/min, 30°C	SCIEX UHPLC X500B QTOF, ESI	TOF-MS and MS/MS	[17]
Protocol 3	Mediterranean Sea C18	0.1% formic acid in water and 99.9% methanol	0.7 mL/min, 40°C	Agilent 6460 Triple Quad, ESI	SRM mode	[12]
Protocol 4	Kinetex PFP C18	0.1% formic acid in water and 99.9% acetonitrile	0.4 mL/min, 40°C	Thermo Q Orbitrap MS/MS, ESI	DDA MS/MS	[13]
Protocol 5	Xterra MS C18	Acetonitrile and water + 5%–10% formic acid	0.5 mL/min, 30°C	Waters SQD, ESI	Full scan (50–150 V)	[18]

Table 4
NMR methods for the detection of plant glycosides (*see* Note 12)

NMR	Solvent	Sample amount	Record NMR spectra on	Spectra acquired	References
Protocol 1	Pyridine-d_5	5 mg in 700 μL or 2 mg in 180 μL	Bruker Advance 500 MHz with 5 mm BB and 2.5 mm inverse detection at 300 K	^{1}H, ^{13}C, 2D NMR	[19]
Protocol 2	MeOD/D_2O (3:2, v/v)	30 mg	Bruker Advance III 400 MHz with BBFO probe using PIR sequence	^{1}H, qNMR	[14]
Protocol 3	Pyridine-d_5/DMSO-d_6	5–10 mg	500 MHz NMR spectrometer	^{1}H, ^{13}C, HMBC	[20]
Protocol 4	DMSO-d_6/CD_3OD	10 mg	Varian INOVA 500 MHz and Bruker 600 MHz with CryoProbe	^{1}H, ^{13}C, APT/DEPTQ, COSY, HSQC, HMBC, NOESY, ROESY	[21]
Protocol 5	DMSO-d_{6+} D_2O	5–10 mg	Bruker 500 MHz NMR spectrometer	^{1}H, HSQC, NOE	[22]

7. Establish calibration curves using multiple concentrations and assess linearity (r^2), repeatability (intraday RSD), and intermediate precision (interday RSD) [23].
8. Calculate LOD (Limit of Detection) and LOQ (Limit of Quantitation) based on the standard deviation of the intercept and slope of the regression line.
9. Formulas for LOD = 3.3 σ/S, LOQ = 10 σ/S [14] (Table 4).

4 Notes

1. To avoid column clogging and pressure problems, filter the sample through a 0.22 μm membrane filter before RP-HPLC injection.
2. To prevent bubble formation, the mobile phase used in HPLC must be prepared freshly and degassed.
3. Based on glycoside polarity and type, select an HPLC column. C18 columns are ideal and mostly used for flavonoids and glycosides
4. Maintain constant column temperature and flow rate for sharp chromatographic patterns.
5. Before GC-MS analysis, verify the sample is completely dry, as any residual moisture can obstruct with conversion efficiency.
6. Keep track on GC–MS ion source and temperature routinely to ensure optimal ionization.
7. Identification of glycosides in GC–MS should be supported by correlating mass fragmentation patterns with renowned spectral databases like NIST or EPA.
8. For LC–MS, use ultrapure water and HPLC-grade solvents to minimize background noise and increase sensitivity.
9. Make sure consistency of injection volume for reproducibility and precise peak area identification and comparison in LC–MS run.
10. Do not use phosphate buffers or high salt solvents during LC–MS analysis as they can inhibit ionization in ESI mode.
11. Use deuterated solvents for NMR spectroscopy for clear spectra and accurate chemical shifts.
12. Pre-saturation should be applied when using aqueous solvent system to minimize interference during the NMR process.

References

1. Azwandia (2015) A review on the extraction methods use in medicinal plants, principle, strength and limitation. Med Aromat Plants 04:3–8
2. Blunder M, Orthaber A, Bauer R, Bucar F, Kunert O (2017) Efficient identification of flavones, flavanones and their glycosides in routine analysis via off-line combination of sensitive NMR and HPLC experiments. Food Chem 218:600–609
3. Sun B, Santos CPR, Leandro MC, De Freitas V, Spranger MI (2007) High-performance liquid chromatography/electrospray ionization mass spectrometric characterization of new products formed by the reaction between flavanols and malvidin 3-glucoside in the presence of acetaldehyde. Rapid Commun Mass Spectrom 21:2227–2236
4. Wang Y, Xing L, Zhang J, Chen Y, Lu S (2025) Simultaneous determination of 32 polyphenolic compounds in Berries via HPLC–MS/MS. Molecules 30:2008
5. Tzanova M, Atanasov V, Yaneva Z, Ivanova D, Dinev T (2020) Selectivity of current extraction techniques for flavonoids from plant materials. Processes 8:1–30
6. Liu D, Kikuchi T, Li W (2023) Characterization and comparison of steroidal glycosides from polygonatum species by high-performance liquid chromatography–electrospray ionization mass spectrometry. Molecules 28:705
7. Liu EH, Liu Q, Chu C, Li P (2011) Characterization and identification of multiple constituents in Yinhuang granules by high-performance liquid chromatography with diode-array and time-of-flight mass spectrometry detection. J Sep Sci 34:2566–2575
8. Swargiary A, Daimari M (2020) Identification of bioactive compounds by GC-MS and α -amylase and α-glucosidase inhibitory activity of *Rauvolfia tetraphylla* L. and *Oroxylum indicum* (L.) Kurz: an in vitro and in silico approach. Clin Phytosci 6:1–11
9. Nokhala A, Siddiqui MJ, Ahmed QU, Bustamam MSA, Zakaria ZA (2020) Investigation of α -glucosidase inhibitory metabolites from Tetracera scandens leaves by GC–MS metabolite profiling and docking studies. Biomolecules 10:1–17
10. Wang XJie, Song HCong, Yang Y, Tao Y Sheng (2020) Chemical profile of terpene glycosides from Meili grape detected by GC–MS and UPLC–Q-TOF-MS. Eur Food Res Technol 246:2323–2333
11. Abdulhafiz F, Reduan MFH, Hisam AH, Mohammad I, Abdul Wahab IR, Abdul Hamid FF, Mohammed A, Nordin ML, Shaari R, Bakar LA, Kari ZA, Wei LS, Goh KW, Ahmad Mohd Zain MR (2022) LC–TOF-MS/MS and GC-MS based phytochemical profiling and evaluation of wound healing activity of Oroxylum Indicum (L.) Kurz (Beka). Front Pharmacol 13:1–14
12. Molina-Calle M, Sánchez De Medina V, Delgado De La Torre MP, Priego-Capote F, Luque De Castro MD (2016) Development and application of a quantitative method based on LC-QqQ MS/MS for determination of steviol glycosides in Stevia leaves. Talanta 154:263–269
13. Rodríguez EP, Li Y, Vaniya A, Shih PM, Fiehn O (2023) Alternative identification of glycosides using MS/MS matching with an in silico-modified aglycone mass spectra library. Anal Chem 95:10618–10624
14. Fernandes JM, Termentzi A, Mandova T, Hammad K, Machera K, Magiatis P, Michel S, Zucolotto SM, Grougnet R (2021) Detection, isolation, and 1H NMR quantitation of the nitrile glycoside sarmentosin from a bryophyllum pinnatum hydro-ethanolic extract. J Agric Food Chem 69:8081–8089
15. Yılmaz FM, Görgüç A, Uygun Ö, Bircan C (2021) Steviol glycosides and polyphenols extraction from Stevia rebaudiana Bertoni leaves using maceration, microwave-, and ultrasound-assisted techniques. Sep Sci Technol 56:936–948
16. Jahan I, Tona MR, Sharmin S, Sayeed MA, Tania FZ, Paul A, Chy MNU, Rakib A, Emran T Bin, Simal-Gandara J (2020) GC-MS phytochemical profiling, pharmacological properties, and in silico studies of chukrasia velutina leaves: a novel source for bioactive agents. Molecules 25:3536
17. Du T, Wang Y, Xie H, Liang D, Gao S (2024) Fragmentation patterns of phenolic C-glycosides in mass spectrometry analysis. Molecules 29:2953
18. Szymborska K, Frański R, Gierczyk B, Beszterda-Buszczak M (2025) Extremely rare flavonoid glycosides identified in the stems of ephedra gerardiana by HPLC-MS and their antioxidant activity. Int J Mol Sci 26:3097

19. Prakash I, Ma G, Bunders C, Charan RD, Ramirez C, Devkota KP, Snyder TM (2017) A novel diterpene glycoside with nine glucose units from Stevia rebaudiana Bertoni. Biomolecules 7:2–11
20. Lee JM, Lee J, Cho SH, Lee SW, Kim YO, Cho IH, Kim HJ, Lee S (2016) Chiisanenside, a new triterpene glycoside from the fruits of Acanthopanax chiisanensis. Appl Biol Chem 59:255–258
21. Afzan A, Marcourt L, Abd Karim HA, Kasim N, Queiroz EF, Osman CP, Ismail NH, Wolfender JL (2024) Chemical characterizations dataset of flavonoid glycoside isomers and other constituents from Ficus deltoidea Jack. Data Br 54:110414
22. Cuny E, Klingler FD (2022) Efficient Isolation and Structure Analysis of (+)-Ranuncoside, a Unique Tricyclic Spiroacetal Glycoside, from Christmas Rose (Helleborus niger L.). Natural Product Communications 17(1):1–9.
23. Rodrigues C, Alves C, Santos-neto AJ, Fernandes C, Lan FM (2007) Analysis of tricyclic antidepressant drugs in plasma by means of solid-phase microextraction-liquid chromatography-mass spectrometry. J mass Spectrom 42:1342–1347

Check for updates

Chapter 16

Polyketides: Comprehensive Chromatographic Analysis

Mansi P. Mehta, Bhikhu S. More and Gaurav S. Shah

Abstract

Polyketides are a diverse group of natural products that are biosynthesized via short acyl-CoA units through polyketide syntheses. They have notable variety in their structure and function. Naturally, they are present in many bacteria, plants, fungi, insects, protists, mollusks, and sponges. Biosynthesis of polyketides consists of a group of enzyme activities known as polyketide synthases (PKSs), whose classification depends on structural organization of their functional domains. Chromatographic analysis of polyketides is vital for separation, identification, and quantification. It ensures high selectivity, sensitivity, and reproducibility in detecting bioactive compounds from natural resources. Basically, HPLC, UPLC, LC-MS, and GC are widely used chromatographic techniques. Advanced techniques such as chiral chromatography and 2D-LC enhance resolution and selectivity. Sample preparation for chromatographic analysis is done using various solvents, solid-phase extraction (SPE), filtration, and derivatization, with new methods such as ultrasound-assisted extraction (UAE) and microwave-assisted extraction (MAE).

Keywords Polyketides, Polyketide synthases (PKSs), Chromatographic analysis, Advanced techniques, Sample preparation

1 Introduction

Plants have developed the ability to produce defensive chemicals through the process of natural selection. Polyketides are a remarkably diverse category of natural compounds and biologically active secondary metabolites, which are not directly involved in cellular functions but play an important role in plant adaptation in their environment [1, 2]. They are biosynthesized via route that involves successive decarboxylative condensation reactions of simple acyl-CoA precursors [3]. They are present extensively across various biological kingdoms such as animals, plants, bacteria, protozoa, and fungi [4]. Due to processes such as condensation, cyclization, glycosylation, hydroxylation, aromatization, methylation, sulfation, and acylation, plant polyketides exhibit structural diversity.

The structural diversity of the plant polyketides results from the number of starter substrates that can be used by the PKSs and from

Dhruti Amin et al. (eds.), *Phytoconstituent Analysis*, Springer Protocols Handbooks,
https://doi.org/10.1007/978-1-0716-5400-2_16,

subsequent modifications via regio-specific condensation, cyclization, aromatization, hydroxylation, glycosylation, acylation, prenylation, sulfation, and methylation reactions [5, 6]. Some plant polyketides have anticancer, antimicrobial, antiviral, antioxidant, neuroprotective, and estrogenic activities [7–11].

1.1 Classification of Polyketides

1.1.1 Type I Polyketide Synthases

Type I polyketide synthases are huge, multifunctional enzymes that consist of multiple modules, each for one non-iterative condensation cycle. In each module, the acyltransferase (AT), ketosynthase (KS), and acyl carrier protein (ACP) domains are usually present, which create β-keto ester intermediates. Further domains such as ketoreductase (KR), dehydratase (DH), and enoyl reductase (ER) can act on the keto group [12]. Before releasing the final product, polyketide chain is passed from module to module.

1.1.2 Type II Polyketide Synthases

Type II polyketide synthases (PKSs) are generally characterized by seven major structural groups: anthracyclines, aureolic acids, tetracyclines, angucyclines, pradimicin-type polyphenols, tetracenomycins, and benzoisochromanequinones. In contrast to type I PKSs, type II PKSs are iterative and employ monofunctional proteins. They are a "minimal PKS" system: KSα, KSβ, and ACP, where chain elongation is managed by KSα and chain length can be controlled by KSβ. Modification of aromatic core can be carried out by ketoreductases, cyclases, and aromatases, while end structure can be tailored by oxygenases, methyltransferases, and glycosyltransferases [13].

1.1.3 Type III Polyketide Synthases

These types consist of homodimeric enzymes having a relatively small size. They catalyze iterative decarboxylative condensations of malonyl units with a CoA-linked starter molecule [14, 15]. CoA-linked starter substrates such as acetyl CoA, malonyl CoA, methyl malonyl CoA, cinnamyl CoA, *p*-coumaroyl CoA, etc. can be utilized by type III PKSs. The applications of plant polyketides in pharmaceutical industries are limited due to their low availability [16].

1.2 Role of Chromatography in Polyketide Analysis

Chromatography is a sensitive and selective analytical technique for polyketide identification and quantification. It is the prime tool in the separation and purity determination of these intricate molecules [17]. High-performance liquid chromatography (HPLC) is universally used because of its good resolution, reproducibility, and versatility with a range of detectors including UV, MS, and ELSD, as well as C18 columns [18, 19]. High-throughput screening can be possible through ultra-performance liquid chromatography (UPLC), which provides faster analysis and improved resolution [20]. For metabolomic profiling, gas chromatography (GC), though restricted to volatile or derivatized polyketides, is still valuable [21].

Thin-layer chromatography (TLC) is also used to detect and monitor polyketide production, which is quick and inexpensive. LC-MS and LC-MS/MS provide structural elucidation and identification of new polyketides from natural extracts [22]. Quantitation is carried out using HPLC-UV or LC-MS/MS techniques validated to ICH guidelines for accuracy, specificity, linearity, and precision [23].

2 Materials

2.1 HPLC

2.1.1 Sample Preparation for HPLC Analysis of Plant-Based Polyketides (see Note 1)

1. **Drying and Grinding of Samples**
   Plant tissues such as roots, stems, leaves, or microbial biomass are air-dried or lyophilized to preserve thermolabile compounds. Fine grinding enhances solvent penetration, improving recovery of compounds such as anthraquinones from *Rhamnus* bark [24].

2. **Extraction of Target Polyketides**

   Conventional Solid–Liquid Extraction

   Classical Soxhlet or maceration techniques using methanol, acetone, or aqueous ethanol are still widely used. Emodin, physcion, and chrysophanol from *Rheum emodi* roots were extracted using 70% ethanol, yielding high recoveries suitable for HPLC profiling. Erythromycin from *Saccharopolyspora erythraea* biomass is efficiently recovered with methanol under controlled agitation [25].

   Modern Extraction Approaches

   - Ultrasonic-Assisted Extraction (UAE): UAE enhanced the yield of curcumin analogs (diarylheptanoid polyketides) from *Curcuma longa* compared to traditional maceration [26].
   - Microwave-Assisted Extraction (MAE): MAE provided rapid and high-yield extraction of hypericin and pseudohypericin from *Hypericum perforatum* in under 10 min [27].
   - Supercritical CO_2 Extraction: Successfully applied for lipophilic anthraquinones from *Aloe vera* resin, offering clean extracts suitable for LC-MS analysis [28].

3. **Clarification of Extracts**
   Crude extracts are clarified by centrifugation or fine filtration, removing debris and polysaccharides that can interfere with HPLC column performance. Clarification steps significantly improved detection sensitivity of tetracycline-type polyketides in fermentation broths of *Streptomyces aureofaciens* [29].

4. **Concentration and Matrix Cleanup**
   Matrix interference is common when working with polyketides due to co-extracted pigments or lipids.
   Solid-Phase Extraction (SPE): HLB or C18 SPE cartridges were used to clean up naphthoquinone polyketides from *Rubia cordifolia*, improving chromatographic resolution. SPE also improved quantification of macrolide antibiotics such as rapamycin from *Streptomyces hygroscopicus* [30].
5. **QuEChERS-Based Approaches**
   QuEChERS has been adapted for polyketide workflows to achieve rapid cleanup. It improved simultaneous detection of anthraquinone derivatives in complex traditional medicine formulations using LC-MS/MS [31].
6. **Advanced Microextraction Techniques**
   Solid-Phase Microextraction (SPME): Applied for solvent-free extraction of volatile polyketides such as geosmin and 2-methylisoborneol from microbial cultures before direct HPLC or GC-MS analysis [32].
   Other Strategies:

   - Stir-bar sorptive extraction (SBSE) has been used for polyether toxins from marine dinoflagellates.
   - Dispersive liquid–liquid microextraction (DLLME) has been explored for hypericin quantification in *Hypericum* extracts.

7. **Solvent Exchange and Filtration**
   Reconstitution in a weak mobile phase and filtration through 0.22 μm membranes ensures compatibility with RP-HPLC and prevents clogging when analyzing complex polyketide matrices [33].
8. **Internal Standards**
   For consistent quantification, internal standards such as actinorhodin or isotopically labeled analogues are added during extraction of macrolides or tetracycline derivatives.
9. **Derivatization (Optional)**

   Though many polyketides absorb in the UV range or ionize readily in MS, derivatization enhances detection for neutral macrolides or compounds lacking strong chromophores.

2.2 HPLC Columns and Mobile Phases for Polyketide Profiling (see Note 2)

1. **Reversed-Phase HPLC—The Standard for Polyketides**
   Most polyketides are moderately polar, making reversed-phase separations ideal.

- C18 Columns: Widely used for anthraquinone polyketides from *Rheum*, *Rubia*, and *Cassia* species. Gradient elution with aqueous formic acid and acetonitrile enabled baseline separation of hypericin and pseudohypericin from *Hypericum perforatum* in <25 min.
- Isocratic Systems: Used for targeted quantification of erythromycin and its analogs with UV detection at 210 nm in fermentation extracts [34].

2. **Alternative Stationary Phases**
   - Phenyl-Hexyl Columns: Provide improved selectivity for aromatic polyketides such as anthraquinones and naphthoquinones due to π–π interactions.
   - Polar-Embedded RP Columns: Enhance retention of polar tetracyclines or glycosylated macrolides in aqueous systems.
   - Calixarene-Modified Phases: Facilitate separation of isomeric anthraquinones and complex macrolides by inclusion complexation [35].
3. **Normal-Phase and HILIC Modes**
   - Normal-Phase HPLC: Applied for non-polar compounds such as polyether marine polyketides and lipid-soluble quinones.
   - HILIC: Effective for glycosylated polyketides or sugar-bound derivatives that are poorly retained in RP systems [36].

2.3 Ultra-High-Performance Liquid Chromatography (UHPLC)

- Used for rapid profiling of macrolides, tetracyclines, and anthraquinones with superior resolution.
- Shorter run times (5–10 min) combined with high sensitivity make UHPLC-MS an excellent choice for metabolite discovery [37].
- Replacement of acetonitrile with ethanol or water-rich systems for anthraquinone analysis.
- Miniaturized UHPLC methods reduce solvent consumption for routine screening of macrolide polyketides in industrial processes [38]. Table 1 presents a comparative analysis of various plant-based polyketides, detailing their sources, structural classes, biological activities, and analytical detection parameters.

Tandem Mass Spectrometry (MS/MS)

Table 1
Comparative study of different plant-based polyketides

Polyketide/class	Column type and dimensions	Flow rate (mL/min)	Mobile phase and gradient	Notes/application	References
Anthraquinones (emodin, chrysophanol, rhein)	C18 reverse-phase, 250 × 4.6 mm, 5 μm	1.0	A: water + 0.1% formic acid; B: acetonitrile; gradient 20%–70% B over 30 min	Quantification in *Rheum palmatum* and *Cassia occidentalis* extracts	[52, 40]
Hypericin and pseudohypericin	C18, 150 × 4.6 mm, 3.5 μm	0.8	A: water + 0.1% formic acid; B: methanol; gradient 40%–90% B over 25 min	UHPLC-MS/MS profiling in *Hypericum perforatum* extracts	[39]
Naphthoquinones (alizarin, purpurin, rubiadin)	Phenyl-hexyl, 250 × 4.6 mm, 5 μm	1.0	A: water + 0.1% TFA; B: acetonitrile; gradient 10%–80% B over 40 min	Separation of closely related naphthoquinones in *Rubia cordifolia*	[30]
Glycosylated anthraquinones (aloe glycosides)	C18, 250 × 4.6 mm, 5 μm	1.0	A: water; B: acetonitrile; isocratic 30% B over 40 min	ELSD/CAD detection for polar derivatives in *Aloe vera*	[51]
Flavonoid-type polyketides (catechin, quercetin derivatives)	C18, 150 × 4.6 mm, 5 μm	1.0	A: 0.1% formic acid in water; B: 0.1% formic acid in acetonitrile; gradient 5%–60% B over 35 min	DAD detection at 278–370 nm for profiling in plant extracts	[39]
Coumarin-type polyketides	C18, 250 × 4.6 mm, 5 μm	1.0	A: water + 0.1% formic acid; B: acetonitrile; gradient 20%–80% B over 30 min	FLD quantification (Ex 320 nm, Em 420 nm) in *Mammea americana*	[41]

3 Methods

3.1 Detection of Plant-Derived Polyketides by HPLC (see Note 3)

1. **UV-Visible and Diode-Array Detection (DAD/PDA)**
Polyketides such as anthraquinones, naphthoquinones, stilbenes, and flavonoid-type polyketides exhibit strong UV/visible absorption due to their conjugated aromatic systems [39].
   - **Fixed-Wavelength UV Detection:**
   Common absorption maxima include:
   - 254 nm for anthraquinones such as emodin and chrysophanol.
   - 278 nm for polyketide flavonoids such as catechins.
   - 350–370 nm for flavonol-derived polyketides such as quercetin derivatives.
   - **Diode-Array Detection (DAD/PDA):**
   Spectral acquisition across 190–600 nm supports both quantification and tentative identification by comparing UV profiles with reference standards or literature data. Anthraquinones from *Cassia occidentalis* were quantified at 254 nm, while spectral libraries confirmed compound identity and purity [40].
2. **Fluorescence Detection (FLD)**
Although fewer polyketides are naturally fluorescent, FLD offers 10–100× higher sensitivity compared to UV detection when applicable.
   - **Natural Fluorophores:**
   Some coumarin-type polyketides and aromatic glycosides show intrinsic fluorescence and can be quantified at trace levels. Coumarin polyketides in *Mammea americana* were detected at excitation 320 nm/emission 420 nm, enabling sensitive quantification in complex matrices [41].
   - **Derivatization for Non-fluorescent Polyketides:**
   Non-fluorescent compounds such as aloe-emodin can be derivatized with reagents such as NDA (naphthalene-2,3-dicarboxaldehyde) or OPA (o-phthaldialdehyde) to form fluorogenic derivatives, enhancing detection sensitivity [42, 43].
3. **Universal Detection: ELSD and CAD**
Evaporative Light-Scattering Detection (ELSD)
Detects non-volatile and weakly chromophoric polyketides, such as glycosylated anthraquinones or macrolide derivatives, by nebulizing and evaporating solvent, then detecting light scatter.

Table 2
Various detectors used to identify polyketides

Detector	Best for	Example applications	References
PDA/DAD	UV-active polyketides	Emodin, chrysophanol, and hypericin profiling in *Cassia* and *Hypericum* species	[49]
FLD	Natural fluorescent polyketides or derivatives	Coumarin-type polyketides in *Mammea americana*; derivatized aloe-emodin	[41]
ELSD	Non-volatile, weakly UV-absorbing polyketides	Glycosylated naphthoquinones in *Rubia cordifolia*	[46]
CAD	Polar/glycosylated polyketides with uniform response	Aloe glycosides and anthraquinone derivatives	[6, 51]
LC–MS/MS	Trace-level quantification and structural confirmation	Hypericin and pseudohypericin in *Hypericum perforatum*; anthraquinones in *Rheum palmatum*	[34, 52]
HRMS (TOF/Orbitrap)	Structural elucidation and metabolite discovery	Novel naphthoquinone glycosides in *Rubia* species	[30, 53]
PDA + CAD or MS	Comprehensive profiling	Simultaneous quantification of anthraquinone aglycones and glycosides in *Aloe vera*	[51]

Glycosylated naphthoquinone polyketides from *Rubia cordifolia* roots were quantified using ELSD, providing a uniform response across derivatives [44].

Charged Aerosol Detection (CAD)

Offers a broader dynamic range and greater sensitivity compared to ELSD, particularly for polar and glycosylated polyketides. CAD enabled robust quantification of glycosidic anthraquinones in *Aloe vera* extracts, even when UV absorption was minimal [45].

4. **Refractive Index (RI) Detection**
   - Occasionally applied for highly concentrated crude extracts of polyketides where simple quantification is needed. Due to low sensitivity and incompatibility with gradient elution, RI detection is rarely used in modern polyketide workflows [46].
5. **Mass Spectrometric Detection (LC-MS and LC-MS/MS)**
   Mass spectrometry provides unparalleled sensitivity and structural insight, making it the preferred detection method for complex polyketide mixtures.

5.1. Ionization Techniques

- Electrospray Ionization (ESI): Effective for polar anthraquinones, flavonoids, and glycosides.
- Atmospheric Pressure Chemical Ionization (APCI): Better for moderately non-polar polyketides such as macrolides and aglycones [47].

5.2. Detection Modes

- Full-Scan MS: Supports untargeted profiling of polyketide mixtures in plant extracts.
- Selected Ion Monitoring (SIM): Enhances sensitivity for targeted compounds such as hypericin or alizarin [48]. Table 2 summarizes the various detectors employed for the identification and quantification of polyketides in analytical studies.

3.2 Tandem Mass Spectrometry (MS/MS)

SRM/MRM Modes: Monitor precursor-to-product ion transitions for exceptional selectivity and trace-level detection. UHPLC-MS/MS detected hypericin and pseudohypericin in *Hypericum perforatum* at sub-ppb concentrations. Targeted MRM transitions quantified rhein and emodin in *Rheum palmatum*, reducing matrix interference [48, 49].

3.2.1 High-Resolution MS (HRMS)

1. TOF and Orbitrap platforms provide exact mass and MS^n fragmentation, enabling characterization of novel or unknown polyketides. HRMS profiling identified new naphthoquinone glycosides in *Rubia* species, expanding chemotaxonomic markers for the genus [50].
2. **Multi-Detector Strategies**

   Combining detectors provides a more comprehensive profile of polyketides.

   - PDA + MS: Simultaneously quantifies UV-absorbing polyketides and identifies non-chromophore compounds.
   - PDA + ELSD or CAD: Ensures accurate detection of glycosylated and non-UV-absorbing derivatives. Comparative profiling of *Aloe vera* revealed that PDA quantified major anthraquinones, while CAD captured minor glycosidic derivatives that lacked UV response [51].

4 Notes

1. Accurate and reproducible HPLC profiling of polyketides depends on proper sample preparation. Since polyketides are often present in complex plant or microbial matrices, optimized

workflows are essential to ensure selective extraction, efficient cleanup, and stability before analysis.

2. HPLC remains the primary tool for resolving and quantifying structurally diverse polyketides. Selecting the right stationary phase and mobile phase is crucial for achieving reproducible, high-resolution separations.
3. Detection of plant-derived polyketides requires careful selection of detectors and analytical modes to match their structural properties, concentration levels, and matrix complexity. The following detection strategies are commonly applied for qualitative and quantitative analysis.

References

1. Pichersky E, Lewinsohn E (2011) Convergent evolution in plant specialized metabolism. Annu Rev Plant Biol 62:549–566. https://doi.org/10.1146/annurev-arplant-042110-103814
2. Wink M (2003) Evolution of secondary metabolites from an ecological and molecular phylogenetic perspective. Phytochemistry 64:3–19. https://doi.org/10.1016/S0031-9422(03)00300-5
3. Baerson SR, Rimando AM (2007) A plethora of polyketides: structures, biological activities, and enzymes. In: Osbourn AE, Lanzotti V (eds) Plant-derived natural products: synthesis, function, and application. ACS symposium series, vol 955. American Chemical Society, Washington, DC, pp 2–14. https://doi.org/10.1021/bk-2007-0955.ch001
4. Hong F, Hopwood DA, Khosla C (1993) Engineered biosynthesis of novel polyketides. Science 262:1546–1550. https://doi.org/10.1126/science.8248802
5. Austin MB, Noel JP (2003) The chalcone synthase superfamily of type III polyketide synthases. Nat Prod Rep 20:79–110. https://doi.org/10.1039/B100917F
6. Kumar A, Ellis BE (2003) A family of polyketide synthase genes expressed in ripening Rubus fruits. Phytochemistry 62:513–526. https://doi.org/10.1016/s0031-9422(02)00572-1
7. Clere N, Faure S, Martinez MC, Andriantsitohaina R (2011) Anticancer properties of flavonoids: roles in various stages of carcinogenesis. Cardiovasc Hematol Agents Med Chem 9:62–77. https://doi.org/10.2174/187152511796196498
8. Gresele P, Cerletti C, Guglielmini G, Pignatelli P, de Gaetano G et al (2011) Effects of resveratrol and other wine polyphenols on vascular function: an update. J Nutr Biochem 22:201–211. https://doi.org/10.1016/j.jnutbio.2010.07.004
9. Yen TL, Hsu CK, Lu WJ, Hsieh CY, Hsiao G et al (2012) Neuroprotective effects of xanthohumol, a prenylated flavonoid from hops (Humulus lupulus), in ischemic stroke of rats. J Agric Food Chem 60:1937–1944. https://doi.org/10.1021/jf204909p
10. Milligan SR, Kalita JC, Heyerick A, Rong H, De Cooman L et al (1999) Identification of a potent phytoestrogen in hops (Humulus lupulus L.) and beer. J Clin Endocrinol Metab 84:2249–2252. https://doi.org/10.1210/jcem.84.6.5887
11. Harborne JB, Williams CA (2000) Advances in flavonoid research since 1992. Phytochemistry 55:481–504. https://doi.org/10.1016/S0031-9422(00)00235-1
12. Staunton J, Weissman KJ (2001) Polyketide biosynthesis: a millennium review. Nat Prod Rep 18:380–416. https://doi.org/10.1039/A909079G
13. Zhang Z, Pan HX, Tang GL (2017) New insights into bacterial type II polyketide biosynthesis. F1000Res 6:172. https://doi.org/10.12688/f1000research.10466.1
14. Abe I, Morita H (2010) Structure and function of the chalcone synthase superfamily of plant type III polyketide synthases. Nat Prod Rep 27:809–838. https://doi.org/10.1039/B909988N
15. Abe I (2008) Engineering of plant polyketide biosynthesis. Chem Pharm Bull (Tokyo) 56:1505–1514. https://doi.org/10.1248/cpb.56.1505
16. Santos CN, Koffas M, Stephanopoulos G (2011) Optimization of a heterologous pathway for the production of flavonoids from glucose. Metab Eng 13:392–40. https://doi.org/10.1016/j.ymben.2011.02.002

17. Hertweck C (2009) Hidden biosynthetic treasures brought to light. Nat Chem Biol 5(7):450–452. https://doi.org/10.1038/nchembio0709-450
18. Yin J, Zhang J, Qu D et al (2014) Simultaneous determination of five polyketide antibiotics in fermentation broth by HPLC. J Sep Sci 37:3140–3146. https://doi.org/10.1002/jssc.201400428
19. Zhou H, He Z, Liang Y et al (2010) Optimization of reversed-phase HPLC separation of macrolide polyketides. J Chromatogr B 878 (5–6):439–444. https://doi.org/10.1016/j.jchromb.2009.12.020
20. Li Y, Xu X, Sun H et al (2015) Application of UPLC in natural product analysis. Fitoterapia 100:208–220. https://doi.org/10.1016/j.fitote.2014.10.003
21. Bhatt P, Vaidya H (2021) GC–MS based identification of polyketides in microbial extracts. J Chromatogr Sci 59(3):257–263. https://doi.org/10.1093/chromsci/bmaa103
22. Zhang L, An H, Wang Y et al (2018) Advances in LC–MS/MS techniques for structural elucidation of natural products. J Pharm Biomed Anal 150:326–341. https://doi.org/10.1016/j.jpba.2017.12.008
23. Gao J, Shi Y, Liu J et al (2020) Validation of an HPLC-UV method for the quantitation of polyketide antibiotics. Chromatographia 83(10):1221–1228. https://doi.org/10.1007/s10337-020-03900-z
24. Gião MS, Pereira CI, Fonseca SC, Malcata F (2001) Effects of drying method and particle size on the antioxidant properties of leaves and teas of Morus alba, Lagerstroemia speciosa and Thunbergia laurifolia. Food Chem 74:289–294. https://doi.org/10.1016/S0308-8146(01)00135-5
25. He Y, Abdin MZ, Shakeel SM, Dhar AK (2015) Assessment of conventional and novel extraction techniques on extraction efficiency of five anthraquinones from Rheum emodi. J Food Sci Technol 52(5):3016–3024. https://doi.org/10.1007/s13197-014-1366-6
26. Mandal V, Mohan Y, Hemalatha S (2009) Design and optimization of ultrasound assisted extraction of curcumin as an effective alternative for conventional solid–liquid extraction of natural products. Nat Prod Commun 4(1):105–112. https://doi.org/10.1177/1934578X0900400117
27. Punegov VV, Kostromin VI, Fomina MG, Zaynullin VG, Yushkova EA, Belykh DV, Chukicheva IY, Zaynullin GG (2015) Microwave-assisted extraction of hypericin and pseudohypericin from Hypericum perforatum. Russ J Bioorg Chem 41:757–761. https://doi.org/10.1134/S1068162015070122
28. Hu Q, Hu Y, Xu J (2005) Free radical-scavenging activity of Aloe vera extracts by supercritical carbon dioxide extraction. Food Chem 91:85–90. https://doi.org/10.1016/j.foodchem.2004.05.052
29. Lyu X, Deng Z, Xiang L, Zhang T, Li X (2014) Enhanced detection of tetracycline-type polyketides in Streptomyces aureofaciens fermentation broths via clarification by centrifugation and fine filtration followed by HPLC. Anal Bioanal Chem 406:4569–4578. https://doi.org/10.1007/s00216-014-7781-9
30. Kashiwada Y, Ikeshiro Y, Onishi A, Nonaka GI, Nishioka I (1986) A novel naphthoquinone glycoside, rubiasin, and a new anthraquinone triglycoside from Rubia cordifolia roots. Chem Pharm Bull 34(3):1146–1153. https://doi.org/10.1248/cpb.34.1146
31. Reddy CS et al (2018) Modified QuEChERS method for determination of multi-pesticide residues in Indian tea samples by LC–MS/MS. Rasayan J Chem 11(2):582–588. https://doi.org/10.7324/RJC.2018.1121855
32. Conrady S et al (2021) Solid-phase microextraction (SPME) for determination of geosmin and 2-methylisoborneol in volatile emissions from soil disturbance. Chemosphere 281:131005. https://doi.org/10.1016/j.chemosphere.2021.131005
33. Ahmed S, Brown R, Nguyen K, Kumar M, Lee S, Davis V (2018) Application of high-performance liquid chromatography for the analysis of polyketides. In Polyketide biosynthesis: methods and protocols, vol 1776. pp 247–260. Humana Press, New York, NY, USA. https://doi.org/10.1007/978-1-4939-7815-4_12
34. Ghasemi R et al (2014) Simultaneous determination of hypericin and pseudohypericin in Hypericum perforatum L. extracts by HPLC. Iran J Pharm Res 13(2):651–657. https://doi.org/10.22037/ijpr.2014.1678
35. Sun F et al (2012) Evaluation of a phenyl-hexyl stationary phase for the separation of polycyclic aromatic hydrocarbons. J Liq Chromatogr Relat Technol 35(19):2735–2747. https://doi.org/10.1080/10826076.2012.678783
36. D'Orazio G et al (2018) Hydrophilic interaction liquid chromatography (HILIC) for the separation and analysis of carbohydrates and glycans. Molecules 23(1):180. https://doi.org/10.3390/molecules23010180
37. Wurtz P et al (2016) Metabolic profiling of human blood by ultra-high-performance liquid chromatography-tandem mass spectrometry. J Proteome Res 15(1):173–185. https://doi.org/10.1021/acs.jproteome.5b00806
38. Yang X et al (2019) A green and rapid method for the determination of macrolide antibiotics in animal-derived foods by

ultra-high-performance liquid chromatography coupled with tandem mass spectrometry. Food Chem 273:218–226. https://doi.org/10.1016/j.foodchem.2018.08.067

39. Farag MA, Wessjohann LA (2012) Metabolome classification of commercial Hypericum perforatum (St. John's Wort) preparations via UPLC-qTOF-MS and chemometrics. Planta Medica 78(5):488–496. https://doi.org/10.1055/s-0031-1298170
40. Kumar N et al (2018) High-performance liquid chromatographic method for the determination of emodin in Cassia occidentalis Linn. J Pharmacogn Phytochem 7(5):1801–1803
41. Braca A et al (2009) Phytochemical analysis of the cytotoxic constituents from the bark of Mammea americana L. J Agric Food Chem 57(11):4496–4501. https://doi.org/10.1021/jf8038317
42. Lindroth P, Mopper K (1979) High performance liquid chromatographic determination of subpicomole amounts of amino acids by precolumn fluorescence derivatization with o-phthaldialdehyde. Anal Chem 51(11):1667–1674. https://doi.org/10.1021/ac50047a016
43. Pan T et al (2018) Naphthalene-2,3-dicarboxaldehyde as a precolumn derivatization reagent for sensitive determination of amino acids by high-performance liquid chromatography with fluorescence detection. Talanta 184:148–154. https://doi.org/10.1016/j.talanta.2018.03.007
44. Kulkarni R et al (2018) Development of a validated HPLC-ELSD method for the quantitative determination of glycosidic compounds in Rubia cordifolia. J Anal Sci Technol 9(1):1–8. https://doi.org/10.1186/s40543-018-0158-7
45. Zong S et al (2018) An efficient method for the separation and purification of rapamycin from a fermentation broth of Streptomyces hygroscopicus. J Ind Microbiol Biotechnol 45(10):833–841. https://doi.org/10.1007/s10295-018-2070-6
46. Kucera P, Guiochon G (1983) The use of the refractive index detector in high-performance liquid chromatography. J Chromatogr A 281:21–40. https://doi.org/10.1016/S0021-9673(01)88390-3
47. Wang Y et al (2018) High-resolution mass spectrometry and electrospray ionization tandem mass spectrometry for phytochemical analysis. TrAC Trends Anal Chem 106:18–27. https://doi.org/10.1016/j.trac.2018.06.012
48. Korf J et al (2018) Fast and efficient determination of antibiotics in wastewater and surface water using a polar-embedded reversed-phase column and UPLC-MS/MS. J Chromatogr A 1552:120–129. https://doi.org/10.1016/j.chroma.2018.04.004
49. Erdoğan A, Orhan IE (2018) LC-DAD-MS-assisted quantification of marker compounds in Hypericum perforatum L. (St. John's wort) and its antioxidant activities. J Agric Food Chem 66(45):11985–11996. https://doi.org/10.1021/acs.jafc.8b04930
50. Wolfender JL et al (2019) Full-scan mass spectrometry for the discovery of new natural products. TrAC Trends Anal Chem 116:165–174. https://doi.org/10.1016/j.trac.2018.10.012
51. Ma ZQ, Liu HM, Ma ZZ, Wang QL (2014) An improved HPLC–ELSD method for the simultaneous determination of eight major saccharides in Aloe vera L. gel. J Agric Food Chem 62(12):2636–2643. https://doi.org/10.1021/jf405367j
52. Zhao D et al (2012) Simultaneous quantification of six anthraquinones in Rheum palmatum L. by high-performance liquid chromatography. Molecules 17(10):11985–11996. https://doi.org/10.3390/molecules171011985
53. Özgen U, Kazaz C, Seçen H, Çaliş İ, Coşkun M, Houghton PJ (2009) A novel naphthoquinone glycoside from Rubia peregrina L. Turk J Chem 33(4):561–568. https://doi.org/10.3906/kim-0806-28

Chapter 17

Non-Ribosomal Peptides: Advanced Profiling Techniques

Krunal Dholiya, Shruti Satashia and Nileshkumar Pandya

Abstract

Non-ribosomal peptides (NRPs) are a diverse class of secondary metabolites with significant ecological and pharmacological importance, produced by multifunctional enzyme complexes in plants and associated microorganisms. Their structural complexity, low abundance, and matrix interference pose substantial challenges for analysis, necessitating advanced profiling techniques. This chapter explores cutting-edge methods for NRP characterization in phytoconstituent analysis, emphasizing high-resolution mass spectrometry (HRMS), nuclear magnetic resonance (NMR) spectroscopy, ultra-high-performance liquid chromatography (UHPLC), and bioinformatics tools. HRMS, including ESI-MS and Orbitrap systems, provides unparalleled sensitivity for detecting trace NRPs, while tandem MS/MS elucidates complex structures through diagnostic fragmentation. NMR, particularly 2D techniques like COSY and NOESY, offers detailed structural insights, complementing MS for resolving stereochemistry. UHPLC and multidimensional chromatography enhance separation of NRPs from interfering phytoconstituents, improving detection accuracy. Bioinformatics platforms, such as NORINE and AntiSMASH, streamline dereplication and predict biosynthetic pathways, integrating metabolomics and genomics data. Emerging techniques, including imaging mass spectrometry for spatial mapping and cryo-electron microscopy for NRPS enzyme visualization, further advance NRP profiling by revealing their ecological roles and biosynthetic mechanisms. Optimized sample preparation, involving tailored solvent extraction and solid-phase enrichment, ensures reliable analysis. Applications span drug discovery, with NRPs such as cyclosporin A showing therapeutic promise, to plant defense studies, such as cyanogenic glycosides in *Manihot esculenta*. Future directions include multi-omics integration, automated high-throughput platforms, and synthetic biology to engineer novel NRPs. These advancements promise to deepen our understanding of NRP diversity and unlock their potential in pharmacology, agriculture, and biotechnology, addressing critical challenges in sensitivity, scalability, and cost-effectiveness.

Keywords Non-ribosomal peptides, Phytoconstituent analysis, High-resolution mass spectrometry, NMR spectroscopy, Bioinformatics tools

1 Introduction

Non-ribosomal peptides (NRPs) represent a diverse class of secondary metabolites characterized by their complex structures and significant biological activities. Unlike peptides synthesized via ribosomal machinery, NRPs are produced by large, multifunctional enzyme complexes known as non-ribosomal peptide synthetases (NRPSs).

Dhruti Amin et al. (eds.), *Phytoconstituent Analysis*, Springer Protocols Handbooks,
https://doi.org/10.1007/978-1-0716-5400-2_17,

These compounds are prevalent in various organisms, including bacteria, fungi, and plants, and play critical roles in ecological interactions and human health applications. In the context of phytoconstituent analysis, NRPs are of particular interest due to their structural diversity, bioactivity, and potential as lead compounds in drug discovery. This chapter aims to elucidate advanced profiling techniques for NRPs, addressing their identification, characterization, and applications in plant science and pharmacology.

NRPs are oligopeptides or polypeptides synthesized independently of the ribosomal translation process, typically ranging from 2 to over 20 amino acid residues. These molecules are distinguished by their incorporation of non-proteinogenic amino acids, modified side chains, and diverse structural motifs, such as cyclic, branched, or glycosylated forms [1]. In phytoconstituent analysis, NRPs are significant because they contribute to the chemical diversity of plant secondary metabolites, which are critical for plant defense, signaling, and adaptation to environmental stresses [2]. Their structural complexity poses challenges for identification and quantification, necessitating advanced analytical techniques to fully explore their chemical and biological profiles. Moreover, NRPs from plants and associated microorganisms (e.g., endophytes) have shown promising therapeutic properties, including antimicrobial, anticancer, and immunosuppressive activities, making them valuable targets for phytochemical research [3].

The biosynthesis of NRPs is mediated by NRPSs, which are large, modular enzyme complexes that operate independently of the ribosomal machinery used for protein synthesis. Each NRPS module is responsible for the incorporation of a specific amino acid or other building block into the growing peptide chain, following a thiotemplate mechanism [4]. NRPSs consist of distinct domains, including adenylation (A), thiolation (T), and condensation (C) domains, which collectively activate, tether, and link amino acids or other substrates. Additional domains, such as epimerization, methylation, or cyclization domains, introduce structural modifications, contributing to the diversity of NRPs [5]. In contrast, ribosomal synthesis relies on messenger RNA (mRNA) templates and the ribosome to produce linear polypeptides composed of the 20 standard amino acids. Unlike ribosomal peptides, which are limited by the genetic code, NRPs incorporate non-proteinogenic amino acids (e.g., D-amino acids, β-amino acids) and other building blocks, such as fatty acids or heterocyclic moieties, enabling a broader chemical repertoire [6]. In plants, NRPS-like enzymes are less common but have been identified in certain species, often in association with endophytic or symbiotic microorganisms that contribute to NRP production [7].

The structural diversity of NRPs arises from the modular nature of NRPSs, which allows for the incorporation of a wide range of substrates and modifications. NRPs may be linear, cyclic, or branched and often feature non-standard amino acids, such as ornithine, hydroxyproline, or β-alanine, as well as post-synthetic modifications such as glycosylation, acylation, or halogenation [8]. For example, cyclic NRPs such as cyclosporin A, produced by fungi but also relevant in plant-associated microbiomes, exhibit remarkable stability and bioactivity due to their constrained structures [9]. In plants, NRPs such as cyanogenic glycosides or cyclic peptides (e.g., orbitides) contribute to defense against herbivores and pathogens [10]. This structural complexity, while biologically advantageous, complicates analytical efforts, as NRPs often co-occur with other secondary metabolites, requiring sophisticated separation and detection methods for accurate profiling.

NRPs play multifaceted roles in plant biology and ecology. In plants and their associated microorganisms, NRPs function as defense compounds, signaling molecules, or mediators of symbiotic interactions. For instance, NRPs are produced by endophytic fungi in plants such as grasses [10]. Their antimicrobial, antifungal, and allelopathic properties contribute to plant resilience against biotic stresses [11]. In ecological contexts, NRPs facilitate interactions between plants and microorganisms, such as nitrogen-fixing bacteria or mycorrhizal fungi, influencing nutrient acquisition and stress tolerance [7]. From a pharmaceutical perspective, NRPs are highly valued for their bioactive properties. Notable examples include vancomycin and daptomycin, bacterial NRPs with potent antibacterial activity, and plant-derived NRPs with potential anticancer or immunosuppressive effects [12]. The ability to profile and characterize these compounds accurately is crucial for harnessing their therapeutic potential.

This chapter aims to provide a comprehensive overview of advanced analytical techniques for the profiling of NRPs in phytoconstituent analysis. By addressing the challenges posed by their structural complexity and low abundance, we explore cutting-edge methods, including high-resolution mass spectrometry, NMR spectroscopy, liquid chromatography, and bioinformatics tools, to enable precise identification and characterization. The chapter also discusses practical applications in drug discovery, plant ecology, and biotechnology, supported by case studies and future directions for NRP research. Through this exploration, we seek to equip researchers with the knowledge and tools necessary to advance the study of these critical secondary metabolites.

1.1 Challenges in NRP Analysis

The analysis of non-ribosomal peptides (NRPs) in phytoconstituent research presents unique challenges due to their inherent chemical complexity, low abundance in plant tissues, interference from co-occurring metabolites, and the limitations of traditional analytical methods. These obstacles necessitate the development and application of advanced profiling techniques to achieve accurate identification, structural elucidation, and quantification of NRPs. This section discusses these challenges in detail, providing a foundation for understanding the need for sophisticated analytical approaches in NRP research (Fig. 1).

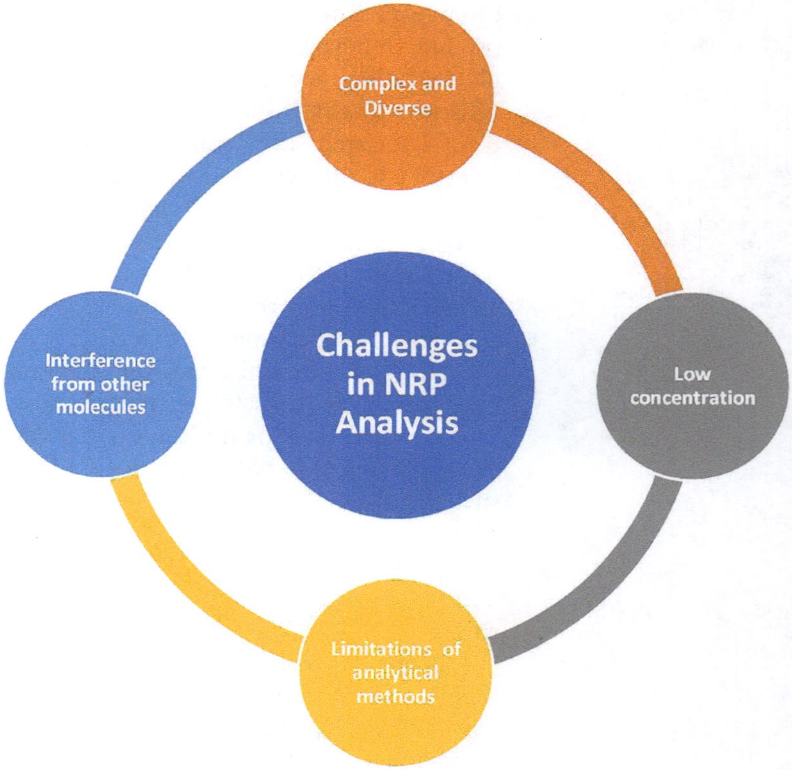

Fig. 1 Major challenges in NRP analysis

1.1.1 Structural Complexity and Diversity

Non-ribosomal peptides exhibit remarkable structural diversity, which poses significant challenges for their analysis. Unlike ribosomally synthesized peptides, NRPs are produced by non-ribosomal peptide synthetases (NRPSs), which incorporate a wide range of building blocks, including non-proteinogenic amino acids (e.g., D-amino acids, β-amino acids, ornithine), fatty acids, and heterocyclic moieties [1]. This diversity results in linear, cyclic, branched, or macrocyclic structures, often with additional modifications such as glycosylation, acylation, or halogenation [8]. For instance, cyclic NRPs such as cyclosporin A or plant-derived orbitides possess constrained conformations that enhance stability but complicate spectroscopic analysis due to overlapping signals in nuclear magnetic resonance (NMR) spectra [9, 11]. The presence of non-standard amino acids and diverse functional groups further hinders mass spectrometry (MS)-based fragmentation, as standard peptide fragmentation patterns do not apply, necessitating customized MS/MS workflows [3]. This structural complexity requires high-resolution analytical platforms and computational tools to deconvolute and accurately characterize NRP structures.

1.1.2 Low Abundance in Plant Tissues

NRPs are often present in trace amounts in plant tissues, which significantly complicates their detection and quantification. Unlike primary metabolites, which are abundant and ubiquitous, NRPs are secondary metabolites typically produced in response to specific environmental or biotic stimuli, resulting in low concentrations [2]. For example, cyclic peptides in plants such as *Kalanchoe* species or cyanogenic glycosides in *Manihot esculenta* are often found at concentrations below 1 μg/g of plant material, making their detection challenging without sensitive techniques [10]. Low abundance necessitates highly sensitive detection methods, such as high-resolution mass spectrometry (HRMS) or quantitative NMR (qNMR), to achieve adequate signal-to-noise ratios. Additionally, the extraction of NRPs from plant matrices must be optimized to maximize yield while minimizing degradation, as these compounds are often chemically labile [7]. The low abundance of NRPs underscores the need for enrichment techniques, such as solid-phase extraction or preparative chromatography, to isolate sufficient quantities for analysis.

1.1.3 Interference from Other Phytoconstituents

Plant extracts are complex mixtures containing a diverse array of phytoconstituents, including alkaloids, flavonoids, terpenoids, and phenolic compounds, which can interfere with NRP analysis. These co-occurring metabolites often have similar physicochemical properties, such as polarity or molecular weight, leading to co-elution during chromatographic separation [13]. For instance, phenolic compounds may produce overlapping mass spectra with NRPs in liquid chromatography-mass spectrometry (LC-MS) analyses, complicating peak assignment and quantification [14]. Additionally, matrix effects in MS can suppress or enhance ion signals, leading to inaccurate quantification of NRPs [15]. The presence of isobaric compounds—molecules with identical nominal masses but different structures—further complicates dereplication efforts, as distinguishing NRPs from other metabolites requires high-resolution MS or complementary techniques such as NMR [6]. Effective separation strategies, such as ultra-high-performance liquid chromatography (UHPLC) or multidimensional chromatography, are essential to mitigate interference and ensure accurate NRP profiling.

1.1.4 Limitations of Traditional Qualitative and Quantitative Methods

Traditional analytical methods, such as thin-layer chromatography (TLC), ultraviolet-visible (UV-Vis) spectroscopy, or low-resolution MS, are often inadequate for NRP analysis due to their limited sensitivity, selectivity, and resolution. TLC, while useful for preliminary screening, lacks the specificity to distinguish NRPs from other phytoconstituents with similar Rf values [16]. UV-Vis spectroscopy is limited by the fact that many NRPs lack strong chromophores, rendering them undetectable without derivatization [17]. Low-resolution MS struggles to resolve isobaric compounds or provide detailed structural information, particularly for NRPs with complex modifications [3]. Furthermore, traditional quantitative methods, such as gravimetric analysis or colorimetric assays, are prone to interference from matrix components and lack the precision required for low-abundance NRPs [14]. These limitations highlight the need for advanced techniques, such as HRMS, tandem MS (MS/MS), and two-dimensional NMR, which offer higher sensitivity and structural resolution. Additionally, the integration of bioinformatics tools, such as the NORINE database or AntiSMASH, is critical for overcoming the shortcomings of traditional methods by enabling rapid dereplication and structural prediction [6, 18].

1.2 Advanced Profiling Techniques for NRPs

Advanced profiling of non-ribosomal peptides (NRPs) uses various analytical techniques for identification and characterization. Table 1 provides a comparative overview of these methods, including their principles, advantages, limitations, and applications.

2 Materials

2.1 Mass Spectrometry-Based Approaches for Profiling Non-Ribosomal Peptides (NRPs)

Mass spectrometry-based profiling of non-ribosomal peptides (NRPs) requires a combination of advanced analytical instrumentation, high-purity reagents, curated databases, and computational tools to address the inherent structural diversity and low abundance of these metabolites.

1. **Biological materials**: Plant tissues (leaves, roots, seeds, bark) or plant-associated microorganisms such as endophytic fungi and bacteria, which are known producers of NRPs [18, 25]. Samples may be processed in fresh or dried form depending on the experimental design.
2. **Chemicals and reagents:** HPLC- or LC-MS-grade solvents including methanol, acetonitrile, and ultrapure water for extraction and chromatographic separation. Volatile additives such as formic acid or ammonium acetate to enhance ionization efficiency during electrospray ionization [13]. Stable isotope-labeled precursors such as ^{13}C-, ^{15}N-, or ^{2}H-labeled amino acids or fatty acids for biosynthetic investigations [5, 7].
3. **Instrumentation:** High-resolution mass spectrometry (HRMS)) systems including electrospray ionization mass spectrometry (ESI-MS), matrix-assisted laser desorption/ionization time-of-flight (MALDI-TOF), and Orbitrap-based HRMS platforms [13, 21]. Tandem mass spectrometry (MS/MS) systems such as quadrupole time-of-flight (Q-TOF) and ion trap–Orbitrap hybrid instruments [8, 23]. Liquid chromatography systems coupled to HRMS (LC-HRMS) for separation of complex phytochemical matrices.
4. **Databases and software:** NORINE database) for known non-ribosomal peptides [6]. AntiSMASH for prediction of

Table 1
Advanced profiling techniques for NRPS

Technique	Principle	Advantages	Limitations	Applications
ESI-MS (HRMS)	Soft ionization for polar/ionic NRPs, generating molecular ions	Preserves structural integrity, high sensitivity	Limited to polar compounds	Detection of intact peptides in plant extracts
MALDI-TOF (HRMS)	Ionization and time-of-flight for high molecular weight NRPs	Rapid screening, spatial profiling via imaging	Matrix interference possible	Mapping NRP distribution in plant tissues
Orbitrap (HRMS)	High-resolution mass analysis with accurate mass measurement	Resolves isobaric compounds, high mass accuracy (<2 ppm)	High cost, complex operation	Differentiation of NRPs from co-eluting phytoconstituents
MS/MS	Fragmentation of precursor ions to produce diagnostic product ions	Detailed structural elucidation, high sensitivity	Non-standard fragmentation rules for NRPs	Amino acid sequence and modification analysis
Dereplication (NORINE, AntiSMASH)	Matching MS/NMR data to databases for known NRPs	Rapid identification, reduces redundant analyses	Requires curated databases	Identifying known NRPs, prioritizing novel compounds
Isotopic labeling with MS	Tracing stable isotopes (^{13}C, ^{15}N, ^{2}H) in NRP biosynthesis	Elucidates biosynthetic pathways, high precision	Requires isotope precursors, complex data analysis	Studying NRP biosynthetic origins
1D and 2D NMR	Magnetic resonance for proton and carbon signals	Resolves complex structures, confirms stereochemistry	Low sensitivity, requires larger samples	Structural characterization of cyclic NRPs
qNMR	Quantification via signal intensity with internal standards	Non-destructive, no calibration standards needed	Needs pure samples, signal overlap possible	Precise quantification of NRPs in extracts

(continued)

Table 1
(continued)

Technique	Principle	Advantages	Limitations	Applications
UHPLC	High-pressure LC with sub-2 μm columns	Superior resolution, speed, and sensitivity	Expensive equipment	Separation of NRPs from complex matrices
LC-MS/MS	LC separation with tandem MS fragmentation	High-throughput, sensitive detection	Requires skilled operation	Profiling and quantifying NRPs in plant extracts
Multidimensional chromatography (2D-LC)	Orthogonal separation mechanisms (e.g., RP and HILIC)	Enhanced peak capacity, resolves co-eluting compounds	Complex optimization, potential sample loss	Untargeted profiling of complex mixtures
Molecular networking	Clustering MS/MS spectra by structural similarity	Visualizes NRP relationships, aids dereplication	Needs computational expertise, curated libraries	Mapping NRP families, discovering novel compounds
IMS (MALDI/DESI)	Spatial mapping via MS with high-resolution imaging	Tissue-specific localization, high sensitivity	Matrix effects, requires high-resolution instruments	Mapping NRP distribution in plant tissues
Cryo-EM	Near-atomic resolution imaging of NRPS enzymes	Reveals NRPS structure and dynamics	Needs purified enzymes, high technical expertise	Studying NRPS mechanisms for synthetic biology
Metabolite fingerprinting with chemometrics	Untargeted profiling with statistical analysis (PCA, PLS-DA)	Holistic profiling, detects low-abundance NRPs	Complex data interpretation, needs reference standards	Identifying novel NRPs, correlating with stressors

NRPS gene clusters and structure–function relationships [18]. Molecular networking platforms for MS/MS spectral clustering and dereplication [26].

2.2 Nuclear Magnetic Resonance (NMR) Spectroscopy for Profiling Non-Ribosomal Peptides

Nuclear magnetic resonance (NMR) spectroscopy represents one of the most authoritative analytical tools for the comprehensive structural elucidation and quantification of non-ribosomal peptides (NRPs). Unlike mass spectrometry, which primarily provides molecular mass and fragmentation patterns, NMR delivers atom-level structural information, making it indispensable for confirming complete molecular architectures, stereochemistry, and conformational features of NRPs.

1. **Biological materials:** Plant tissues, plant extracts, or purified fractions enriched in NRPs. Isolated NRPs obtained through chromatographic purification (e.g., preparative HPLC or LC-MS-guided fractionation). Plant-associated microorganisms such as endophytic fungi or bacteria producing NRPs. Because NRPs often occur in trace concentrations within complex phytochemical matrices, careful selection and pre-concentration of biological material are essential prior to NMR analysis [7, 28].
2. **Chemicals and reagents:** Deuterated solvents selected based on solubility and stability of NRPs, including: CDCl3 for relatively non-polar peptides, CD3OD and D2O for polar and ionic NRPs, DMSO-d6 for poorly soluble or highly modified peptides. Internal standards for qNMR analysis with chemically inert and well-resolved signals. Buffer additives (where required) to maintain peptide integrity during prolonged acquisitions.
3. **Instrumentation:** High-field NMR spectrometers operating at 400, 500, 600 MHz or higher magnetic field strengths. Cryogenically cooled probes (cryoprobes) to enhance sensitivity for low-abundance NRPs. Variable-temperature control systems for conformational studies of cyclic or flexible peptides.
4. **Software and databases:** Advanced NMR data processing and spectral assignment software. Structural databases and literature repositories for peptide reference comparison. Complementary metabolomic and dereplication databases such as NORINE [6].

2.3 Liquid Chromatography and Hyphenated Techniques for Profiling Non-Ribosomal Peptides

Liquid chromatography (LC) and its hyphenated techniques form a critical component of non-ribosomal peptide (NRP) analysis, as efficient chromatographic separation is a prerequisite for accurate detection, identification, and structural characterization of these compounds in complex phytochemical matrices. Given the structural diversity, polarity variation, and frequent co-elution of NRPs with other secondary metabolites, robust chromatographic systems and optimized analytical conditions are essential.

1. **Biological materials:** Plant tissues (leaves, roots, seeds, bark) containing endogenous NRPs, plant extracts or partially purified fractions enriched in NRPs, and endophyte-derived cultures or plant-associated microorganisms producing NRPs.
2. **Chemicals and reagents:** LC-MS-grade solvents such as water, acetonitrile, and methanol. Mobile phase additives such as formic acid and ammonium acetate or ammonium formate.
3. Sample preparation: solvents for extraction and dilution. Calibration standards and reference compounds where available.
4. **Chromatographic columns:** Reversed-phase columns such as C18 (sub-2 μm particle size for UHPLC), polar-embedded or mixed-mode columns for enhanced selectivity, hydrophilic interaction liquid chromatography (HILIC) columns for orthogonal separations, and guard columns to protect analytical columns from matrix contamination.
5. **Instrumentation:** Ultra-high-performance liquid chromatography (UHPLC) systems capable of operating at pressures up to 15,000 psi [36]. Liquid chromatography–tandem mass spectrometry (LC-MS/MS) platforms including quadrupole time-of-flight (Q-TOF) and Orbitrap-based systems [21]. Multidimensional chromatography setups such as two-dimensional LC (2D-LC) or comprehensive two-dimensional LC (LC×LC).
6. **Software and databases:** LC-MS data acquisition and processing software, molecular networking and chemometric analysis tools, and dereplication databases such as NORINE [6].

3 Methods

3.1 Mass Spectrometry-Based Approaches for Profiling Non-Ribosomal Peptides (NRPs)

3.1.1 Sample Preparation and Extraction of NRPs

1. Plant material or microbial biomass is cleaned, dried (if required), and finely powdered.
2. Extraction is carried out using methanol or methanol–water mixtures under controlled conditions to ensure efficient recovery of NRPs.

3. The extract is centrifuged to remove insoluble debris and filtered through membrane filters.
4. The clarified extract is subjected to LC-HRMS analysis for initial profiling of NRP constituents. This step is critical, as NRPs are often present at low concentrations and coexist with a wide range of primary and secondary metabolites that may interfere with detection [13].

3.1.2 Tandem Mass Spectrometry (MS/MS) for Structural Elucidation

Structural elucidation of NRPs relies heavily on tandem mass spectrometry due to their non-standard amino acid composition and extensive post-synthetic modifications.

1. Precursor ions corresponding to putative NRPs are selected from HRMS full-scan data.
2. Fragmentation is induced using collision-induced dissociation (CID) or higher-energy collisional dissociation (HCD) [23].
3. Fragment ions are analyzed to deduce amino acid sequence, cyclization patterns, and structural modifications [3].
4. Linear NRPs yield characteristic b- and y-ions, whereas cyclic NRPs produce diagnostic fragmentation patterns unique to macrocyclic peptides.
5. Advanced platforms such as Q-TOF and ion trap–Orbitrap hybrids provide high sensitivity and resolution, allowing detection of low-abundance NRPs with methylated or hydroxylated residues [8]. MS/MS-based structural analysis has been effectively applied to orbitides, cyclic NRPs found in *Linum usitatissimum*, where diagnostic cleavage patterns enabled complete structural characterization [11].
6. Data acquisition strategies such as data-dependent acquisition (DDA) and data-independent acquisition (DIA) further enhance analytical coverage by systematically fragmenting detectable ions in complex samples [24].

3.1.3 High-Resolution Mass Spectrometry (HRMS) Analysis

High-resolution mass spectrometry forms the foundation of NRP analysis due to its ability to provide accurate mass measurements and resolve isobaric compounds.

3.1.4 Electrospray Ionization Mass Spectrometry (ESI-MS)

ESI-MS is widely employed for NRP profiling owing to its soft ionization characteristics, which allow intact molecular ions to be generated with minimal fragmentation [13]. This technique is particularly effective for polar, ionic, and cyclic NRPs, including glycosylated or otherwise labile peptides. Previous studies have demonstrated the utility of ESI-MS in detecting intact NRPs in complex plant extracts, such as glycosylated NRPs reported in *Ginkgo biloba* [19].

3.1.5 Matrix-Assisted Laser Desorption/Ionization Time-of-Flight (MALDI-TOF)

MALDI-TOF MS is advantageous for the analysis of higher molecular weight NRPs and is less affected by salt contamination. Additionally, MALDI imaging mass spectrometry enables spatial localization of NRPs within plant tissues, providing insight into their physiological roles [20].

3.1.6 Orbitrap-Based HRMS

Orbitrap mass analyzers provide ultra-high mass resolution (> 100,000) and exceptional mass accuracy (<2 ppm), enabling the discrimination of NRPs from co-eluting phytoconstituents with similar nominal masses [21]. Orbitrap systems have been successfully applied to identify novel NRPs produced by endophytic fungi associated with medicinal plants, resolving isobaric interferences from flavonoids and alkaloids [22].

3.1.7 Dereplication Strategies for Known NRP Identification

Dereplication is an essential step to avoid re-isolation of previously reported NRPs and to prioritize novel compounds.

1. Accurate mass and MS/MS data are matched against the NORINE database, which contains curated information on over 1700 known NRPs [6].
2. LC-HRMS data are cross-referenced with predicted NRPS gene clusters using AntiSMASH, facilitating correlation between genomic and metabolomic datasets [18].
3. Molecular networking is employed to cluster MS/MS spectra based on structural similarity, enabling visualization of relationships among known and unknown NRPs [26].

Such dereplication strategies have been successfully utilized in studies involving endophytic fungi isolated from *Taxus baccata*, where known cyclic NRPs were rapidly identified, allowing focused investigation of novel metabolites [25].

3.1.8 Integration of Isotopic Labeling for Biosynthetic Elucidation

Isotopic labeling combined with HRMS provides critical insights into the biosynthesis and metabolic flux of NRPs.

1. Stable isotope-labeled precursors (^{13}C, ^{1}⁵N, or ^{2}H) are incorporated into plant systems or microbial cultures [5].
2. Labeled substrates are metabolized and incorporated into NRPs through NRPS-mediated pathways.
3. HRMS analysis detects characteristic mass shifts corresponding to isotope incorporation with high precision [27].
4. Orbitrap or FT-ICR MS platforms enable identification of biosynthetic intermediates and pathway branching points.
5. Isotopic flux analysis quantifies precursor contribution under varying environmental or physiological conditions [7].

This approach has been successfully applied to elucidate the biosynthesis of cyanogenic glycosides in *Manihot esculenta*, revealing the involvement of NRPS-like enzymes and precursor utilization patterns [10]. Such studies enhance understanding of the ecological and physiological roles of NRPs in plant defense and symbiotic interactions.

3.2 Nuclear Magnetic Resonance (NMR) Spectroscopy for Profiling Non-Ribosomal Peptides

3.2.1 Sample Preparation for NMR Analysis

1. NRP-containing extracts or fractions are concentrated under reduced pressure to remove volatile solvents.
2. Samples are dried completely to eliminate residual protic solvents that may interfere with NMR signals.
3. The dried residue is reconstituted in an appropriate deuterated solvent based on polarity and solubility.
4. Solutions are sonicated or gently heated, if necessary, to ensure complete dissolution.
5. Insoluble particulates are removed by filtration or centrifugation.
6. For quantitative NMR experiments, a precisely weighed internal standard is added prior to acquisition [32]. Meticulous sample preparation is particularly important for NRPs, as impurities, solvent residues, or peptide aggregation can significantly compromise spectral resolution and quantitative accuracy [28].

3.2.2 One-Dimensional (1D) NMR for Preliminary Structural Assessment

One-dimensional NMR experiments serve as the foundation for structural characterization of NRPs.

^{1}H NMR Spectroscopy

1. Provides information on proton environments within the NRP molecule.
2. Enables identification of amide protons indicative of peptide bonds, aromatic protons associated with phenylalanine, tyrosine, or heterocyclic residues, and aliphatic protons corresponding to side chains and lipid moieties.
3. Signal multiplicity and chemical shift dispersion offer preliminary insights into molecular complexity [28].

^{13}C NMR Spectroscopy

1. Facilitates characterization of the carbon framework of NRPs. Allows detection of carbonyl carbons of peptide bonds, quaternary carbons in modified residues, and unusual carbons associated with non-proteinogenic amino acids.
2. Particularly useful for identifying backbone and side-chain modifications common in NRPs [28].

However, due to the dense proton and carbon environments of NRPs—especially in cyclic peptides—1D spectra frequently exhibit extensive signal overlap, limiting unambiguous structural interpretation [11].

3.2.3 Two-Dimensional (2D) NMR for Comprehensive Structural Elucidation

To overcome spectral congestion and resolve complex connectivity, a combination of 2D NMR experiments is systematically employed.

1. **COSY (Correlation Spectroscopy)**: Identifies scalar (through-bond) proton–proton couplings. Enables assignment of neighboring protons within individual amino acid residues. Serves as the initial step in defining amino acid spin systems [29].
2. **TOCSY (Total Correlation Spectroscopy):** Extends COSY correlations across entire spin systems. Facilitates complete identification of individual amino acid residues within NRPs. Particularly useful for peptides containing non-standard or heavily modified residues [29].
3. **NOESY (Nuclear Overhauser Effect Spectroscopy)**: Provides through-space correlations between protons in close spatial proximity. Essential for confirming cyclic peptide conformations, determining folding patterns and secondary structures, and assigning relative stereochemistry [29].
4. **HSQC (Heteronuclear Single Quantum Coherence)**: Correlates proton signals with directly attached carbon atoms. Enables accurate proton–carbon assignments. Aids in distinguishing overlapping proton signals by carbon dimension separation [30].
5. **HMBC (Heteronuclear Multiple Bond Correlation)**: Detects long-range proton–carbon correlations (2–3 bonds). Crucial for establishing backbone connectivity and confirming sequence order. Particularly valuable for identifying post-synthetic modifications such as methylation, hydroxylation, or glycosylation [30].

The combined application of these techniques has enabled elucidation of highly complex NRPs, such as kalata B1 from *Oldenlandia affinis*, where NMR revealed its cyclic cystine knot topology and disulfide connectivity [31].

3.2.4 Quantitative NMR (qNMR) for Accurate Determination of NRPs

Quantitative NMR (qNMR) offers a universal and calibration-free approach for determining NRP concentrations.

1. A known quantity of internal standard is added to the NRP sample.
2. ^{1}H NMR spectra are acquired using optimized relaxation delays to ensure full signal relaxation.
3. Well-resolved, non-overlapping proton signals are selected for integration.

4. Concentration is calculated based on the proportional relationship between signal area and number of nuclei [32].
5. Solvent suppression and baseline correction techniques are applied to enhance accuracy [33].

qNMR is particularly advantageous for NRPs lacking chromophores, which are unsuitable for UV-based detection [34]. Additionally, it allows simultaneous quantification of multiple components in a single experiment, reducing analytical time and variability.

This approach has been successfully applied to the quantification of cyanogenic glycosides in *Manihot esculenta*, yielding precise and reproducible results despite matrix complexity [10].

Such integrated strategies have been successfully applied in the structural characterization of cyclic peptides from *Viola* species, where MS-guided isolation followed by NMR resolved stereochemical and conformational ambiguities [35]. Furthermore, bioinformatic platforms such as AntiSMASH and curated repositories such as NORINE support NMR-based structural assignments by linking spectral data with biosynthetic predictions [6, 18].

3.3 Liquid Chromatography and Hyphenated Techniques for Profiling Non-Ribosomal Peptides

3.3.1 Sample Preparation Prior to LC Analysis

1. Plant material or microbial biomass is extracted using suitable organic solvents.
2. Extracts are filtered or centrifuged to remove insoluble debris.
3. Samples are diluted in initial mobile phase composition to prevent peak distortion.
4. Where required, solid-phase extraction (SPE) or pre-fractionation is performed to reduce matrix complexity.
5. Samples are transferred to LC vials and maintained at controlled temperature prior to analysis.

Effective sample preparation minimizes column fouling, reduces matrix effects, and improves chromatographic reproducibility, which is particularly important for NRPs present at trace levels [13].

3.3.2 Ultra-High-Performance Liquid Chromatography (UHPLC) for NRP Separation

UHPLC has become the preferred chromatographic platform for NRP analysis due to its superior resolving power, speed, and sensitivity compared to conventional HPLC.

1. UHPLC separation is performed using sub-2 μm particle-size columns operated under high-pressure conditions (up to 15,000 psi) [36].

2. Reversed-phase C18 columns are commonly employed owing to their broad applicability for peptides with varying polarity.
3. Gradient elution is optimized using acetonitrile or methanol in combination with aqueous mobile phases containing formic acid to enhance ionization efficiency [38].
4. Flow rates, column temperature, and gradient slope are carefully optimized to resolve NRPs from co-eluting phytoconstituents such as flavonoids, alkaloids, and terpenoids [13].

UHPLC has been successfully applied to the separation of cyclic NRPs, such as orbitides from *Linum usitatissimum*, achieving baseline resolution from structurally similar phenolic compounds within short run times [11]. The enhanced peak capacity of UHPLC also minimizes matrix effects, improving the accuracy of subsequent MS-based detection.

Polar-embedded and mixed-mode stationary phases are employed when additional selectivity is required, particularly for NRPs containing glycosylated, acylated, or highly polar modifications [37].

3.3.3 Liquid Chromatography–Tandem Mass Spectrometry (LC-MS/MS) for High-Throughput Profiling

LC-MS/MS combines chromatographic separation with sensitive and selective detection, making it a cornerstone technique for large-scale NRP profiling.

1. UHPLC is coupled directly to a tandem mass spectrometer using an electrospray ionization interface.
2. Separated NRP ions are subjected to MS/MS fragmentation using collision-induced dissociation (CID) and higher-energy collisional dissociation (HCD) [3].
3. Fragmentation patterns are analyzed to determine amino acid sequence information, cyclization motifs, and post-synthetic modifications such as methylation or glycosylation.
4. LC-MS/MS has been extensively used for profiling cyanogenic glycosides in *Manihot esculenta*, where characteristic fragment ions corresponding to non-standard amino acid residues enabled confident identification [10].
5. Advanced LC-MS/MS platforms, such as Q-TOF and Orbitrap systems, provide exceptional mass accuracy (<2 ppm) and high resolution (>100,000), allowing discrimination of isobaric NRPs from co-eluting phytoconstituents [21].
6. Data acquisition is performed using data-dependent acquisition (DDA) for targeted fragmentation and data-independent acquisition (DIA) for comprehensive, unbiased profiling [24].
7. MS/MS spectra are matched against dereplication databases such as NORINE to rapidly identify known NRPs and prioritize novel compounds [6]. The high sensitivity of LC-MS/MS enables detection of NRPs at nanogram levels,

making it particularly suitable for metabolomic studies involving low-abundance peptides [26].

3.3.4 Multi-Dimensional Chromatography for Highly Complex Plant Matrices

Single-dimension LC may be insufficient for resolving NRPs in highly complex extracts. Multidimensional chromatography provides enhanced separation by combining orthogonal mechanisms.

1. Two-dimensional LC (2D-LC) employs sequential separations using complementary modes, such as:
   - Reversed-phase chromatography in the first dimension
   - Hydrophilic interaction liquid chromatography (HILIC) in the second dimension [39]
2. Fractions from the first dimension are transferred to the second, increasing peak capacity and resolving power.
3. Comprehensive two-dimensional LC (LC×LC) transfers all fractions between dimensions, enabling exhaustive separation of complex mixtures [40].

2D-LC has been successfully applied to isolate cyclic NRPs from *Viola* species, effectively separating them from interfering flavonoids and alkaloids using orthogonal C18–HILIC combinations [35].

3.3.5 Hyphenation of Multi-Dimensional LC with Mass Spectrometry

The integration of multi-dimensional chromatography with MS detection (2D-LC-MS) provides a powerful platform for comprehensive NRP profiling.

1. Orthogonally separated fractions are analyzed by MS or MS/MS.
2. Multidimensional datasets are generated, capturing retention time, accurate mass, and fragmentation information.
3. Advanced data analysis tools, including molecular networking and chemometric approaches, are used to interpret complex spectral relationships [13].

Such integrated workflows are particularly valuable for plants with diverse secondary metabolite profiles or endophyte-rich systems, such as *Taxus baccata*, where NRPs coexist with numerous structurally related compounds [25].

4 Notes

1. **Limitations of NMR and complementary integration with mass spectrometry:** Despite its strengths, NMR presents several limitations in NRP analysis, including lower sensitivity compared to MS, necessitating relatively large sample

quantities [28]; extensive spectral overlap in complex or cyclic NRPs, increasing acquisition and interpretation time [29]; high capital and maintenance costs of high-field instruments; and limited throughput compared to MS-based screening methods [13]. To address these challenges, NMR is routinely integrated with mass spectrometry in NRP workflows. LC-MS is employed for rapid detection, molecular weight determination, and prioritization of candidate NRPs [13]. Target compounds identified by MS are isolated and subjected to NMR for definitive structural confirmation [3]. Hyphenated techniques such as LC-NMR-MS combine chromatographic separation with complementary detection capabilities, enabling comprehensive profiling of NRPs in complex phytochemical systems [13].

2. **Method optimization and practical considerations:** Mobile phase compatibility between dimensions must be carefully optimized to prevent peak distortion or sample loss [39]. Column selection and gradient design are critical for maximizing orthogonality and resolution. Transfer modulation parameters in 2D-LC systems must be optimized to balance resolution and analysis time. Despite increased complexity and operational demands, multidimensional chromatography provides unparalleled separation capability for challenging NRP matrices.

References

1. Finking R, Marahiel MA (2004) Biosynthesis of nonribosomal peptides. Annu Rev Microbiol 58:453–488. https://doi.org/10.1146/annurev.micro.58.030603.123615
2. Demain AL, Fang A (2000) The natural functions of secondary metabolites. Adv Biochem Eng Biotechnol 69:1–39. https://doi.org/10.1007/3-540-44964-7_1
3. Schwarzer D, Finking R, Marahiel MA (2003) Nonribosomal peptides: from genes to products. Nat Prod Rep 20(3):275–287. https://doi.org/10.1039/b111145k
4. Sieber SA, Marahiel MA (2005) Molecular mechanisms underlying nonribosomal peptide synthesis: approaches to new antibiotics. Chem Rev 105(2):715–738. https://doi.org/10.1021/cr0301191
5. Koglin A, Walsh CT (2009) Structural insights into nonribosomal peptide enzymatic assembly lines. Nat Prod Rep 26(8):987–1000. https://doi.org/10.1039/b904543k
6. Caboche S, Pupin M, Leclère V, Fontaine A, Jacques P, Kergomard A (2008) NORINE: a database of nonribosomal peptides. Nucleic Acids Res 36(Database issue):D970–D974. https://doi.org/10.1093/nar/gkm792
7. Rai M, Varma A, Pandey R (2005) Endophytes and their secondary metabolites: a review. Crit Rev Biotechnol 25(4):193–205. https://doi.org/10.1080/07388550500346757
8. Hur GH, Vickery CR, Burkart MD (2012) Explorations of catalytic domains of the nonribosomal peptide synthetases. Chem Rev 112(7):3939–3959. https://doi.org/10.1021/cr200307w
9. Borel JF, Kis ZL (1991) The discovery and development of cyclosporine. Transplant Proc 23(2):1867–1870.
10. Gleadow RM, Møller BL (2014) Cyanogenic glycosides: synthesis, physiology, and phenotypic plasticity. Annu Rev Plant Biol 65:155–185. https://doi.org/10.1146/annurev-arplant-050213-040027
11. Tan NH, Zhou J (2006) Plant cyclopeptides. Chem Rev 106(3):840–895. https://doi.org/10.1021/cr040699h
12. Newman DJ, Cragg GM (2007) Natural products as sources of new drugs over the last 25 years. J Nat Prod 70(3):461–477. https://doi.org/10.1021/np068054v
13. Wolfender JL, Marti G, Thomas A, Bertrand S (2015) Current approaches and challenges for the

metabolite profiling of complex natural extracts. J Chromatogr A 1382:136–164. https://doi.org/10.1016/j.chroma.2014.10.091
14. Sumner LW, Mendes P, Dixon RA (2003) Plant metabolomics: large-scale phytochemistry in the functional genomics era. Phytochemistry 62(6):817–836. https://doi.org/10.1016/S0031-9422(02)00708-2
15. Annesley TM (2003) Ion suppression in mass spectrometry. Clin Chem 49(7):1041–1044. https://doi.org/10.1373/49.7.1041
16. Wagner H, Bladt S (1996) Plant drug analysis: a thin layer chromatography atlas, 2nd ed. Springer
17. Hostettmann, K, Marston, A, 1997. Preparative chromatography techniques: applications in natural product isolation, 2nd ed.. Springer
18. Blin K, Shaw S, Steinke K et al (2019) AntiSMASH 5.0: updates to the secondary metabolite genome mining pipeline. Nucleic Acids Re 47(W1):W81–W87. https://doi.org/10.1093/nar/gkz310
19. Beckwith JR, Li X, Chen Y (2015) Advances in the chemical analysis and biological activities of ginkgolic acids. Nat Prod Rep 32(7):1048–1062. https://doi.org/10.1039/c5np00015a
20. Spraker JE, Luu GT, Sanchez LM (2020) Imaging mass spectrometry for natural products discovery: a review of new approaches. Nat Prod Rep 37(2):150–162. https://doi.org/10.1039/c9np00040k
21. Zubarev RA, Makarov A (2013) Orbitrap mass spectrometry. Anal Chem 85(11):5288–5296. https://doi.org/10.1021/ac4001223
22. Kusari S, Hertweck C, Spiteller M (2012) Chemical ecology of endophytic fungi: origins of secondary metabolites. Chem Biodivers 9(7):1231–1247. https://doi.org/10.1002/cbdv.201100296
23. Paizs B, Suhai S (2005) Fragmentation pathways of protonated peptides. Mass Spectrom Rev 24(4):508–548. https://doi.org/10.1002/mas.20024
24. Gillet LC, Leitner A, Aebersold R (2016) Mass spectrometry applied to bottom-up proteomics: entering the high-throughput era for hypothesis testing. Annu Rev Anal Chem 9:449–472. https://doi.org/10.1146/annurev-anchem-071015-041535
25. Soliman SSM, Trobaugh KP, Raizada MN (2018) Endophytic fungi as a source of novel metabolites. In: Fungal biology. Springer, pp 139–160.https://doi.org/10.1007/978-3-319-77317-9_6
26. Watrous J, Roach P, Alexandrov T et al (2012) Mass spectral molecular networking of living microbial colonies. Proc Natl Acad Sci 109(26):E1743–E1752. https://doi.org/10.1073/pnas.1203689109
27. Marshall AG, Hendrickson CL (2008) High-resolution mass spectrometers. Annu Rev Anal Chem 1:579–599. https://doi.org/10.1146/annurev.anchem.1.031207.112945
28. Keeler J (2010) Understanding NMR spectroscopy, 2nd ed. Wiley
29. Wüthrich K (1986) NMR of proteins and nucleic acids. Wiley
30. Claridge TDW (2016) High-resolution NMR techniques in organic chemistry, 3rd ed. Elsevier
31. Craik DJ, Daly NL, Bond T, Waine C (1999) Plant cyclotides: a unique family of cyclic and knotted peptides that have multiple biological activities. J Mol Biol 294(5):1327–1336. https://doi.org/10.1006/jmbi.1999.3388
32. Pauli GF, Gödecke T, Jaki BU, Lankin DC (2012) Quantitative ^{1}H NMR: development and potential of an analytical method. J Nat Prod 75(4):834–851. https://doi.org/10.1021/np200993k
33. Ardenkjær-Larsen JH, Boebinger GS, Comment A et al (2015) Facing and overcoming sensitivity challenges in biomolecular NMR spectroscopy. Angew Chem Int Ed 54(32):9162–9185.https://doi.org/10.1002/anie.201410653
34. Simmler C, Napolitano JG, McAlpine JB, Chen SN, Pauli GF (2014) Universal quantitative NMR analysis of complex natural samples. Curr Opin Biotechnol 25:51–59. https://doi.org/10.1016/j.copbio.2013.08.004
35. Göransson U, Luijendijk T, Johansson S, Bohlin L, Claeson P (1999) Seven novel macrocyclic polypeptides from Viola arvensis. J Nat Prod 62(2):283–286. https://doi.org/10.1021/np980380e
36. Guillarme D, Nguyen DT, Rudaz S, Veuthey JL (2007) Recent developments in liquid chromatography: impact on qualitative and quantitative performance. J Chromatogr A 1149(1):20–29. https://doi.org/10.1016/j.chroma.2007.02.001
37. Snyder LR, Kirkland JJ, Dolan JW (2010) Introduction to modern liquid chromatography, 3rd ed. Wiley
38. Fiehn O (2002) Metabolomics: the link between genotypes and phenotypes. Plant Mol Biol 48(1–2):155–171. https://doi.org/10.1023/A:1013713905833
39. Stoll DR, Li X, Wang X, Carr PW, Porter SEG, Rutan SC (2007) Fast, comprehensive two-dimensional liquid chromatography. J Chromatogr A 1168(1–2):3–43. https://doi.org/10.1016/j.chroma.2007.08.054
40. Guiochon G, Marchetti N, Mriziq K, Vaughn MA (2008) Comprehensive two-dimensional liquid chromatography. Anal Chem 80(16):6111–6121. https://doi.org/10.1021/ac8004663

Check for updates

Chapter 18

Steroids—High-Precision Metabolic Profiling

Hirangi Vansia and Ami Naik

Abstract

Phytosteroids, also known as plant steroids, are naturally occurring steroids that are found in plants. Among the plant steroids, phytosterols are ubiquitous and diverse in the plant kingdom. High-precision metabolic profiling of steroids is essential for understanding plant biosynthetic pathways, bioactivity, and the standardization of herbal materials. The present protocol chapter emphasizes on an integrated workflow for the quantitative and qualitative analysis of secondary metabolites, specifically focusing on steroids and antimicrobial leads. Advanced chromatographic separation of phytosteroids is achieved via ultra-high-performance liquid chromatography (UHPLC) or high-performance thin-layer chromatography (HPTLC). Detection and identification are performed using high-resolution mass spectrometry (HRMS) and tandem mass spectrometry (MS/MS), leveraging high mass accuracy and rich fragmentation patterns for both targeted quantification and untargeted metabolite discovery. Furthermore, these experimental findings are complemented by computational studies, including molecular docking and molecular dynamics (MD) simulations, to predict binding affinities and interactions with microbial drug-target proteins. This comprehensive approach provides scalable platform for identifying novel bioactive steroids for wide range of applications.

Keywords Steroidomics, LC–MS/MS, GC–MS, Metabolic profiling, Internal standards, Solid-phase extraction, Liquid–liquid extraction, High-performance thin-layer chromatography (HPTLC)

1 Introduction

Plant steroid analysis involves identifying and quantifying phytosterols (e.g., sitosterol, stigmasterol) using advanced chromatographic techniques, primarily GC-MS, HPLC,, and UHPLC [1]. These compounds are crucial for plant growth and stress adaptation, often analyzed using methods that include extraction, purification, and detection with mass spectrometry (MS).

Plant-derived materials constitute a structurally diverse and functionally critical class of natural products that underpin key aspects of plant physiology, including membrane integrity, growth regulation, and stress responses. Major groups of secondary metabolites found within these materials include alkaloids, flavonoids, tannins, and steroids, which are of growing interest in pharmacology and

Dhruti Amin et al. (eds.), *Phytoconstituent Analysis*, Springer Protocols Handbooks,
https://doi.org/10.1007/978-1-0716-5400-2_18,

biotechnology due to their antimicrobial, immunomodulatory, and anticancer properties. These bioactive constituents are distributed across various plant organs, including rhizomes, stems, leaves, flowers, and bark. However, the high market demand for these medicinal resources often leads to the overexploitation and endangerment of natural populations, necessitating the search for sustainable harvesting methods and alternative plant parts that provide similar therapeutic benefits without destroying the whole organism [2, 3].

The protocol accompanying this introduction presents an integrated, high-precision workflow for steroid metabolic profiling in plants and plant-derived phytoconstituents. It describes standardized procedures for sample collection and handling to preserve native metabolite states; optimized extraction strategies to efficiently recover free and conjugated steroids while minimizing artifacts; and optional enzymatic or chemical hydrolysis steps to enable comprehensive coverage of aglycones and conjugates. The method further details liquid–liquid and solid-phase extraction approaches for matrix cleanup and enrichment, UHPLC conditions for resolving closely related steroid structures and isomers, and mass spectrometric settings suitable for both targeted and untargeted analyses.

A strong emphasis is placed on the use of internal standards, matrix-matched calibration, system suitability, and quality control samples to ensure high analytical precision, accuracy, and reproducibility [4]. Finally, the workflow incorporates structured data processing and annotation using curated libraries of exact masses, retention times, and MS/MS spectra, enabling confident identification and quantification of plant steroids and their metabolites. Together, this protocol provides a robust and adaptable platform for high-precision steroid metabolic profiling that can be tailored to different plant species, steroid classes, and instrumental setups.

Advanced chromatographic separation is achieved through ultra-high-performance liquid chromatography (UHPLC) or high-performance thin-ayer chromatography (HPTLC), coupled with high-resolution mass spectrometry (HRMS). These platforms allow for the simultaneous targeted quantification of known markers and the untargeted discovery of novel compounds by leveraging high mass accuracy and rich fragmentation patterns [5]. Establishing these precise analytical fingerprints is like creating a high-resolution map of a dense forest; it allows researchers to identify every individual "tree" or metabolite, ensuring that the final medicinal product is both authentic and consistently potent.

2 Materials

1. Extraction solvents (HPLC/LC-MS grade): methanol, acetonitrile, ethyl acetate, petroleum ether, chloroform, n-hexane, or MTBE (for lipid removal).

2. Mobile phase additives: formic acid (≥99%, 0.1% v/v), ammonium formate, or ammonium acetate (5–10 mM).
3. Steroid standards (analytical grade): β-sitosterol, stigmasterol, campesterol, diosgenin, cholesterol (for reference).
4. Internal standards (isotopically labeled): d7-cholesterol, d4-cortisol/d5-testosterone (optional, depending on target panel).
5. Reagents: anhydrous sodium sulfate, ultrapure water (18.2 MΩ).
6. Media: Mueller–Hinton agar, nutrient broth, potato dextrose agar (PDA).
7. Sample processing: liquid nitrogen container, cryomill or mortar–pestle, lyophilizer (freeze-dryer), analytical balance (±0.1 mg).
8. Extraction and cleanup: Soxhlet apparatus, ultrasonic bath, refrigerated centrifuge (≥15,000 *g*), rotary evaporator, nitrogen evaporator, SPE manifold with C18 cartridges.
9. Analytical platform.
10. UHPLC system.
11. C18 reversed-phase column (2.1 × 100 mm, 1.7 μm).
12. High-resolution mass spectrometer (Orbitrap/Q-TOF) or triple quadrupole MS.
13. Computational tools: AutoDock/AutoDock Vina, Schrodinger (Desmond), or GROMACS.

3 Method

3.1 Plant Sample Collection and Stabilization

1. Fresh plant material (leaves/rhizomes) is collected and authenticated.
2. Samples are harvested using clean, sterile tools to avoid microbial contamination.
3. Immediately after collection, the plant material is immersed in liquid nitrogen (−196°C) to achieve rapid freezing (flash-frozen in liquid nitrogen to arrest enzymatic degradation) [3] (Note 1).
4. This process instantaneously halts cellular metabolism and enzymatic activity within the tissue.
5. The flash-frozen samples are transferred to airtight, labeled containers and stored at −80°C until further processing [6].

3.2 Lyophilization and Powder Preparation

1. Remove plant samples from −80°C storage and immediately place them into the precooled lyophilizer sample chamber without allowing thawing.

2. Arrange samples in a single layer to ensure uniform drying.
3. Set the lyophilizer to: condenser temperature: $\leq -50°C$ and vacuum pressure: ≤ 0.1 mbar [7].
4. Run the freeze-drying cycle for 48–72 h until constant weight is achieved.
5. Once drying is complete, slowly vent the chamber with dry inert gas or air.
6. Immediately cap the samples to prevent atmospheric moisture absorption.
7. Transfer dried samples to a cryomill or mortar and add liquid nitrogen to maintain cryogenic temperature [8].
8. Grind the material in uniform powder and transfer quickly into prelabeled glass vials.
9. Add a silica gel desiccant to each vial, then seal vials tightly and store at −20°C until extraction.

3.3 Steroid Extraction

1. Accurately weigh 50–100 mg of the dried, powdered plant material into clean, labeled centrifuge tubes using an analytical balance.
2. A known concentration of an isotopically labeled internal standard (e.g., d7-cholesterol) is added directly to each sample before extraction.
3. Add 1–2 mL of cold 80% methanol or acetonitrile to each tube containing the plant powder.
4. Place the tubes in an ultrasonic water bath and sonicate for 15 min at room temperature.
5. After sonication, samples are subjected to orbital shaking for 30–60 min at 4°C.
6. Centrifuge the extracts at 15,000 *g* for 15 min at 4°C.
7. Carefully transfer the clear supernatant to a fresh tube without disturbing the pellet [9] (Note 2 and 3).

3.4 Sample Cleanup and Fraction Enrichment

3.4.1 Liquid–Liquid Extraction (LLE)

1. The crude solvent extract obtained after centrifugation is diluted with ultrapure water in a 1:1 (v/v) ratio [4].
2. The diluted extract is transferred to a separatory tube and extracted with n-hexane or methyl tert-butyl ether (MTBE).
3. The extraction is performed three consecutive times (3×) with vigorous vortexing followed by phase separation.
4. The organic layers are carefully pooled and passed through anhydrous sodium sulfate to remove residual moisture [10].

3.4.2 Solid-Phase Extraction (SPE)

1. Conditioning of SPE cartridge: A C18 reversed-phase SPE cartridge is sequentially conditioned with methanol (to activate the hydrophobic sorbent) and water (to equilibrate the stationary phase).
2. Sample loading: The dried organic extract is reconstituted in a suitable solvent and loaded onto the conditioned SPE cartridge at a controlled flow rate.
3. The cartridge is washed with 10%–30% methanol in water.
4. Elution of steroids: Steroidal metabolites are eluted using 100% methanol [11].

3.5 UHPLC–HRMS/MS Steroid Profiling

3.5.1 Chromatographic Conditions

Proper chromatographic conditions are essential to achieve high resolution, sensitivity, and reproducibility, especially for structurally similar steroidal compounds.

- A reversed-phase C18 column with 1.7 μm particle size is used for steroid separation.
- Flow-rate optimization: The mobile phase is delivered at a flow rate of 0.3 mL/min.
- An injection volume of 5 μL is used [12, 13].

Mobile phase: Two mobile phases are employed:

- Mobile phase A: water with 0.1% formic acid.
- Mobile phase B: acetonitrile or methanol with 0.1% formic acid.

1. Gradient elution Program: The chromatographic separation is achieved using a linear gradient:

- Initial: 5% mobile phase B.
- Final: 95% mobile phase B.
- Duration: 20 min.

3.5.2 Mass Spectrometric Conditions

1. The mass spectrometer is operated using electrospray ionization in positive-ion mode (ESI+).
2. Ionization: ESI(+) is preferred for steroids.
3. Scan mode: To comprehensively characterize the steroid profile, both untargeted and targeted acquisition modes are employed.
4. Full-scan acquisition (m/z 100–1000): In full-scan mode, the instrument continuously records all ions within the m/z range of 100–1000.
5. Targeted MS/MS (PRM/MRM): For precise quantification and confirmation, targeted MS/MS acquisition is performed using:

6. Parallel Reaction Monitoring (PRM)—commonly used on HRMS systems.
7. Multiple Reaction Monitoring (MRM)—used on triple quadrupole instruments.
8. The mass spectrometer is operated at a resolution of 30,000–70,000 FWHM (full width at half maximum) [14, 15].

3.6 Data Processing and Quantification

After chromatographic separation and mass spectrometric detection, the acquired data are processed using dedicated LC-MS software.

1. Accurate mass measurement (±5 ppm): Each detected peak is screened based on its exact mass, with a tolerance of ±5 parts per million (ppm).
2. Retention time matching: The retention time (RT) of each detected compound is compared with that of authentic reference standards analyzed under identical chromatographic conditions [16].
3. MS/MS fragmentation pattern analysis: The MS/MS spectra of detected steroids are compared with reference standard spectra, spectral libraries, and published literature data.
4. Quantification is performed using internal-standard calibration curves.
5. Known concentrations of steroid standards are analyzed together with a fixed amount of isotopically labeled internal standard [17].
6. Calibration curves are generated by plotting:
   a. Peak area ratio (analyte/internal standard)
   b. Versus analyte concentration [18].

4 Notes

1. Adapt good laboratory practices while working with liquid nitrogen or any hazardous chemicals.
2. Plant extracts can be stored at 4°C for up to a week.
3. The sample needs to be passed through filter paper (Whatman No. 1) prior to chromatography. It should be free from suspended particulates.

References

1. Nystrom L (2012) Analysis methods of phytosterols. In: Analysis of antioxidant-rich phytochemicals. pp 313–351
2. Yang B, Zhao T, Liu Y, Lu B (2023) Guidelines for extraction and quantitative analysis of phytosterols and oxidation products. eFood 4(4):e108
3. Pandey V, Parihar S, Saxena HO, Yadav AK, Pawar G (2025) A developed thin-layer chromatography method for the quantification of stigmasterol in leaves, stem and root of Sida cordifolia Linn., assessment of secondary metabolites and in vitro antioxidant activity. J Planar Chromatogr Mod TLC 38(6):507–523
4. Yoon BH, Truong VL, Jeong WS (2025) Phytosterols: extraction methods, analytical techniques, and biological activity. Molecules 30(12):2488
5. Porte S, Joshi V, Shah K, Chauhan NS (2022) Plants' steroidal saponins-a review on its pharmacology properties and analytical techniques. World J Tradit Chin Med 8(3):350–385
6. Sarkar R, Kumar S, Singh R (2025) Development of a supercritical fluid extraction process for preparation of phytosteroids rich extract from Tinospora cordifolia stem and determination of content of three phytosteroids using a validated ultra-high performance liquid chromatography method. J Sep Sci 48(8):e70246
7. Maisl C, Doppler M, Seidl B, Bueschl C, Schuhmacher R (2023) Untargeted plant metabolomics: evaluation of lyophilization as a sample preparation technique. Metabolites 13(6):686
8. Patel KS, Singh A, Litoriya NS, Sharma AK (2024) Impact of different drying method on phytosterols and proximate biochemical contents in leaf, stem and root of Jivanti [Leptadenia reticulata (Retz.) Wight and Arn.]
9. Song Y, Feng XS (2023) Sample preparation and analytical methods for steroid hormones in environmental and food samples: an update since 2012. Crit Rev Anal Chem 53(1):69–87
10. Shiju SG, Augustyniak A, Alaydi H, Rajauria G, Gaffey J, McMahon H (2025) Green liquid–liquid extraction for environmental chemicals
11. Li Y, Wang T, Qin S, Li Y, Hu W, Cheng W (2023) Magnetic solid-phase extraction pretreatment method for quantitative analysis of 12 steroids in human serum based on LC–MS/MS. Chromatographia 86(4):333–347
12. Zhang X, Zheng F, Zhao C, Li Z, Li C, Xia Y, Xu G (2022) Novel method for comprehensive annotation of plant glycosides based on untargeted LC-HRMS/MS metabolomics. Anal Chem 94(48):16604–16613
13. Chakraborty A, Bhattacharjee A, Mondal B, Chakraborty M, Mukhopadhyay G, Majumder A, Ghosh M (2023) A validated HPTLC method for quantification of linoleic acid and beta-sitosterol in Solanum nigrum extract. J Nat Remedies 23:871–878
14. Gachumi G, El-Aneed A (2017) Mass spectrometric approaches for the analysis of phytosterols in biological samples. J Agric Food Chem 65(47):10141–10156
15. Fang X, Szołtysik R, Tang J, Bajkacz S (2022) Efficient extraction and sensitive HPLC-MS/MS quantification of selected ecdysteroids in plants. J Food Compost Anal 110:104580
16. Tripathi A, Sahni P, Pandey S, Kumar A (2025) Extraction and quantification of health-promoting phytosterols. In: Phytochemicals: extraction, quantification, isolation, and characterization. Springer US, New York, NY, pp 81–93
17. Visconti G, De Figueiredo M, Salamin O, Boccard J, Vuilleumier N, Nicoli R et al (2023) Straightforward quantification of endogenous steroids with liquid chromatography–tandem mass spectrometry: comparing calibration approaches. J Chromatogr B 1226:123778
18. Batool M, Fatima A, Khan U, Jahangir M, Nazir M, Shakoor A et al (2025) Detection and quantification of steroid adulteration in herbal medicines using TLC and UV-spectroscopy

Check for updates

Chapter 19

Flavonoids: Chromatographic Profiling Approaches

Priyang Kanojiya and Sanket Ray

Abstract

Flavonoids are among the most abundant and bioactive secondary metabolites in plants, necessitating accurate analytical procedures for separation and identification. In this chapter, different chromatographic and mass-spectrophotometric methods are combined to accomplish thorough profiling. TLC was initially employed to obtain a quick qualitative overview, allowing for the identification of key flavonoid groups based on polarity, UV response, and reagent-induced fluorescence. HPTLC enhance fingerprint resolution and reproducibility by automating application controlling development and evaluating densitometric results. GC–MS allowed for the characterization of volatile and derivatized compounds by defined EI fragmentation patterns. LC–MS provided highly sensitive detection of non-volatile flavonoids, glycosides, and phenolic acids via reversed-phase C18 gradients with ESI in both positive and negative modes. MS/MS fragmentation also improved structural confirmation. Together, these complementary approaches form a strong pipeline that progresses from simple screening to comprehensive structural elucidation, resulting in accurate and reproducible profiling.

Keywords Flavonoids, TLC, HPTLC, GC–MS, LC–MS, ESI

1 Introduction

Flavonoids are a key family of plant secondary metabolites known for their protective, ecological, and nutritional properties. They have a distinctive C6–C3–C6 carbon skeleton and are frequently found in leaves, fruits, flowers, and medicinal herbs [1]. Their biological significance, including antioxidant, anti-inflammatory, and antibacterial capabilities, has generated a rising interest in their extraction and analytical process [2]. Traditional extraction procedures, such as maceration and Soxhlet remain widespread, but may involve extensive processing periods and significant solvent consumption as well as the risk of thermolabile flavonoid detection [3]. Recent innovations like ultrasound and microwave-assisted extraction aim to improve efficiency and match with eco-friendly approaches to phytochemical recovery. The analytical profiling of flavonoids is highly dependent on chromatographic techniques. Thin Layer Chromatography (TLC)

Dhruti Amin et al. (eds.), *Phytoconstituent Analysis*, Springer Protocols Handbooks,
https://doi.org/10.1007/978-1-0716-5400-2_19,

continues to function as a quick and inexpensive preliminary method for detecting flavonoid classes utilizing UV and derivatization chemicals [4]. High-Performance Thin Layer Chromatography (HPTLC) offers greater resolution, reproducible fingerprints, and densitometric quantifications, making it ideal for comparative plant analysis and quality monitoring [5]. Gas Chromatography-Mass Spectrometry (GC-MS) is effective for profiling volatile compounds, It provides robust EI fragmentation patterns for chemical identification [6]. The most effective method for non-volatile flavonoids, particularly glycosides and phenolic acids, is Liquid Chromatography-Mass Spectrometry (LC–MS), which allows for sensitive detection and MS/MS based structural elucidation under mild ionization circumstances [7].

2 Materials

2.1 Sample Preparation for Flavonoid Extraction [3]

1. Plant materials (leaves/flowers/roots)
2. Dark glass bottle or Amber glass sealed container
3. Solvent: 70% (v/v) methanol/ethanol
4. Mortar and pestle
5. Rotary evaporator
6. Whatman No.1 filter paper

2.2 Thin Layer Chromatography [8]

1. Sample extract (Methanol/Ethanol)
2. TLC plates: Silica gel 60 F254
3. Capillary tubes/Micropipettes
4. Developing chamber with lid
5. Solvent systems: (Chloroform: Ethanol: Glacial acetic acid), (Tetrahydrofuran: Water: Formic acid), (Methanol: Chloroform: Hexane).
6. Visualizing agents: 1% $AlCl_3$ Solution, H_2SO_4 (1% in methanol)
7. Hot air dryer
8. Glass vials

2.3 High Performance Thin Layer Chromatography (HP-TLC) [9]

1. Pre-coated HPTLC plates—Silica gel 60 F254 (10×10 or 10×20 cm)
2. Solvents (analytical grade): ethyl acetate, toluene, chloroform, methanol, acetonitrile, formic acid, glacial acetic acid.
3. HPTLC components—Automatic applicator (Linomat 5), developing chamber, TLC scanner/Chronology Analysis Measurement and Grading (CAMAG) scanner, Windows based Chromatography, and TLC software (winCATS) software

4. Visualization reagents: Natural Product (NP) reagent, Poly ethylene glycol (PEG-400), 1% $AlCl_3$, 1% H_2SO_4 in methanol
5. Filters and Syringes: 0.22 μm syringe filters
6. Rotary evaporator, glass vials for storage

2.4 Gas Chromatography–Mass Spectroscopy (GC–MS)

1. Plant extracts or dried plant residues.
2. GC reagents: Methanol, *n*-hexane, chloroform.
3. GC-MSGC-MS instrument: Agilent 7890/5975 [5, 10, 11].
4. GC Column: HP-5MS or DB-5MS fused silica capillary column, GC Column dimension (30 m × 0.25 mm × 0.25 μm).
5. Carrier gas: Helium 99.99% purity.
6. Micro syringes and clean glass vials.

2.5 Liquid Chromatography–Mass Spectrometry (LC–MS)

1. Plant extracts or dried plant residues.
2. HPLC grade solvents: water, acetonitrile (MeCN), methanol (MeOH), formic acid.
3. Solid phase extraction (SPE) cartridges (C18).
4. LC system: HPLC coupled Quadrupole-Time of Flight (Q-TOF)/Triple Quad.
5. 0.22 μm syringe filters. 1.5 mL amber glass vials.

3 Methods

3.1 Flavonoids Extraction Method [1–3, 12]

1. Weigh the necessary amount of dried plant material and grind it into a fine powder using a mortar and pestle
2. To prevent flavonoid deterioration due to light, transfer the powdered sample to a clean, dry, amber-colored glass container [3].
3. Add 70% methanol/ethanol solvent to fully submerge the plant material [12].
4. Seal the jar and store it at room temperature for 24–48 h at 25°C–30°C.
5. After maceration period, sieve the mixture through Whatman No. filter paper into a clean, dry flask [1].
6. Concentrate the filter by using rotary evaporator at 40°C at low pressure.
7. Filter the final extract and store it in labeled amber color glass bottle at 4°C (short term) and 20°C (long term) [2].

3.2 Thin Layer Chromatography

1. Plate preparation: Heat the silica gel TLC plate for 10 min at 100°C to pre-activate it, then let it cool at room temperature.
2. Sample loading: Draw a baseline of 1–1.5 cm from lower edge with a pencil and spot 2–5 μL extract using a capillary tube (keep the spot small and visible).
3. Preparation of development chamber.
   - Pour 10–15 mL mobile phase (Table 1) and pre-saturate it for 20–30 min.
   - Place the TLC plate inside the chamber without perturbing the solvent front.
   - Allow the solvent to rise up to 6–8 cm depending on the plate size.
4. Drying and Visualization (Table 1)
   - Air-dry the TLC plates or use warm airflow to dry it until all solvent evaporates
   - After drying, visualize or examine under UV-254 nm and UV-366 nm (flavonoids)
   - Spray with 1% $AlCl_3$ and dry it to observe fluorescence—yellow, blue, pink spots indicate flavonoids
5. Record results
   - For preparative TLC and TLC—scrape silica gently with scrapper and extract it with methanol, centrifuge, and collect it or store supernatant in glass vials.

Table 1
TLC solvent systems for flavonoid separation

TLC	Mobile phase	Stationary phase	Visualization	References
TLC 1	Tetrahydrofuran: water: formic acid (40:60:6)	RP-18	UV 366 nm	[13]
TLC 2	Chloroform: ethanol: acetic acid (9.4:0.5:0.1)	Silica gel 60 F24	UV 254 and 366 nm and H_2SO_4 spray	[8]
TLC 3	Methanol: chloroform: hexane (7:2:1)	Silica gel	UV 365 nm, $AlCl_3$ spray	[14]
TLC 4	EtOH: formic acid: acetic acid: water (100:11:11:27)	Silica gel	Natural product (NP)/ polyethylene glycol (PEG) reagent	[15]

3.3 High Performance Thin Layer Chromatography (HP-TLC) [7, 9, 16, 17]

1. Sample and standard preparation
   - Dissolve the dried plant extract (10 mg) in 1 mL methanol and calibrate it with known ng/μL standards, such as rutin, quercetin, eugenol
   - Filter it using a 0.22 μm filter before use [9]
2. Application
   - Use an automated applicator (Linomat 5) or a microsyringe.
   - Apply as bands of approximately 10–15 mm from the lower border, with 8–10 mm between bands, and typical application rate should be maintained between 0.1 and 0.3 μL/s.
3. Developing chamber
   - Fill the chamber with mobile phase about 10–15 mL and line the chamber walls with filter paper, and saturate it for 15–30 min at room temperature. This enhances Rf repeatability
4. Plate Development
   - Insert plate gently and develop until the solvent front reaches up to the desired distance. Slowly, remove the plate and quickly mark the solvent front
   - To remove the solvent, air-dry it or use warm mild air (do not overheat) and visualize at UV 254 nm.
   - To detect flavonoids, spray NP reagent and dry it, and then spray PEG-400 and let it dry and examine at 366 nm
5. Scanning, calculation, and documentation (Table 2)

Table 2
HPTLC analytical conditions for the separation and detection of flavonoids

HPTLC	Mobile phase	Stationary phase	Visualization	References
HPTLC 1	Ethyl acetate: formic acid: glacial acetic acid: water (100:11:11:26)	Silica gel 60 F254	NP/PEG (366 nm)	[9]
HPTLC 2	Acetone solution (development up to 80 mm)	Silica gel 60 F254	UV 254 nm densitometry	[17]
HPTLC 3	Ethyl acetate: formic acid: water (100:11:26)	HPTLC plates (CAMAG system)	NP/PEG (366 nm) densitometric scanning	[7]
HPTLC 4	Ethyl acetate mixtures/ toluene	Silica gel 60 F254	NP/PEG, ATR-FTIR	[16]

Table 3
Comparative analysis of GC–MS, oven programs and MS conditions

GC–MS	Column	Injector temperature and carrier gas	Oven program	MS mode	Scan range	References
GC–MS 1	HP-5MS	250°C He 1 mL/min	60°C–180–250°C		40–550 m/z	[10]
GC–MS 2	DB-5MS	250°C He 1 mL/min	80°C–250°C		40–500 m/z	[5]
GC–MS 3	HP-5MS	260°C He 1 mL/min	60°C–280°C	EI 70 eV	50–550 m/z	[18]
GC–MS 4	DB-5	250°C He 1 mL/min	80°C–280°C		40–450 m/z	[6]
GC–MS 5	HP-5MS	250°C He 1 mL/min	70°C–300°C		50–600 m/z	[11]

- Scan plates with CAMAG scanner at 254 nm and record densitograms.
- Use winCATS for peak integration, overlaying standards and quantification.
- DetermineRf values,Limit of detection (LOD),Limit of quantitation (LOQ) linearity, regression, precision, and robustness
- Lastly, scrap the silica from the target band, extract it with methanol/ethanol, centrifuge it, and evaporate it before re-analyzing or storing it [7].

3.4 Gas Chromatography–Mass Spectroscopy [5, 6, 10, 11, 18] (Table 3)

1. Evaporate the solvent under low pressure to prepare the dry plant extract and reconstitute in hexane or methanol, gently vortex, then filter through 0.22 μm filter
2. Fill an autosampler vial with 1 μL of the sample.
3. Let the column execute the temperature program after injecting in split mode (1:10 or 1:20).
4. Start at 60°C and then hold for 120 s, then ramp at 5°C/min to 250°C and hold for 10 min.
5. Obtain mass spectra by scanning m/z ranging between 40 and 550 with Electron Ionization (EI) at 70 eV and quadrupole temperature 150°C, Ion source temperature 200°C–230°C.

Table 4
Comparative overview of LC–MS chromatographic and mass spectrometric

LC–MS	LC column	Mobile phases	Gradient	Flowrate and injection volume	Ionization and mass analyzer	Scan range	References
LC–MS 1	RP-C18 column	water + 0.1% ACN	5%–95% B over 30 min	0.2–0.4 mL/min and 1–5 µL	ESI (±) and Q-TOF	50–1000 m/z	[19]
LC–MS 2	C18 column	water + ACN + formic acid	10%–95% B	0.3–0.5 mL/min and 1–10 µL	ESI (±) and triple quad	50–1000 m/z	[20]
LC–MS 3	C18 column	Water + acetic acid	10%–100% B	0.4 mL/min and 1 µL	ESI (±) and Q-TOF	50–1200 m/z	[21]
LC–MS 4	C18 column	Water + 1% acetic acid	4%–33% B then to 100%	0.5 mL/min and 5 µL	ESI (−) and triple quad (MS/MS)	50–1000 m/z	[22]
LC–MS 5	Agilent C18	Water + 0.005% formic acid	Linear gradients to 100% B	0.3–0.5 mL/min and 1–5 µL	ESI (−) and Q-TOF	50–1000 m/z	[4]

6. Use % peak area normalization to quantify and identify main peaks.
7. Use NIST library and Wiley MS library matching to identify components based on retention time and mass fragmentation pattern.

3.4.1 Liquid Chromatography–Mass Spectroscopy (Table 4) [4, 19–22]

1. To eliminate extraction of solvents, evaporate the crude plant extract under low pressure at 40°C–45°C.
2. Mix 5–20 mg of dry weight extract with 1–2 mL of HPLC-grade methanol, vortex, and sonicate for 5 min to guarantee complete breakdown.
3. For heavy matrix samples, load them into a conditioned C18 SPE cartridge, wash with water and elute with methanol or MeOH/Acetonitrile (ACN).
4. Evaporate the eluate and reconstitute in 200–500 μL injection solvent. To eliminate particles, filter the solution with 0.22 μm syringe filter and transfer into LC vial.
5. Prepare mobile phases
   - Solvent A: water with 0.1% formic acid.
   - Solvent B: Acetonitrile with formic acid.
6. Run a gradient from 5% B to 95% B for 35 min and flowrate should be 0.2–0.5 mL/min. Maintain the column temperature between 25°C and 40°C.
7. Use electrospray ionization in both positive and negative modes. Set the spray voltage to 3.5–5.0 kV and capillary temperature to 200°C–350°C, with nitrogen as nebulizer.
8. Obtain mass spectra in the m/z 50–1000 range and process the chromatograms.
9. Compare retention time, exact mass and MS/MS fragments to NIST/Wiley databases and validate with legitimate standards when available.

4 Notes

1. Always use fresh, high-purity solvents. Old or contaminated mobile phase shift Rf/RT values in TLC, HPTLC, LC–MS, and GC-MS.
2. Keep all samples and solvents totally moisture free. Water degrades GC-MS columns and inhibits EI ionization, which leads to reduce results.

3. Filter all samples using a 0.22 μm membrane. Unfiltered samples produce streaking in TLC/HPTLC.
4. Do not touch HPLC/TLC plates with bare hands. Fingerprints inject oils, resulting in misleading spots.
5. Maintain adequate chamber saturation in TLC/HPTLC.
6. Use new visualization ranges. NP/PEG and other reagents decay quickly, resulting in weak bands.
7. Clean or replace liners on a regular basis to prevent GC inlet contamination.
8. Use LC–MS grade solvents, and avoid buffer salts.
9. Keep the ESI source and needle very clean.
10. Avoid overloading or large sample volumes or heavy spots, which may result in tailing or merged peaks.
11. Conduct regular instrument calibration and system appropriateness assessments.

References

1. Chávez-González ML, Sepúlveda L, Verma DK, Luna-García HA, Rodríguez-Durán L V., Ilina A, Aguilar CN (2020) Conventional and emerging extraction processes of flavonoids. Processes 8:434
2. Tzanova M, Atanasov V, Yaneva Z, Ivanova D, Dinev T (2020) Selectivity of current extraction techniques for flavonoids from plant materials. Processes 8:1–30
3. Rodríguez De Luna SL, Ramírez-Garza RE, Serna Saldívar SO (2020) Environmentally friendly methods for flavonoid extraction from plant material: impact of their operating conditions on yield and antioxidant properties. Sci World J 2020:6792069.
4. Salih EYA, Fyhrquist P, Abdalla AMA, Abdelgadir AY (2017) LC-MS/MS Tandem Mass Spectrometry for Analysis of Phenolic Compounds and Pentacyclic Triterpenes in Antifungal Extracts of *Terminalia brownii* (Fresen.). Antibiotics 6:37.
5. Bhardwaj P, Naryal A, Thakur MS, Aggarwal NK, Saxena S, Chaurasia OP, Kumar R (2020) Comparative antioxidant, antibacterial, and GC-MS analysis of methanol extract's fractions and isolation of luteolin from leaves of trans-Himalayan *Codonopsis clematidea*. Ind Crops Prod 144:112046
6. Karimi E, Jaafar HZE (2011) HPLC and GC-MS determination of bioactive compounds in microwave obtained extracts of three varieties of *Labisia pumila* benth. Molecules 16:6791–6805
7. Jesionek W, Majer-Dziedzic B, Choma IM (2015) Separation, Identification, and Investigation of antioxidant ability of plant extract components using TLC, LC-MS, and TLC-DPPH•. J Liq Chromatogr Relat Technol 38:1147–1153
8. Syarifah AL, Retnowati R, Soebiantoro S (2019) Characterization of secondary metabolites profile of flavonoid from salam leaves (*Eugenia polyantha*) using TLC and UVSpectrophotometry. Pharm Sci Res 6:155–163
9. Singh, MK, Bindhu, OS, 2019. Plant latex: A rich source of haemostatic proteases, Herbal medicine in India: indigenous knowledge, practice, innovation and its value. Springer, Singapore, pp. 143–153
10. Al-Owaisi M, Al-Hadiwi N, Khan SA (2014) GC-MS analysis, determination of total phenolics, flavonoid content and free radical scavenging activities of various crude extracts of *Moringa peregrina* (Forssk.) Fiori leaves. Asian Pac J Trop Biomed 4:964–970
11. Proestos C, Komaitis M (2013) Analysis of naturally occurring phenolic compounds in aromatic plants by RP-HPLC coupled to diode array detector (DAD) and GC-MS after silylation. Foods 2:90–99
12. Liga S, Paul C, Péter F (2023) Flavonoids: overview of biosynthesis, biological activity, and current extraction techniques. Plants 12:2732
13. Pobłocka-Olech L, Głód D, Zebrowska ME, Sznitowska M, Krauze-Baranowska M (2016) TLC determination of flavonoids from different

cultivars of *Allium cepa* and *Allium ascalonicum*. Acta Pharm 66:543–554

14. Gwatidzo L, Dzomba P, Mangena M (2018) TLC separation and antioxidant activity of flavonoids from *Carissa bispinosa*, *Ficus sycomorus*, and *Grewia bicolar* fruits. Nutrire 43:1–7
15. Kowalska T (2022) Botanicals—its versatile potential and selected applications. Molecules 27:6607
16. Kustrin SA, Gegechkori V, Petrovich DS, Ilinichna KT, Morton DW (2021) ATR-FTIR characterisation of major flavonoids and polyphenolics. Molecules 26:1–15
17. Das S, Ray A, Nasim N, Nayak S, Mohanty S (2019) Effect of different extraction techniques on total phenolic and flavonoid contents, and antioxidant activity of betelvine and quantification of its phenolic constituents by validated HPTLC method. 3 Biotech 9:37
18. Alkhatib RQ, Almasarweh AB, Abdo NM, Mayyas AS, Al-Qudah MA, Abu-Orabi ST (2022) Chromatographic analysis (LC-MS and GC-MS), antioxidant activity, antibacterial activity, total phenol, and total flavonoid determination of *Cleome arabica* L. growing in Jordan. Int J Food Prop 25:1920–1933
19. Rahim NAF, Roslan MNF, Muhamad M, Seeni A (2022) Antioxidant Activity, Total Phenolic and Flavonoid Content and LC-MS Profiling of Leaves Extracts of *Alstonia angustiloba*. Separations 9:234.
20. Al-Dalahmeh Y, Al-Bataineh N, Al-Balawi SS, Lahham JN, Al-Momani IF, Al-Sheraideh MS, Mayyas AS, Abu-Orabi ST, Al-Qudah MA (2022) LC-MS/MS Screening, Total Phenolic, Flavonoid and Antioxidant Contents of Crude Extracts from Three *Asclepiadaceae* Species Growing in Jordan. Molecules 27(3):859.
21. Keskes H, Belhadj S, Jlail L, El Feki A, Sayadi S, Allouche N (2017) LC–MS–MS and GC–MS analyses of biologically active extracts of Tunisian Fenugreek (*Trigonella foenum-graecum* L.) Seeds. J Food Meas Charact 11:177–189.
22. Quan H, Sun N, Liu S, Li M, Wang H, Wang Z (2021) The analysis of flavonoids and triterpenes extracted from Urtica by LC-MS and the antimicrobial activity of the extracts. J Food Process Preserv 45, e15533.

Check for updates

Chapter 20

Saponins: Advanced Profiling Techniques for Saponins

Amit Gamit, Rishita Trivedi and Dhruti Amin

Abstract

Saponins are structurally diverse amphiphilic glycosides found in plants, marine organisms, and microbes, valued for ecological functions and broad applications in food, medicine, and industry. Their structural complexity comes from variations in aglycone backbones, sugar chains, and chemical modifications, necessitating robust, standardized analytical workflows for accurate profiling. This chapter offers an integrated protocol for sample preparation, extraction, enrichment, qualitative screening, and advanced analysis of saponins. It covers classical methods like Soxhlet, maceration, ultrasound-assisted extraction, and rapid screening tools, such as foam test, TLC, and FTIR. High-resolution chromatographic and spectrometric techniques, including HPLC, UHPLC–HRMS, ion mobility MS, and targeted MRM, enable sensitive detection and differentiation of isomers and low-abundance saponins. For complex mixtures, LC–MS–NMR platforms support comprehensive characterization. The protocol highlights key parameters, optimization, and reproducibility, guiding researchers in natural product chemistry, plant biochemistry, and pharmacognosy for reliable saponin profiling.

Keywords Saponins extraction, UHPLC–HRMS, Metabolomics, Structural elucidation, Quantification

1 Introduction

Saponins are a diverse class of naturally occurring amphiphilic glycosides widely distributed across higher plants, selected marine organisms, and certain microorganisms [1–3]. Structurally, they consist of a hydrophobic aglycone linked to one or more hydrophilic sugar moieties, imparting distinctive surface-active properties, such as foaming, emulsification, and membrane interactions [4]. Based on the aglycone core, saponins are broadly classified into triterpenoid and steroidal types, each with unique biosynthetic origins, chemical complexity, and biological functions [5]. Their sugar chains vary in composition, length, branching, and linkage, creating high structural diversity, while further modifications, such as acylation, sulfation, and oxidation, add to the overall chemical complexity of these metabolites.

Dhruti Amin et al. (eds.), *Phytoconstituent Analysis*, Springer Protocols Handbooks,
https://doi.org/10.1007/978-1-0716-5400-2_20,

In plants, saponins serve as key defensive phytochemicals that protect against insects, pathogens, and herbivores through membrane-disruptive and deterrent activities. They also regulate ecological interactions including allelopathy, stress tolerance, and symbiosis. For humans, saponins possess significant pharmacological, nutraceutical, cosmetic, and industrial relevance. Reported activities include anti-inflammatory, antimicrobial, immunomodulatory, and cholesterol-lowering properties, along with applications as vaccine adjuvants (e.g., QS-21), bio-surfactants, and quality markers in foods and herbal preparations [1, 6–16]. Given these diverse bioactivities, accurate identification and quantification of saponins are essential for quality control, standardization, and mechanistic research.

Despite their importance, saponin profiling remains analytically challenging. Their large molecular weights (600–2000+ Da), structural isomerism, variable polarity, and inconsistent ionization behavior complicate detection and separation. Many saponins lack commercial reference standards, hindering quantification, while in-source fragmentation and formation of multiple adducts during mass spectrometry further complicate data interpretation [17, 18].

Recent innovations in chromatography, mass spectrometry, sample preparation, and data analysis have markedly improved saponin profiling. Modern workflows integrate multidimensional approaches, such as UHPLC–HRMS [19], ion mobility spectrometry (IMS),

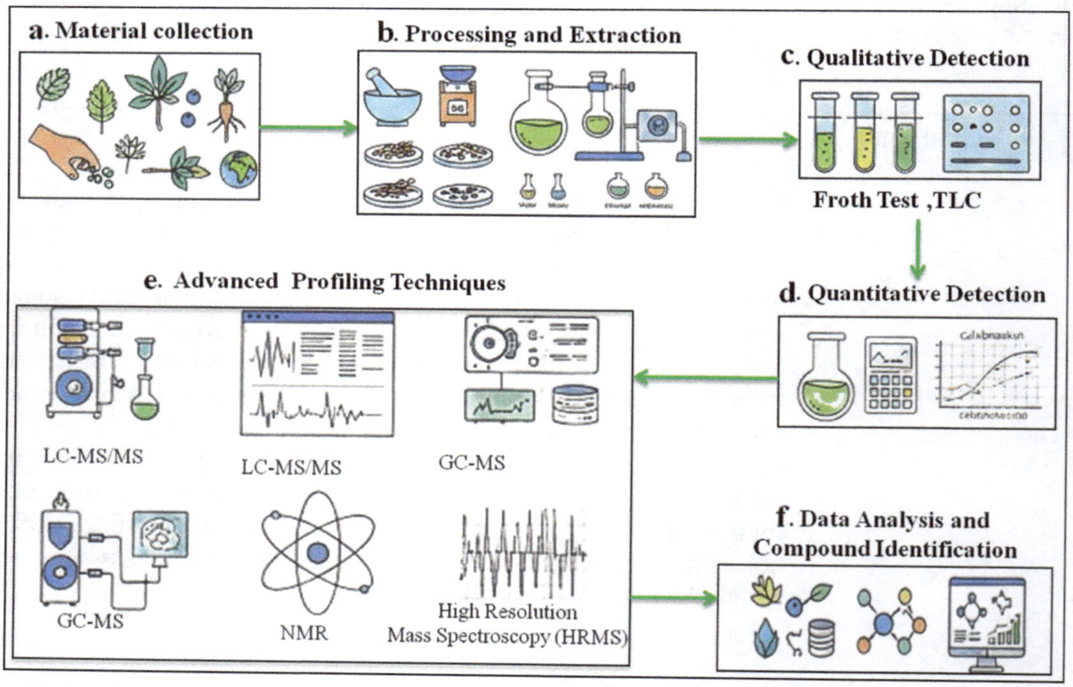

Fig. 1 Workflow of advanced saponin profiling: plant material collection, extraction, qualitative and quantitative detection, advanced profiling techniques, and compound identification

MS/MS molecular networking, data-independent acquisition (DIA), and NMR-guided structural elucidation. High-resolution MS platforms, including Q-TOF, Orbitrap, and triple quadrupole systems, enable accurate mass measurement, detailed fragmentation analysis, and enhanced detection sensitivity [20]. Advanced methods like HILIC and 2D-LC enable better separation of polar and isomeric saponins, and when combined with chemometric tools, such as PCA and OPLS-DA, they support comparative profiling across different tissues, cultivars, treatments, and conditions (Fig. 1) [21].

Together, these advancements mark a transition from traditional colorimetric assays to high-fidelity, metabolomics-oriented workflows. A modern protocol must, therefore, encompass optimized extraction, enrichment, chromatographic separation, high-resolution detection, structural annotation, and quantitative validation. Such standardized frameworks enhance reproducibility and enable deeper insights into saponin biosynthesis, ecological significance, and therapeutic potential across diverse biological systems.

2 Materials

2.1 Sample Collection and Extraction

2.1.1 Sample Collection and Pre-Processing

1. Plant sample (*Hedera helix*)
2. Air-flow heating oven (60°C capability)
3. Electric grinder/mechanical mill
4. Stainless steel sieve set (<315 μm mesh size)
5. Analytical balance (±0.001 g accuracy)
6. Stainless steel trays for drying
7. Forceps, scissors, or pruning shears
8. Sealed plastic containers/airtight sample jars
9. Refrigerator (4–5°C for storage)

2.1.2 Extraction of Saponins

A. **For Soxhlet Extraction**
   1. Soxhlet extractor (with cellulose thimbles)
   2. Condenser and connecting tubes
   3. Round-bottom flasks (150–250 mL)
   4. Heating mantle or water/oil bath
   5. Retort stand with clamps
   6. Rotary evaporator with vacuum pump
   7. Amber collection vials

8. 70% aqueous methanol (analytical grade)
9. Distilled or deionized water

B. **For Ultrasound-Assisted Extraction (UAE)**

1. Jacketed glass reactor (250–500 mL)
2. Water circulator (temperature-controlled)
3. Ultrasonic probe system (20%, 40%, 60% amplitude options)
4. Magnetic stirrer with stir bars (900 rpm capability)
5. Centrifuge (2500 rpm)
6. Centrifuge tubes (15 and 50 mL)
7. Ethanol (0%–96%, analytical grade)
8. Distilled water
9. *n*-Butanol (analytical grade)
10. Anhydrous sodium sulfate (Na_2SO_4)
11. Separatory funnel (250–500 mL)
12. Glass funnels and beakers
13. Filter paper (Whatman No. 1)
14. C18 SPE cartridges (1–3 g)
15. SPE vacuum manifold (optional)
16. Methanol (HPLC grade for SPE conditioning)
17. Milli-Q or deionized water

2.2 Qualitative Profiling of Saponins

2.2.1 Primary Phytochemical Tests for Saponins [22]

1. Plant extract (aqueous, alcoholic, or hydroalcoholic)
2. Distilled water
3. Test tubes, pipettes
4. Water bath/heating source
5. Vegetable oil/olive oil (for emulsification test)
6. Sodium bicarbonate ($NaHCO_3$)
7. Optional: fresh RBC suspension (2% in isotonic saline) for hemolysis test
8. Timer for observing foam persistence

2.2.2 Thin Layer Chromatography (TLC) for Saponins

1. Pre-coated silica gel 60 F_{254} TLC plates (0.25–1 mm thickness; analytical or preparative)
2. Capillary tubes or micropipettes (for spotting samples and standards)
3. TLC chamber or development tank with tight lid

4. Mobile phase solvents (any of the following systems):
    - *n*-Butanol: Water: Acetic acid = 84: 14: 7 (v/v/v)
    - Chloroform: Methanol: Water = 65: 35: 10 (v/v/v)
    - Chloroform: Glacial acetic acid: Methanol: Water = 16: 8: 3: 2
5. Spraying/derivatization reagent components:
    - *p*-Anisaldehyde, glacial acetic acid, ethanol, concentrated sulfuric acid
6. Hot air oven or TLC plate heater (100–110°C)
7. Plant extract (crude or saponin-rich fraction)
8. Optional saponin standard (for Rf comparison)
9. Pencil, ruler, forceps, atomizer/spray bottle

2.2.3 FTIR Analysis of Saponins

1. Crude or partially purified plant extract containing saponins (dried powder form preferred).
2. Drying oven or desiccator (for moisture removal).
3. Potassium bromide (KBr) IR-grade, analytical purity.
4. Mortar and pestle/ball mill (sample grinding and mixing).
5. Pellet die set and hydraulic pellet press (for KBr pellet preparation).
6. FTIR spectrometer (mid-IR range: 4000–650 cm^{-1}).

2.3 Quantitative Profiling of Saponins

1. **Total Saponins Content Determination**
    - Plant extract (fresh extract)
2. Methanol (80% v/v aqueous methanol)
3. Ethanol (for preparing vanillin reagent)
4. Vanillin (for 8% w/v vanillin solution)
5. Sulfuric acid (72% w/v)
6. Diosgenin standard (for calibration curve, 40–550 $mg{\cdot}L^{-1}$ range)
7. Distilled water
8. Ice (for ice–water bath)
9. Round-bottom vials (small size)
10. Clean glass vials/test tubes
11. Pipettes (0.5, 1, 5 mL capacity) and pipette tips
12. Volumetric flasks (10, 100 mL)
13. Measuring cylinders
14. Sealing caps/stoppers for vials
15. Blank cuvettes for UV–Vis

16. Rotary evaporator (for evaporation under reduced pressure)
17. Water bath (maintained at 60°C)
18. Ice–water bath
19. UV–Vis spectrophotometerUV–Vis spectrophotometer (capable of reading at 544 nm)
20. Analytical balance

2.3.1 HPLC Profiling

1. Methanol
2. Acetonitrile
3. Water
4. Formic acid
5. Saponin standards (hederacoside C, α-hederin, hederagenin)
6. API 3200 Triple Quadrupole Mass Spectrometer—*Sciex (Toronto, Canada)*
7. Infinity 1260 Binary Pump + Autosampler – *Agilent (Santa Clara, CA, USA)*
8. Luna NH_2 Column (150 × 2 mm, 3 μm, 100 Å)—*Phenomenex (Torrance, CA, USA)*
9. IKA Vortex 3—*IKA (Staufen, Germany)*
10. Analyst 1.5.2—*Sciex*

2.4 Advanced Profiling Techniques

2.4.1 Metabolomic Based Profiling

a. **Untargeted Liquid Chromatography–Mass Spectrometry (LC–MS)**

1. Methanol(LC-MS grade)
2. Acetonitrile (LC–MS grade)
3. Water (LC–MS grade, Milli-Q)
4. Formic acid, ammonium acetate/ammonium formate (LC–MS grade)
5. *n*-Butanol (HPLC grade) for enrichment (optional)
6. SPE cartridges: C18 (various capacities)
7. 0.22 μm syringe filters (PTFE or PVDF)
8. HPLC vials, caps, microcentrifuge tubes, pipette tips, volumetric flasks
9. Authentic saponin standards (where available), e.g., hederacoside C, α-hederin, hederagenin, diosgenin (for TSC calibration)
10. Internal standards (isotopically labeled standards if available, or structurally related surrogate)

11. Quality control (QC) pooled sample (aliquot of combined extracts)
12. UHPLC system capable of binary/ternary gradients and low-dispersion plumbing
13. High-resolution mass spectrometer (HRMS) Q-TOF or Orbitrap (for untargeted work)
14. Triple quadrupole (QQQ) MS (for targeted MRM quantification)
15. Ion mobility spectrometer (optional; for IMS-HRMS workflows)
16. LC–NMR hyphenation system or offline NMR (400–800 MHz) for structure confirmation
17. Autosampler with temperature control, column oven, degasser
18. Software: vendor acquisition software + data processing tools (e.g., XCMS/MzMine/Compound Discoverer), GNPS for molecular networking, and chemometrics packages (SIMCA, MetaboAnalyst, R packages)
19. Reversed-phase UHPLC column (C18, sub-2 μm)
20. HILIC column (e.g., amino, zwitterionic) for very polar glycosides
21. Guard cartridges, frits, tubing compatible with low flow (≤ 0.5 mL/min)
22. Centrifuge (capable of ≥5000 × *g*)
23. Vortex mixer, sonicator (bath or probe)
24. Rotary evaporator or nitrogen evaporator
25. Analytical balance, pH meter, refrigerated storage (4°C/−20°C)

b. **Targeted Multiple Reaction Monitoring (MRM)**

1. Hederacoside C (authentic standard)
2. α-Hederin (authentic standard)
3. Hederagenin (authentic standard)
4. Internal standard (stable-isotope labeled saponin or structurally related surrogate not present in samples)
5. Methanol (LC–MS grade)
6. Acetonitrile (LC–MS grade)
7. Water (Milli-Q/LC–MS grade)
8. Formic acid (LC–MS grade)
9. *n*-Butanol (HPLC grade) for partitioning
10. Reagent grade methanol for SPE conditioning (if different from LC–MS grade)

11. C18 SPE cartridges (e.g., 500 mg bed)
12. Anhydrous sodium sulfate (Na_2SO_4) if drying organic fractions is required
13. 0.22 μm syringe filters (PTFE or PVDF)
14. LC–MS vials with low-volume inserts and caps (2 mL/suitable for autosampler)
15. Autosampler vials, microcentrifuge tubes (1.5–2.0 mL)
16. Volumetric flasks, pipette tips (filtered), glass Pasteur pipettes
17. Syringes for SPE and infusion
18. Luna NH_2 (Amino) HILIC column, 150 × 2.0 mm, 3 μm *Phenomenex*) column validated for saponins)
19. Triple-quadrupole mass spectrometer (for MRM), e.g., API 3200 *Sciex* (or other triple-quadrupole QqQ)
20. UHPLC system with autosampler and column oven, e.g., Infinity 1260 binary pump + autosampler *Agilent* (or equivalent UHPLC)
21. Vortex mixer (e.g., IKA Vortex 3) for sample mixing
22. Centrifuge capable of 10,000 rpm (microcentrifuge) and bench centrifuge for lower speeds
23. Nitrogen evaporator or rotary evaporator (for solvent removal), e.g., Büchi Rotavapor (or equivalent)
24. Analytical balance (0.1 mg readability)
25. Sonicator (bath or probe) for extraction
26. Glassware: round-bottom flasks, separatory funnel (for partitioning)
27. Acquisition/control software compatible with QqQ (e.g., Analyst v1.5.2 for Sciex)
28. Data processing/quantification software (vendor MRM quantitation module or third-party tools)
29. Spreadsheet software for calibration and calculations

2.4.2 Hyphenated Technique

a. **LC–MS and NMR Analysis of Purified Saponin Fractions**

1. Ultimate 3000 RSLC System (Dionex, Germering, Germany)
2. LTQ XL Linear Ion Trap Mass Spectrometer with ESI source (Thermo Scientific, San José, CA, USA)
3. Bruker Avance III 500 MHz NMR spectrometer (5 mm TXI probe, Z-gradient)
4. Bruker Avance III 400 MHz NMR spectrometer (5 mm BBO probe, Z-gradient)
5. Xcalibur v3.0.63 software (Thermo Scientific)

6. TopSpin 4.0.7 (Bruker)
7. MestReNova v14.2.0 (Mestrelab Research)
8. LC–MS grade Methanol (MeOH), Water (H_2O), Acetonitrile (ACN)
9. Formic acid (LC–MS grade)
10. Deuterated Methanol ($CD_3OD \geq 99.8\%$ D; Sigma-Aldrich)
11. MPLC/HPLC fractions (e.g., B1, B3)
12. Purified saponins: Qb1, S13, commercial QS-21
13. 2 mL LC vials with caps
14. 0.22 μm PTFE syringe filters
15. 5 and 3 mm NMR tubes
16. Micropipettes and tips
17. Glass vials for drying

3 Methods

3.1 Sample Collection and Extraction

3.1.1 Sample Collection and Pre-Processing [23]

1. Collect fresh leaves of selected plant and dry them in an air-flow heating oven at 60°C until constant weight is achieved (final moisture content ~5.8%).
2. Grind the dried leaves using an electric grinder and sieve the powder to obtain particles <315 μm.
3. Weigh 25 g of the sieved leaf powder into sealed plastic containers.
4. Store the prepared samples at 4–5°C until further extraction of bioactive compounds.

3.1.2 Extraction of Saponins

a. **Soxhlet extraction**Soxhlet extraction **of saponins** [24]

1. Weigh 5–10 g of plant powder and transfer it into a cellulose extraction thimble, tapping gently to distribute powder uniformly.
2. Insert the loaded thimble into the main chamber of the Soxhlet extractor, ensuring it sits properly and does not obstruct solvent siphoning.
3. Add the extraction solvent 70% aqueous methanol into a round-bottom flask (typically 150–250 mL, enough to allow 8–12 siphon cycles per hour).

4. Attach the round-bottom flask, Soxhlet chamber, and condenser securely. Connect the condenser to running cold water to ensure efficient solvent condensation.
5. Place the setup on a heating mantle or controlled water/oil bath, adjusting the temperature to keep the solvent gently refluxing.
6. Allow the system to run for 6–8 h, during which the Soxhlet apparatus performs continuous cyclic extraction (approximately 8–12 extraction cycles per hour).
7. After extraction, turn off the heat, allow the system to cool, then carefully detach the apparatus. Remove the solvent-rich extract from the round-bottom flask.
8. Concentrate the extract under reduced pressure using a rotary evaporator until most solvent is removed, obtaining a thick crude extract.
9. Store the concentrated extract in a sealed vial at 4–5°C if further processing or purification steps are needed.

b. **Ultrasound-Assisted Extraction of saponins (UAE)** [4] (*see* **Note 1**)

1. Prepare the jacketed glass reactor and maintain the extraction temperature (30°C, 40°C, or 50°C) by circulating temperature-controlled water.
2. Mix leaf powder with solvent (0%–96% ethanol) at plant-to-solvent ratios of 1:10, 1:20, or 1:30 (w:v).
3. Perform extraction either by inserting the ultrasonic probe (20%, 40%, or 60% amplitude = 5.5, 27.9, 58.9 W) or by conventional stirring (900 rpm) for selected times (1–100 min).
4. Centrifuge the mixture at 2500 rpm for 5 min and collect the supernatant.
5. Evaporate the supernatant to dryness using a rotary evaporator and store the dried extract in a freezer until HPLC analysis; repeat all experiments in triplicate.
6. Combine all extracts and evaporate under reduced pressure at 40–45°C using a rotary evaporator until a concentrated aqueous residue remains.
7. Dissolve the residue in 50–100 mL water and transfer to a separatory funnel for cleanup.
8. Perform liquid–liquid extraction with *n*-butanol (3 × equal volumes, e.g., 3 × 50 mL), collect and combine the *n*-butanol layers, dry over anhydrous Na_2SO_4, filter, and evaporate to obtain a saponin-enriched fraction.

3.2 Qualitative Profiling of Saponins

3.2.1 Primary Phytochemical Tests for Saponins [22]

1. For the froth/foam test, take 5 mL of plant extract in a clean test tube, add 5 mL distilled water, shake vigorously for 15–30 s and allow to stand; stable foam ≥1 cm that persists for 10–15 min indicates saponins.
2. For the heated froth test, take 2 mL extract, boil gently for 5 min, cool to room temperature, shake with 2 mL distilled water; persistent froth confirms saponins.
3. For the emulsification test, add 1 mL plant extract to 2 mL vegetable/olive oil, shake vigorously for 1 min; presence of a stable emulsion indicates saponins due to their natural surfactant properties.
4. For the hemolysis test (advanced), prepare 2% RBC suspension in isotonic saline, add a few drops of plant extract, incubate, and observe; hemolysis producing a red supernatant denotes the presence of strong saponins and requires biosafety precautions.
5. For the $NaHCO_3$ foam test, add a pinch of sodium bicarbonate to 5 mL extract, shake vigorously for ~15 s; formation of persistent foam further supports the presence of saponins.

3.2.2 Thin Layer Chromatography (TLC) [25]

1. Dissolve the crude extract or purified saponin fraction in a small volume of methanol, ethanol, or water–methanol mixture.
2. Mark a straight origin line 1 cm above the bottom of the TLC plate using a pencil.
3. Spot 2–5 μL of sample on the origin using a capillary/micropipette; avoid spreading.
4. If available, spot a saponin standard on the same plate for reference.
5. Allow the plate to air-dry completely before development.
6. Add selected mobile phase to the TLC chamber (0.5–1 cm depth).
7. Close the chamber and allow vapor saturation for 15–30 min.
8. Place the spotted TLC plate vertically inside the chamber, ensuring spots remain above solvent level.
9. Allow the solvent front to ascend until ~1 cm below the top edge.
10. Remove the plate immediately and mark the solvent front.
11. Let the developed plate air-dry before spraying.
12. Prepare *p*-anisaldehyde–sulfuric acid reagent by mixing:

- 0.5 mL *p*-anisaldehyde
- 10 mL glacial acetic acid
- 85 mL ethanol
- 5 mL concentrated sulfuric acid (add slowly and carefully)

13. Spray the dried TLC plate evenly using a fine atomizer.
14. Heat the plate in an oven or TLC heater at 100°C–110°C for 5–10 min until colored spots appear.
15. Saponins typically develop violet, blue, purple, or pink spots.
16. Mark visible spots lightly with a pencil.
17. Calculate Rf values: $\frac{\text{Distance traveled by saponin spot}}{\text{Distance traveled by solvent front}}$
18. Compare Rf values with standards or literature to classify saponin types.
19. Photograph the TLC plate for documentation and record keeping.

3.2.3 FTIR Analysis [26]

1. Ensure the crude saponin extract is completely dry remove moisture by keeping the sample in a hot air oven (40°C–50°C) or desiccator.
2. Grind the dry sample into a fine, uniform powder.
3. Dry IR-grade KBr powder in a hot air oven (≈110°C for 1 h) to remove adsorbed moisture.
4. Accurately weigh a small amount of sample (typically 1–2 mg).
5. Mix with ≈100–150 mg KBr in a mortar/pestle until homogeneous (*see* **Note 2**).
6. Transfer the powdered mixture to a pellet die.
7. Apply high pressure using a hydraulic press (≈5–8 tons) for 1–2 min.
8. Carefully remove the transparent KBr pellet and preserve it in a desiccator to prevent moisture absorption.
9. Place the KBr pellet into the sample holder of the FTIR spectrometer.
10. Collect the IR spectrum across 4000–650 cm^{-1} (full functional group + fingerprint region).
11. Use air or pure KBr pellet as the background scan before sample measurement.
12. Analyze the spectrum for characteristic saponin peaks, typically:
    - Broad O–H stretching (3300–3400 cm^{-1})
    - C–H stretching (2920–2850 cm^{-1})

- C–O–C and C–O glycosidic bond peaks (1200–1000 cm^{-1})
- Fingerprint region peaks indicating triterpenoid or steroidal aglycone structures

13. Compare peaks with the reference FTIR patterns reported as per [26] for confirmation.
14. Save and document the spectrum for qualitative profiling.

3.3 Quantitative Profiling of Saponins

3.3.1 Total Saponins Content Determination [27]

1. Take 0.5 mL of fresh extract in a small round-bottom vial and evaporate to dryness under reduced pressure (rotary evaporator).
2. Dissolve the dried residue in 10 mL of 80% (v/v) methanol in water and mix thoroughly.
3. Pipette 0.5 mL of this methanolic solution into a clean vial, add 0.5 mL of 8% (w/v) vanillin in ethanol, then add 5.0 mL of 72% (w/v) sulfuric acid (add acid carefully down the vial wall).
4. Incubate the sealed vial in a 60°C water bath for 10 min, then cool immediately in an ice–water bath for 4 min.
5. Measure absorbance of the resulting chromophore at 544 nm against a reagent blank using a UV–Vis spectrophotometer.
6. Quantify total saponin content as mg diosgenin equivalents (DE) per g dry matter (mg DE/g DM) using a diosgenin calibration curve prepared across 40–550 mg·L^{-1}; run samples in triplicate and include blanks and quality controls.

3.3.2 HPLC Profiling [28]

1. Dissolve dried extract in 1.0 mL methanol and vortex 5 min; prepare a 1:10,000 (v:v) dilution in methanol and mix.
2. Set up LC: Luna NH_2 (150 × 2.0 mm, 3 μm, 100 Å) column, mobile phases A = water + 0.1% formic acid, B = acetonitrile + 0.1% formic acid, flow 0.3 mL/min, injection 10 μL.
3. Use HILIC gradient: 5% A (0–0.5 min) → ramp to 50% A (0.5–1.0 min) → hold 50% A (1.0–1.5 min) → ramp to 95% A (1.5–2.0 min) → hold 95% A (2.0–4.0 min) → re-equilibrate to initial conditions.
4. Use triple-quadrupole MS (API3200) with ESI in negative ion mode, acquire in MRM for target saponins; process data with Analyst v1.5.2.
5. Prepare calibration curves in methanol for each analyte (matrix adjustment as required): hederacoside C: 15–1500 ng/mL, α -hederin: 5–500 ng/mL, hederagenin: 2.5–250 ng/mL.
6. Inject standards and samples; expected RTs: hederagenin ~1.41 min, hederacoside C ~3.97 mincalibration curve.)

7. Run samples and standards in triplicate, include blanks and QC samples to monitor carryover and matrix effects.
8. Report concentrations as ng/mL in the injected solution and convert to sample basis (e.g., ng analyte per g dry extract) using dilution factors and sample weights.

3.4 Advanced Profiling Techniques

3.4.1 Metabolomic-Based Profiling [29–34]

a. **Untargeted Liquid Chromatography–Mass Spectrometry (LC–MS)**

1. Weigh 100 mg dried plant powder and extract with 2.0 mL of 70% methanol by sonication for 20 min.
2. Centrifuge at 10,000 rpm, 10 min and collect supernatant; perform *n*-butanol partition (1 mL extract + 1 mL BuOH, vortex 30 s, settle 2 min, centrifuge 8000 rpm, 5 min).
3. Load BuOH layer onto C18 SPE (500 mg) pre-conditioned with 5 mL MeOH → 5 mL water; wash with 5 mL water, elute saponin-rich fraction with 5 mL 80% MeOH.
4. Evaporate eluate at 40°C under nitrogen, reconstitute in 1.0 mL of 80% MeOH, filter through 0.22 μm PTFE filter, and keep at 1 mg/mL in LC vials.
5. Prepare pooled QC by mixing 50 μL from each extract; vortex and aliquot into 200 μL vials.
6. Prepare blank (80% MeOH) and set run order: 5–7 QC injections for conditioning, randomized samples, with QC every 6–10 injections.
7. Run RP-UHPLC using Hypersil GOLD C18 (150 × 2.1 mm, 3 μm), 0.40 mL/min, 25°C, 5 μL injection.
8. Apply RP gradient: 2% B (0 min) → 20% (4 min) → 68% (30 min) → 98% (32–40 min), then re-equilibrate to 2%.
9. Run HILIC-UHPLC using Luna NH_2(150 × 2 mm, 3 μ m), 0.25 mL/min, 30°C, 2 μL injection in ≥70% ACN.
10. HILIC gradient: 90% B → 60% (0.5–4 min) → 40% (6–7 min), re-equilibrate to 90% by 13 min.
11. Acquire on UHPLC–HRMS (Orbitrap/Q-TOF) in both positive and negative ESI.
12. Use full scan m/z 100–2000, resolution 60,000–120,000 (Orbitrap) or high-res TOF mode.
13. Use typical ion source settings: 3.5 kV (pos), −3.0 kV (neg), 380°C–420°C source, 40 psi nebulizer, 600 L/h desolvation gas.
14. Acquire MS/MS using DDA Top-5/Top-10, CE ramp 20–40 eV, with 10–15 s dynamic exclusion.

15. (Optional) Perform DIA/SWATH acquisition using 20–25 variable windows.
16. Inject QC every 6–10 samples; ensure <20% CV in peak areas.
17. Evaluate RT stability (<0.1–0.2 min variation) and signal drift; apply QC-based LOESS correction.
18. Convert raw files to mzML (ProteoWizard).
19. Process with MZmine/XCMS/Compound Discoverer: noise filtering → peak detection → RT alignment → isotope & adduct grouping → feature alignment across all samples.
20. Perform blank subtraction and remove duplicate adducts/fragments.
21. Annotate features using accurate mass (±5 ppm), MS/MS matching (GNPS, MassBank), and in-silico tools (SIRIUS, CFM-ID, MetFrag).
22. Build GNPS Feature-Based Molecular Network to group saponin clusters and identify structural families.
23. Prioritize features based on intensity, clustering, unique MS/MS patterns, and sample-specific enrichment.
24. Isolate candidate saponins by preparative RP-HPLC (load 1–5 mg per run), dry fractions, and re-check them via LC-HRMS for purity.
25. Confirm structures using 1D/2D NMR (^{1}H, ^{13}C, HSQC, HMBC, COSY), HRMS^n, and CCS values (if using ion mobility), and add confirmed data to the lab's spectral library.

b. **Targeted Multiple Reaction Monitoring (MRM)**

1. Extract sample as per your validated workflow (*n*-BuOH enrichment → C18 SPE → dry → reconstitute), then dilute/exchange into methanol; target final extract concentration ≈ 1 mg·mL^{-1}. Filter 0.22 μm into autosampler vials.
2. Select an appropriate internal standard (stable-isotope-labeled saponin or structurally related compound not present in the matrix). Spike IS at a constant concentration into all standards, QCs and samples (e.g., 100 ng·mL^{-1} final).
3. Prepare calibration standards in solvent or matrix-matched extract over the working ranges: hederacoside C 15–1500 ng·mL^{-1}; α-hederin 5–500 ng·mL^{-1}; hederagenin 2.5–250 ng·mL^{-1} (seven levels are recommended). Add IS to each standard.

4. Prepare three QC levels (Low, Mid, High) within the calibration range (e.g., 3×: 3×LOQ, mid-point, 75% of top) and a blank (solvent) and zero (IS only).
5. LC conditions(HILIC, use Luna $NH_2$150×2 mm, 3 μ m):A = water + 0.1% formic acid; B = ACN + 0.1% formic acid; flow 0.25 mL·min^{-1}; temp 30°C; inj vol 2–5 μ L; sample solvent ≥70% ACN. Gradient: 0–0.5 min 90% B; 0.5–4.0 min 90→60% B; 4–6 min 60→40% B; 6–7 min hold 40% B; 7–13 min re-equilibrate to 90% B.
6. Triple-quad source settings: ESI negative mode; ion spray ≈ −4.0 kV; source temp 450°C; curtain/GS1/GS2 gas per vendor recommendations (e.g., CUR 25, GS1 50, GS2 50).
7. Determine precursor ions: infuse each standard (1–10 μ g·mL^{-1}) to identify the dominant precursor (e.g., $[M–H]^-$ or adduct). Record precursor m/z for each analyte.
8. Acquire product ion spectra for each precursor across a CE range (e.g., 10–40 eV) to identify the most intense and diagnostic fragments (sugar losses and aglycone ions).
9. Select two transitions per analyte: one quantifier (most intense, high S/N) and one qualifier (diagnostic fragment). Note precursor→product m/z, optimized DP, CE, and CXP for each.
10. Create an MRM method with scheduled MRM windows around expected RTs (±0.2–0.5 min) to allow longer dwell times; set dwell times 20–50 ms (longer for low-abundance analytes).
11. Set collision gas to medium and enable unit resolution on Q1/Q3 (unless high-res MRM available). Enable interscan pause minimal.
12. Prepare run sequence: 5× pooled QC injections to condition; calibration standards lowest→highest (triplicate if desired); blanks; QCs (Low/Mid/High); randomized samples; QC every 8–10 injections; mid-level standard at end for stability check.
13. Inject standards, QCs, and samples. Monitor RT, peak shape, peak area, and qualifier/quantifier ion ratio for each analyte.
14. Build calibration curves: plot analyte area ratio (analyte/IS) vs nominal concentration; use weighted linear regression (1/x or $1/x^2$) to minimize low-end bias. Accept $r^2 \geq 0.99$.
15. Determine LOD/LOQ: S/N ≈ 3:1 (LOD) and 10:1 (LOQ) (or statistical approach). Verify LOQ by precision

and accuracy at LOQ (≤20% RSD and within ±20% accuracy).

16. Quantify samples by interpolation from calibration curve using analyte/IS area ratio. Convert to sample basis (mg analyte/g dry sample) using dilution factors and initial sample mass:
    a. mg/g = C_{meas} (ng\cdotpmL−1) × V_{final} (mL)/mass (g) × 10^6
    b. mg/g = $\frac{C_{meas} \text{ (ng\cdotpmL − 1) × } V_{fina}l\text{(mL)}}{\text{mass (g) × 106}}$
17. Validate method performance: assess intra-day ($n \geq 5$) and inter-day precision ($n \geq 3$ days), target CV ≤ 15% (≤20% at LOQ); accuracy (recovery) 80%–120%; matrix effects via post-extraction spike (compare slope solvent vs matrix).
18. Apply identity confirmation criteria per sample: RT within ±0.2–0.3 min of standard, qualifier/quantifier ion ratio within ±30% of standard, S/N ≥ 10 at reporting level.
19. Monitor carryover by injecting a blank after the highest standard and after highest sample; accept ≤1% carryover. If >1%, add wash steps or increase needle wash.
20. Document and report: calibration equation and r^2, LOD/LOQ, precision [%RSD], accuracy [%recovery], matrix effect %, qualifier ratios, and final concentrations (mean ± SD, n replicates). Archive raw data, integration details, and MRM parameters in lab records.

3.4.2 Hyphenated Technique [34]

a. **LC–MS and NMR Analysis of Purified Saponin Fractions**

1. Collect MPLC or HPLC fractions containing target saponins (e.g., B1, B3) based on UV chromatograms.
2. Dry each fraction under nitrogen or vacuum concentrator at ≤40°C.
3. Reconstitute dried fractions in 1 mL methanol or extraction solvent used previously.
4. Filter reconstituted samples through a 0.22 μm PTFE filter into LC vials.
5. Prepare reference standards (e.g., QS-21) at matching concentrations for comparison.
6. Inject samples into an LC–MS system such as Ultimate 3000 RSLC coupled to a Thermo LTQ XL linear ion trap.
7. UseESI negative mode with previously validated source parameters (capillary temp, sheath/aux gas, spray voltage).
8. Employ a C18 analytical column with a water–acetonitrile gradient containing 0.1% formic acid.
9. Run samples under Full Scan mode (m/z ~150–2000) for initial profiling.

10. Acquire MS/MS spectra using data-dependent acquisition to target major saponin ions.
11. Collect SIM (Selected Ion Monitoring) scans for known diagnostic saponin fragments.
12. Review chromatograms to identify peaks corresponding to the expected molecular ions and adducts.
13. Compare retention times and MS/MS fragmentation patterns with commercial QS-21 or literature-reported profiles.
14. Select fractions containing high-purity target ions for structural analysis.
15. Transfer 1–2 mg of each isolated compound into a 3 mm NMR tube.
16. Add ~200 μL CD_3OD (≥99.8% D) to dissolve the sample completely.
17. Vortex gently or sonicate briefly to ensure full dissolution.
18. Acquire ^1H NMR spectra at 500 MHz using a Bruker Avance III system at 298 K.
19. Acquire ^{13}C NMR spectra at 400 MHz when sample amount allows.
20. Reference chemical shifts using residual CD_3OD peaks (δ H 3.31 ppm; δC 49.0 ppm).
21. Record 2D spectra including HSQC, HMBC, TOCSY, and NOESY depending on sample quantity.
22. Adjust mixing times for TOCSY (20–100 ms) and NOESY (200–300 ms) for optimal correlations.
23. Process all NMR data using MestReNova or TopSpin software.
24. Combine LC–MS fragmentation, retention-time data, and 2D NMR correlations to assign aglycone and sugar residues.
25. Confirm final structures by matching MS/MS pathways and NMR chemical shifts with published saponin datasets or reference compounds.

4 Notes

1. Purify further using a C18 SPE cartridge: condition with methanol and water, load sample, wash with water, elute saponins with 50%–80% methanol, and evaporate the eluate for analysis.
2. The sample must be sufficiently diluted so that the final pellet is transparent.

References

1. Sharma K, Kaur R, Kumar S, Saini RK, Sharma S, Pawde S V Kumar V (2023) Saponins: a concise review on food related aspects, applications and health implications. Food Chem Adv 2:100191. https://doi.org/10.1016/j.focha.2023.100191
2. Sparg SG, Light ME, Van Staden J (2004) Biological activities and distribution of plant saponins. J Ethnopharmacol 94:219–243. https://doi.org/10.1016/j.jep.2004.05.016
3. Vincken JP, Heng L, de Groot A, Gruppen H (2007) Saponins, classification and occurrence in the plant kingdom. Phytochemistry 68:275–297. https://doi.org/10.1016/j.phytochem.2006.10.008
4. Timilsena YP, Phosanam A, Stockmann R (2023) Perspectives on saponins: food functionality and applications. Int J Mol Sci 24(17):13538. https://doi.org/10.3390/ijms241713538
5. Moses T, Papadopoulou KK, Osbourn A (2014) Metabolic and functional diversity of saponins, biosynthetic intermediates and semi-synthetic derivatives. Crit Rev in Biochem Mol Biol 49:439–462. https://doi.org/10.3109/10409238.2014.953628
6. Man S, Yujuan W, Yiming L, Sheng L (2018) Advances in pharmacology and clinical application of platycodin. J Shanghai Univ Tradit Chin Med 32:86–91
7. Jun Z, Baikun Y, Xiaoyang H (2019) Research progress in the chemical constituents and modern pharmacology of *Platycodon*. J Liaoning Univ Tradit Chin Med 21:113–116
8. Xiongxiong X, Chi Z, Jinxiang Z, Chenhui Z, Zhu M, Junwei H, Hongling W, Guoyue Z, Shouwen Z, Fengyu H (2018) Advances in research on chemical constituents and pharmacological activity of *Platycodon grandiflorum*. Bull Tradit Chin Med 17(13):66–72
9. Mian T, Xiaofen X (2018) Advances in studies on chemical constituents and pharmacological effects of medicinal *Astragalus*. Herald Tradit Chin Med 24:117–122
10. Lili W, Li C, Ying S, Tianyang X (2018) Chemical composition and pharmacological action of *Rhizoma Anemarrhenae*. *Jilin*. J Tradit Chin Med 38:90–92
11. Yashuang C, Shiwei S (2014) Advances in research on chemical constituents and pharmacological effects of *Bupleurum chinense*. *Heilongjiang*. Medicine 27:630–633
12. Meiling Y, Liu Y, Ajiao H, Xinyue G, Wenjing M, Xudong X, Hua H (2018) Research progress on chemical composition and pharmacological effect of *Bupleurum chinense*. Inform Tradit Chin Med 35:103–109
13. Xiaohui S, Liang Z (2017) Radix *Bupleuri*: research progress of pharmacological effects. China Med Herald 14:52–55
14. Dawei L, Liping K, Baiping M (2012) Chemical constituents and pharmacological activities of *Polygala tenuifolia* Willd.: research advances. Int J Pharm Res 39(32–36):44
15. Yingying S, Yue L, Keji C (2016) Cardiovascular pharmacological effects of ginsenosides: progress and thinking. Sci China Life Sci 46:771–778
16. Juan L, Rufeng W, Li Y, Zhengtao W (2015) Structure and biological action on cardiovascular systems of saponins from *Panax notoginseng*. *Chinese*. J Tradit Chin Med 40:3480–3487
17. Savarino P, Demeyer M, Decroo C, Colson E, Gerbaux P (2023) Mass spectrometry analysis of saponins. Mass Spectrom Rev 42(3):954–983. https://doi.org/10.1002/mas.21728
18. Wang Y, Ma Y, Tao L, Zhang X, Hao F, Zhao S, Han L, Bai C (2022) Recent advances in separation and analysis of saponins in natural products. Separations 9(7):163. https://doi.org/10.3390/separations9070163
19. Bai Y, Cai G, Guo N, Huang X, Gong J, Liu S, Guo Y, Wang W (2023) UHPLC-HRMS based saponins profiling of three morphological regions in American ginseng (*Panax quinquefolium* L.) and correlation with antioxidant activity. Food Sci Biotechnol 33(7):1685–1696. https://doi.org/10.1007/s10068-023-01453-4
20. Zheng W, Li G, Yang G, Lu P, Li Q, Zhang M, Ma B (2023) Two-dimensional liquid chromatography and ion mobility-mass spectrometry for multicomponent characterization of *Gynostemma longipes*. Front Chem 11:1203418
21. Zheng W, Li G, Yang G, Lu P, Li Q, Zhang M, Ma B (2023) Two-dimensional liquid chromatography and ion mobility-mass spectrometry for multicomponent characterization of *Gynostemma longipes*. Front Chem 11:1203418
22. Harborne AJ (1998) Phytochemical methods a guide to modern techniques of plant analysis. Springer Science & Business Media
23. Gavrila AI, Tatia R, Seciu-Grama AM, Tarcomnicu I, Negrea C, Calinescu I, Zalaru C, Moldovan L, Raiciu AD, Popa I (2022) Ultrasound-assisted extraction of saponins from *Hedera helix* L. and in vitro biocompatibility evaluation. Pharmaceuticals 15(10):1197. https://doi.org/10.3390/ph15101197
24. Seidel V (2006) Initial and bulk extraction. In: Natural products isolation. Humana Press, pp 27–46
25. Sharma OP, Kumar N, Singh B, Bhat TK (2012) An improved method for thin layer

chromatographic analysis of saponins. Food Chem 132:671–674
26. Almutairi MS, Ali M (2015) Direct detection of saponins in crude extracts of soapnuts by FTIR. Nat Prod Res 29(13):1271–1275
27. Hiai S, Oura H, Nakajima T (1976) Color reaction of some sapogenins and saponins with vanillin and sulfuric acid. Planta Med 29:116–122. https://doi.org/10.1055/s-0028-1097639
28. Tiwari R, Udayabanu M, Chanda S (2017) Quantitative analysis of secondary metabolites in aqueous extract of *Clerodendrum serratum*. Int Res J Pharm 7:61–65. https://doi.org/10.7897/2230-8407.0712148
29. Nord LI, Kenne L (2000) Novel acetylated triterpenoid saponins in a chromatographic fraction from *Quillaja saponaria* Molina. Carbohydr Res 329:817–829
30. Schleucher J, Schwendinger M, Sattler M, Schmidt P, Schedletzky O, Glaser SJ, Sørensen OW, Griesinger C (1994) A general enhancement scheme in heteronuclear multidimensional NMR employing pulsed field gradients. J Biomol NMR 4:301–306
31. Bax A, Donald GD (1969) MLEV-17-based two-dimensional homonuclear magnetization transfer spectroscopy. J Magn Reson 65:355–360
32. Nyberg NT, Duus JØ, Sørensen OW (2005) Heteronuclear two-bond correlation: enhancing 2JCH correlations while suppressing higher-order correlations. J Am Chem Soc 127:6154–6155
33. Claridge TDW, Pérez-Victoria I (2003) Enhanced 13C resolution in semi-selective HMBC for complex organic structure elucidation. Org Biomol Chem 1:3632–3634
34. Broeckling CD, Beger RD, Cheng LL, Cumeras R, Cuthbertson DJ, Dasari S, Davis WC, Dunn WB, Evans AM, Fernández-Ochoa A, Gika H, Goodacre R, Goodman KD, Gouveia GJ, Hsu P-C, Kirwan JA, Kodra D, Kuligowski J, Lan RS-L, Monge ME, Moussa LW, Nair SG, Reisdorph N, Sherrod SD, Ulmer Holland C, Vuckovic D, Yu L-R, Zhang B, Theodoridis G, Mosley JD (2023) Current practices in LC–MS untargeted metabolomics: a scoping review on the use of pooled quality control samples. Anal Chem 95:18645–18654. https://doi.org/10.1021/acs.analchem.3c02924

Chapter 21

Coumarins: Metabolic Profiling by Chromatography

Anil Prajapati, Glory Simon and Ketankumar Panchal

Abstract

Coumarins are naturally occurring compounds majorly sourced from plants having great therapeutic potential. Metabolic profiling of coumarins using chromatographic techniques, such as Thin-Layer Chromatography (TLC), High-Performance Liquid Chromatography (HPLC), and Ultra-Performance Liquid Chromatography coupled with Mass Spectrometry (UPLC-MS), is essential for the qualitative and quantitative analysis of these bioactive molecules. This chapter comprehensively reviews sample preparation methods, chromatographic conditions, detection strategies, and data interpretation approaches for coumarin analysis across various biological and plant matrices. Emphasis is placed on the advances in chromatographic technology that facilitate improved sensitivity, resolution, and metabolite identification, informing drug development, pharmacokinetics, and toxicological assessments of coumarin derivatives.

Keywords Coumarins, Chromatography, Plant metabolites, Biotransformation, Mass spectrometry, Analytical techniques

1 Introduction

Coumarins are a class of natural compounds mainly sourced from various plant families, particularly Rutaceae and Umbelliferae. Structural diversity from simple coumarins to complex furocoumarins have their broad spectrum importance for various biological activities, such as anti-inflammatory, antimicrobial, anticancer, antioxidant, and anticoagulant effects. Profiling of coumarin from biological samples requires utilization of chromarographic techniques, such as ultra-performance liquid chromatography (UPLC) and ultra-high-performance liquid chromatography (UHPLC) interfaced with electrospray ionization-quadrupole time-of-flight mass spectrometry (ESI–QTOF–MS). Coumarin structural modification, such as prenylation or isopentenyl substitutions, by enzymes influence its metabolic rates and pathways highlights the complexity of coumarin biotransformation. Although generally regarded as safe at dietary levels, some coumarins can metabolize into reactive or potentially toxic intermediates, underscoring the importance of detailed metabolic studies for drug

Dhruti Amin et al. (eds.), *Phytoconstituent Analysis*, Springer Protocols Handbooks,
https://doi.org/10.1007/978-1-0716-5400-2_21,

development and toxicological risk evaluations. Chromatographic and mass spectrometric technologies for profiling of coumarins in this chapter advances understanding of coumarin metabolism, fostering novel therapeutic avenues as well as refined risk assessments.

2 Materials

2.1 Extraction of Coumarins from Sample [1–5]

1. Organic solvents (methanol, ethanol, ethyl acetate).
2. Apparatus for maceration/percolation or Soxhlet extraction.
3. Ultrasonic bath for ultrasound-assisted extraction.
4. Microwave unit (if using microwave-assisted extraction).
5. Filtration materials (Whatman filter paper, syringe filters 0.22–0.45 μm).
6. Round-bottom flasks and water bath for concentration.
7. Amber vials for protected storage of extracts.

2.2 Thin-Layer Chromatography (TLC) Protocol for Coumarin Profiling [6–8]

1. Standard coumarins: fraxin, fraxetin, aesculin, aesculetin + synthetic coumarins
2. Silica gel 60 F254 TLC plates (glass or aluminum-backed)
3. Methanol, ethanol, or suitable solvent for sample extraction
4. Developing solvents: Toluene:ether (1:1, saturated with 10% acetic acid) or Cyclohexane:ethyl acetate (13:7)
5. Visualization reagents: 10% ethanolic potassium hydroxide (KOH), 1% methanolic diphenylboryloxyethylamine (DPBA)
6. Capillary tubes or micropipettes for spotting
7. UV lamp (254 and 366 nm)
8. Developing chamber with filter paper for solvent saturation
9. Pencil and ruler for marking plates

2.3 High-Performance Liquid Chromatography (HPLC) Protocol for Coumarin Profiling [9–11]

1. Reversed-phase C18 column (150–250 mm length, 4.6 mm internal diameter, 3–5 μm particle size)
2. Mobile phases: solvent A as water with 0.1% formic acid or 0.1% trifluoroacetic acid (TFA); solvent B as acetonitrile or methanol with 0.1% formic acid or 0.1% TFA
3. Coumarin standards
4. Sample extracts prepared in methanol (optionally in suitable solvent)
5. Filtration unit (0.45 or 0.22 μm syringe filters)
6. Vials for sample and standard injection

2.4 Analysis of Coumarins by Ultra-Performance Liquid Chromatography Coupled with Mass Spectrometry (UPLC-MS) [12–15]

1. Reversed-phase C18 columns, 100 mm × 2.1 mm, sub-2 μm particle size
2. Mobile phases: solvent A as water with 0.1% formic acid or 10 mM ammonium formate (pH adjusted to 3–4); solvent B as acetonitrile or methanol with 0.1% formic acid
3. Reference standards of coumarins and furocoumarins prepared in solvents like methanol and ethyl acetate.
4. Sample extracts prepared in methanol or suitable solvent
5. Syringe filters (0.22 μm)
6. Vials for injections

3 Methods

3.1 Extraction of Coumarins from Sample [1–5]

1. Powder the plant or biological sample uniformly using liquid nitrogen to increase extraction efficiency.
2. Select an appropriate extraction approach depending on sample nature: maceration, percolation, Soxhlet extraction,Soxhlet extraction,ultrasound-assisted extraction, or microwave-assisted extraction.
3. Add a suitable organic solvent, such as methanol, ethanol, or ethyl acetate, to the powdered sample. For 25 g freeze dried powder, use 250 mL 95% methanol (v/v) or proportionately for lower volume. Maintain the sample–solvent mixture for the required duration depending on the method (steady soaking for classical methods or shorter times when using ultrasound/microwave assistance).
4. Filter the extract through Whatman filter paper to remove solid material.
5. Concentrate the filtrate using any of water bath, reduced pressure, rotatory evaporator at 35°C–40°C to obtain a smaller volume of crude extract.
6. Store the concentrated extract in amber vials until analysis.
7. Optionally, the concentrated extract allowed until it dries, collect and store it in amber color glass bottle.

3.2 Thin-Layer Chromatography (TLC) Protocol for Coumarin Profiling [6–8]

1. Sample preparation requires methanolic extract of source 1 g of dry powder of plant or biological sample material in 10 mL methanol refluxed in waterbath for 30 min at 65°C–70°C followed by filtering the extract using Whatman filter paper to remove particulate matter. Alternately follow protocol-1 to obtain plant extract containing coumarins.

2. Concentrate the filtrate by evaporation under reduced pressure or gentle heating to approximately 1 mL.
3. Draw a straight origin line approximately 1–2 cm from the bottom edge of the TLC plate using a pencil.
4. Spot 1–10 μL of concentrated sample on the TLC at line marking of origin using a capillary tube or micropipette or automated syringe on HPTLC. Similarly, spot standard (one or mixture) (0.1–0.5 mg/mL). Allow the spots to air dry thoroughly.
5. Simultaneously, prepare the developing chamber by adding the chosen mobile phase solvent system to a depth of about 0.5 cm and line the chamber with filter paper soaked in the same solvent to saturate the chamber.
6. Place the spotted TLC plate vertically inside the chamber and allow the plate to develop until the solvent reaches about 1–2 cm from the top edge.
7. Remove the plate and mark the solvent front with a pencil. Allow the plate to dry at room temperature.
8. Observe the plate under UV illumination at wavelengths 254 and 366 nm. Coumarins typically fluoresce blue-white or yellow-green depending on their structure. Mark selected spots using pencil.
9. Spray the plate with 10% ethanolic potassium hydroxide (KOH) solution to enhance coumarin fluorescence, then visualize again under UV 366 nm. Optionally, spray with 1% methanolic diphenylboryloxyethylamine (DPBA) for further enhancement of flavonoid and coumarin spots under UV 366 nm.
10. Record the retention factor (Rf) for each spot using the formula (Rf = Distance travelled by the compound/Distance travelled by the solvent front).
11. Compare Rf values and fluorescence colors with known coumarin standards.

3.3 High-Performance Liquid Chromatography (HPLC) Protocol for Coumarin Profiling [9–11]

1. Prepare methanolic extract of 1 g dry plant or biological sample with 10 mL methanol refluxed at 65°C–70°C for 30 min followed by filtering extract through Whatman filter paper. Optionally, directly use dried powder prepared in protocol-1.
2. Concentrate extract to approximately 1 mL and final filter extract through a 0.45 μm syringe filter.
3. Prepare coumarin standard solutions in methanol at various concentrations for calibration and validation. Dissolve standard coumarins in 80% methanol (v/v) to obtain stock (optionally mixed standards) at 0.1 mg/mL concentration.

4. Setup C18 column on the instrument and allow it to equilibrate at 30°C–40°C.
5. Use a gradient elution program with Solvent A (water + 0.1% formic acid) and Solvent B (acetonitrile + 0.1% formic acid) starting at 10% B (hold for 2 min) with linearly increase to 60% B over 20 min and raise to 90% B over 5 min and hold for 3 min. Return to initial conditions (10% B) and re-equilibrate for 5 min.
6. Set the flow rate at 0.5–1.0 mL/min (vary depending on column dimensions).
7. Inject 10–20 μL of concentrated filtered sample or standard solution.
8. Monitor UV absorbance using DAD at wavelengths 320–330 nm for coumarins.
9. Run time for each injected sample has an estimated run time of approximately 30–35 min.
10. Identify coumarin peaks by comparing retention times against standards. Quantify coumarin content by integrating peak areas and using calibration curves derived from standard solutions.

3.4 Analysis of Coumarins by Ultra-Performance Liquid Chromatography Coupled with Mass Spectrometry (UPLC-MS) [12–15]

1. Prepare methanolic extract using 1 g of the powdered sample in 10 mL methanol refluxed at 65°C–70°C for 30 min or ultrasonication (sonication at 40 kHz for 30 min at room temperature).
2. Filter the extract through Whatman No. 1 filter paper followed by concentration of filtrate under reduced pressure at ≤40°C to 1 mL.
3. Pass the concentrated extract through a 0.22 μm syringe filter to remove particulates. Transfer the filtrate to the UPLC vial for injection.
4. Prepare standard stock solutions of coumarin in methanol. Generate calibration standards by serial dilution covering the expected concentration range (0.1–500 ng/mL).
5. While maintaining column oven temperature at 30°C, equilibration C18 UPLC column for at least 15 min with initial mobile phase conditions.
6. Based on column specifications adjust the mobile phase flow rate to 0.3–0.4 mL/min
7. Start gradient at 5%–10% solvent B (organic phase) with linearly increase to 90% solvent B over 10–12 min and maintain 90% solvent B for 2 min to elute strongly retained compounds.
8. Re-equilibrate column by returning to initial conditions over 3 min and hold for 3–5 min.

9. Choose injection volume between 1 and 5 μL to prevent column overload and peak distortion.
10. Use electrospray ionization (ESI) in positive and/or negative mode depending on analyte chemical properties.
11. Set capillary voltage at 3–4 kV, source temperature at 350°C–450°C, desolvation gas flow at 600–800 L/h (nitrogen), cone gas flow: 50–100 L/h.
12. Employ acquisition using Multiple Reaction Monitoring (MRM) or Selected Reaction Monitoring (SRM) for targeted quantification of coumarins.
13. User needs to optimize precursor ion and product ion transitions, collision energies, cone voltages, and dwell times for each analyte.
14. Set the mass range to m/z 100–1000 (adjust based on target compounds).
15. Identify peaks by comparing retention time and characteristic MS/MS fragmentation patterns with known standards. Quantify compounds using internal standard calibration curves, correcting for matrix effects. Validate method for linearity, Limit of Detection (LOD),Limit of Quantification (LOQ), accuracy, precision, and recovery according to standard guidelines (e.g., ICH or FDA bioanalytical method validation).

4 Notes

1. Chamber saturation is essential for separation sample and result reproducibility.
2. For optimal resolution and visualization, use freshly prepared solvents and developing reagents.
3. Avoid overloading the spots as it can result in tailing and poor resolution.
4. Protect coumarin extracts from bright light to reduce degradation.
5. Store extracts at low temperature and preferably in amber vials.
6. Avoid long heating during concentration, as it may alter sensitive compounds.
7. Freshly prepared solvents and mobile phases help maintain uniform detection.
8. Keep samples covered or sealed during preparation to avoid evaporation loss.

9. Mobile phase pH and gradient conditions may need optimization based on sample complexity and coumarins of interest.
10. Maintain consistent column temperature for reproducible retention times.
11. Use fresh prepared, degased mobile phases to prevent noise in baseline.
12. Prepare calibration standards in the same solvent as samples for reliable quantification.
13. Optional sample preparation using solid-phase extraction (SPE) can improve sensitivity for complex samples.
14. Positive mode is commonly preferred for coumarins.
15. For metabolite profiling, use full scan and MS/MS fragmentation data acquisition.
16. Gradient profiles and MS parameters must be optimized during method development specific to your sample matrix and target compounds.
17. Establish strict quality control procedures with controls and blanks for robust data.
18. Use derivatization only when the native form of the coumarin does not give satisfactory detection or separation.
19. Perform the derivatization reaction after extraction and before HPLC or UPLC-MS injection.
20. Mix the extract with the derivatizing reagent and allow reaction for the recommended time.
21. Filter the derivatized sample if required and transfer it to analysis vials.

References

1. Harborne JB (1998) Phytochemical methods: a guide to modern techniques of plant analysis, 3rd edn. Springer
2. Sarker SD, Latif Z, Gray AI (2006) Natural products isolation. Methods in biotechnology. Humana Press
3. Houghton P, Raman A (1998) Laboratory handbook for the fractionation of natural extracts. Chapman and Hall
4. Martín-Esteban A (2013) Sample preparation for chromatographic analysis. Adv Chromatogr 51:215–240
5. Harrison M, Quillien V (2013) Natural product isolation, 2nd edn. Springer
6. Kovac-Besović EE, Durić K (2003) Thin layer chromatography-application in qualitative analysis on presence of coumarins and flavonoids in plant material. Bosn J Basic Med Sci 3(3):19–26. https://doi.org/10.17305/bjbms.2003.3523
7. Kandzia M, Pichler F (2000) Thin-layer chromatography of coumarins of medicinal and aromatic plants. J Chromatogr A 889(1–2):213–228. https://doi.org/10.1016/S0021-9673(00)88357-0
8. Krüger R, et al. (2020) High-performance thin-layer chromatography-fluorescence detection-mass spectrometry/effect-direct analysis (HPTLC-FLD-MS/EDA) for rapid coumarin analysis. Anal Bioanal Chem 219:121306. https://doi.org/10.1007/s00216-020-02741-6
9. Zhang Y, Wang Q, Zhang Z, Li X (2020) HPLC-DAD-MS profiling of coumarins in medicinal herbs. J Chromatogr B 1132:121899. https://doi.org/10.1016/j.jchromb.2019.121899

10. Perkowska A, Kowalska M, Nowak P, Lewandowska J (2019) Analysis of coumarins in plasma by UHPLC-ESI-QTOF-MS. J Pharm Biomed Anal 165:343–350. https://doi.org/10.1016/j.jpba.2019.04.042
11. Duan B, Chen L, Liu Y, Zhang F (2018) Optimized HPLC method for quantifying coumarins in plant extracts. Anal Methods 10(36):4352–4360. https://doi.org/10.1039/C8AY01022D
12. Tine Y, Zengin M, Mahomoodally M et al. (2017) A method for LC-MS/MS profiling of coumarins in Zanthoxylum fruit essential oils and extracts. J Chromatogr B 1044:33–43. https://doi.org/10.1016/j.jchromb.2016.11.028
13. Perkowska I, Hawrylak-Nowak M, Matysiak A et al (2021) Identification and quantification of coumarins by UHPLC-MS in Arabidopsis thaliana root exudates. Metabolites 11(3):148. https://doi.org/10.3390/metabo11030148
14. Wei L, Cai P, Yang L et al. (2016) An UPLC–MS/MS method for simultaneous quantitation of umbelliferone and daphnoretin in rat plasma: application to pharmacokinetic study. Biomed Chromatogr 30(3):361–367. https://doi.org/10.1002/bmc.3509
15. Dugrand A, Pellé L, Girodon C et al. (2013) Coumarin and furanocoumarin quantitation in Citrus peel via ultra-performance liquid chromatography coupled with mass spectrometry (UPLC-MS). J Agric Food Chem 61(43):10434–10442. https://doi.org/10.1021/jf402763t

Check for updates

Chapter 22

Tannins: Advanced Techniques for Tannin Profiling

Hetvi Naik and Natarajan Amaresan

Abstract

Tannins are complex polyphenolic compounds that are widely distributed in plants and have significant biological, nutritional, and industrial importance. Accurate profiling of tannin content, composition, and structure is essential for understanding their bioactivity and optimizing their application. Traditional spectrophotometric methods provide limited structural information, necessitating the development of advanced analytical techniques to overcome these limitations. Here, we describe comprehensive protocols for tannin profiling using multiple complementary methods: High-Performance Liquid Chromatography coupled with Mass Spectrometry (HPLC-MS/MS), matrix-assisted laser desorption/ionization time-of-flight mass spectrometry (MALDI-TOF MS), and depolymerization techniques, including thiolysis and phloroglucinolysis. These advanced methods enable the determination of tannin concentration, oligomer distribution, mean degree of polymerization (mDP), monomer composition, and structural characteristics of both condensed tannins (proanthocyanidins) and hydrolyzable tannins.

Keywords Tannins, Proanthocyanidins, HPLC–MS/MS, MALDI–TOF MS, Thiolysis

1 Introduction

Tannins are high molecular weight plant polyphenols classified into two major groups: hydrolysable tannins (gallotannins and ellagitannins) and condensed tannins (proanthocyanidins). These compounds exhibit diverse biological activities, including antioxidant, antimicrobial, and protein-binding properties, making them valuable in food, pharmaceutical, and agricultural applications [1]. Traditional tannin analysis methods, such as the Folin-Ciocalteu assay and butanol-HCl method, provide quantitative data but lack specificity and structural information. Advanced analytical techniques, including HPLC coupled with UV or MS detection, LC–MS, and MALDI–MS, have revolutionized tannin analysis by enabling detailed structural characterization [2]. Modern liquid chromatography-mass spectrometry methods can obtain two-dimensional tannin fingerprints to better visualize both tannin content and diversity in plants with just one 10-min analysis per sample [3].

Dhruti Amin et al. (eds.), *Phytoconstituent Analysis*, Springer Protocols Handbooks,
https://doi.org/10.1007/978-1-0716-5400-2_22,

For polymeric proanthocyanidins, chemical depolymerization methods coupled with HPLC separation, UV detection, and ESI–MS detection are useful analytical approaches [4]. These reactions include phloroglucinolysis, thiolysis, and thioglycolysis, which are SN1 reactions involving carbocation intermediates under strongly acidic conditions [5]. MALDI–TOF MS is highly suitable for the analysis of highly polydisperse and heterogeneous proanthocyanidins [6]. This chapter provides detailed protocols for comprehensive tannin profiling, including sample preparation, extraction, HPLC–MS/MS analysis, MALDI–TOF MS characterization, and depolymerization. The methods described enable the complete structural characterization of tannins from various plant sources.

2 Materials

Prepare all solutions using ultrapure water (prepared by purifying deionized water to attain a sensitivity of 18 MΩ-cm at 25°C) and analytical grade reagents. Prepare and store all reagents at room temperature (unless indicated otherwise). All waste disposal regulations should be diligently followed when disposing of waste materials.

2.1 Sample Collection and Preparation

1. Plant material (leaves, bark, seeds, or other tissues).
2. Liquid nitrogen for freezing.
3. Freeze dryer.
4. Laboratory mill or grinder.
5. Desiccator with indicating silica gel.
6. 0.5 mm sieve.

2.2 Extraction Reagents

1. Acetone: water (70:30, v/v): Mix 700 mL HPLC-grade acetone with 300 mL ultrapure water. Store at 4°C (*see* **Note 1**).
2. Methanol: water (50:50, v/v): Mix equal volumes of HPLC-grade methanol and ultrapure water. Store at 4°C.
3. Acetone: water:formic acid (70:29.5:0.5, v/v/v): Mix 700 mL acetone, 295 mL water, and 5 mL formic acid. Prepare fresh before use.
4. Ethyl acetate (HPLC grade)
5. Sephadex LH-20 resin
6. Ethanol (95%, HPLC grade)

2.3 HPLC–MS/MS Analysis

1. HPLC systemHPLC system: Agilent 1200 LC system or equivalent, equipped with degasser, binary pump, and autosampler maintained at 6°C.
2. Analytical column: Waters Acquity UPLC BEH C18 column (2.1 mm × 50 mm, 1.7 μm) or Agilent Eclipse XDB C18 column (150 × 4.6 mm, 5 μm).

3. Mobile phase A: Water with 0.1% formic acid.
4. Mobile phase B: Acetonitrile with 0.1% formic acid or methanol with 0.01% TFA.
5. Mass spectrometer: Triple quadrupole or Q-TOF instrument with ESI source.
6. Standards: (+)-catechin, (−)-epicatechin, gallic acid, ellagic acid, procyanidin B1, B2.

2.4 Thiolysis Reagents

1. Benzyl mercaptan (phenylmethane thiol) (*see* **Note 2**).
2. Methanol (HPLC grade).
3. Hydrochloric acid (37% analytical grade).
4. Acidified methanol: Add 3.3 mL of 37% HCl to 1 L of methanol to prepare 0.1 M HCl in methanol. Store at 4°C.
5. (+)-Catechin standard.
6. (−)-Epicatechin standard.
7. (−)-Epigallocatechin standard.

2.5 Phloroglucinolysis Reagents

1. Phloroglucinol (1,3,5-trihydroxybenzene).
2. Phloroglucinol reagent: Dissolve 50 g phloroglucinol in 1 L of acidified methanol (0.2 M HCl in methanol). Prepare fresh before use (*see* **Note 3**).
3. Ascorbic acid solution: 40 mg/mL in water. Prepare fresh daily.
4. Sodium acetate (anhydrous)
5. Sodium acetate solution: 200 mM sodium acetate in water

2.6 MALDI-TOF MS Analysis

1. MALDI-TOF mass spectrometer: Bruker Reflex III, Voyager DE-STM, or equivalent instrument.
2. Matrix solution: 2,5-Dihydroxybenzoic acid (DHB, 10 mg/mL in water or methanol: water 50:50, v/v) (*see* **Note 4**).
3. Alternative matrices: α-Cyano-4-hydroxycinnamic acid (CHCA) or sinapinic acid.
4. Cation-exchange resin: Amberlite IRP-64, equilibrated in deionized water.
5. MALDI target plate (stainless steel).
6. Calibration standards: Angiotensin II (1046.5 Da), bombesin (1619.8 Da), ACTH clip 18-39 (2465.2 Da).

2.7 Spectrophotometric Quantification

1. Folin-Ciocalteu reagent.
2. Sodium carbonate solution: 7.5% (w/v) in water.
3. Gallic acid standard.
4. *n*-Butanol (analytical grade).

5. Ferric reagent: Dissolve 2 g ferric ammonium sulfate in 100 mL of 2 M HCl. Store in dark bottle at 4°C.
6. Butanol-HCl reagent: Mix 950 mL *n*-butanol with 50 mL of 37% HCl. Prepare fresh.

2.8 Additional Equipment

1. Centrifuge capable of 10,000 × *g*.
2. Rotary evaporator.
3. 0.45 μm PTFE or nylon syringe filters.
4. Vortex mixer.
5. Water bath or heating block (up to 95°C).
6. Micropipettes (10–1000 μL).
7. HPLCHPLC vials and inserts.
8. Glass columns for Sephadex chromatography (2.5 × 50 cm).

3 Methods

Carry out all procedures at room temperature unless otherwise specified. Protect tannin extracts from light during storage and analysis to prevent oxidation.

3.1 Sample Preparation and Extraction

3.1.1 Sample Collection and Drying

1. Collect fresh plant material and immediately freeze in liquid nitrogen (*see* **Note 5**).
2. Transfer frozen samples to a freeze dryer and lyophilize for 48–72 h until completely dry.
3. Grind the dried material to a fine powder using a laboratory mill.
4. Pass the ground material through a 0.5 mm sieve.
5. Store the powdered samples in a desiccator at 4°C until extraction (*see* **Note 6**).

3.1.2 Tannin Extraction

1. Weigh 100 mg of ground plant material into a 15 mL centrifuge tube.
2. Add 5 mL of acetone: water (70:30, v/v) extraction solvent.
3. Vortex vigorously for 1 min.
4. Sonicate the mixture for 15 min at room temperature.
5. Centrifuge at 10,000 × *g* for 10 min at 4°C.
6. Carefully transfer the supernatant to a clean tube.
7. Repeat the extraction two more times with the pellet using fresh solvent.

8. Combine all three supernatants.
9. Store extracts at −20°C or proceed immediately to purification (*see* **Note 7**).

3.1.3 Sephadex LH-20 Purification

1. Prepare a Sephadex LH-20 column (2.5×50 cm) by swelling the resin in methanol overnight.
2. Pack the column and equilibrate with three column volumes of methanol.
3. Concentrate the combined extract to approximately 5 mL using a rotary evaporator at 35°C.
4. Load the concentrated extract onto the column.
5. Elute sequentially with:
   - two column volumes of methanol: water (50:50, v/v) to remove sugars and polar compounds.
   - three column volumes of acetone: water (70:30, v/v) to elute tannins.
   - two column volumes of methanol to elute remaining phenolics.
6. Collect fractions (20 mL each) and monitor at 280 nm (*see* **Note 8**).
7. Pool tannin-containing fractions (those eluting with acetone: water).
8. Remove solvent using rotary evaporation at 35°C.
9. Lyophilize to obtain purified tannin powder.
10. Store at −20°C in dark bottles.

3.2 HPLC–MS/MS Analysis

3.2.1 Sample Preparation for HPLC–MS/MS

1. Dissolve 1–5 mg of purified tannin extract in 1 mL methanol (*see* **Note 9**).
2. Filter through a 0.45 μm PTFE syringe filter into an HPLC vial.
3. Prepare a calibration curve using tannin standards at concentrations of 1, 5, 10, 25, 50, and 100 μg/mL.

3.2.2 HPLC-MS/MS Operating Conditions

1. Set autosampler temperature to 6°C.
2. Set column temperature to 25°C.
3. Program the following gradient elution (for UPLC BEH C18 column):
   - 0–2 min: 5% B.
   - 2–15 min: 5%–30% B (linear gradient).
   - 15–20 min: 30%–95% B (linear gradient).
   - 20–22 min: 95% B (isocratic).

- 22–23 min: 95%–5% B (linear gradient).
- 23–28 min: 5% B (re-equilibration).

4. Set flow rate to 0.3 mL/min for UPLC or 0.7 mL/min for conventional HPLC.
5. Set injection volume to 5–10 μL.
6. Monitor at 280 nm for general phenolic detection and 520 nm for anthocyanins.

3.2.3 Mass Spectrometer Settings

1. Configure ESI source in negative ionization mode (*see* **Note 10**).
2. Set source parameters:
   - Nebulizer gask (nitrogen): 30–40 psi.
   - Dry gas flow: 10.0 L/min.
   - Dry temperature: 320–350°C.
   - Capillary voltage: −3500 to −4500 V.
   - Fragmentation voltage: 100–150 V.
3. For Multiple Reaction Monitoring (MRM), set up specific transitions for target compounds based on $[M-H]^-$ ions.
4. Scan range for full scan mode: m/z 100–1800.
5. For MS/MS analysis, use collision energy of 20–40 eV with argon as collision gas.

3.2.4 Data Analysis

1. Identify tannin compounds by comparing retention times and mass spectra with authentic standards.
2. Confirm identity using MS/MS fragmentation patterns.
3. Quantify compounds using peak area integration and external calibration curves..
4. Calculate mDP from oligomer distribution using the formula: mDP = (Sum of all monomers)/(Sum of terminal units).

3.3 Thiolysis Analysis

3.3.1 Thiolysis Reaction

1. Weigh 5 mg of purified tannin sample into a 2 mL screw-cap vial with PTFE-lined cap (*see* **Note 11**).
2. Add 1 mL of acidified methanol containing 5% (v/v) benzyl mercaptan (50 μL benzyl mercaptan in 1 mL of 0.1 M HCl in methanol).
3. Seal the vial tightly and vortex briefly.
4. Incubate at 50°C for 30 min in a heating block or water bath (*see* **Note 12**).
5. Cool the vial immediately in an ice bath to stop the reaction.

6. Add 50 μL of ascorbic acid solution (40 mg/mL) to prevent oxidation.
7. Centrifuge at 10,000 × *g* for 5 min.
8. Transfer supernatant to an HPLC vial.
9. Analyze immediately by HPLC or store at −20°C for up to 24 h (*see* **Note 13**).

3.3.2 HPLC Analysis of Thiolysis Products

1. Use the same HPLC column and mobile phases as described in Sect. 3.2.2.
2. Modify the gradient program for better separation of thiolysis products:
   - 0–5 min: 5% B.
   - 5–35 min: 5%–50% B (linear gradient).
   - 35–40 min: 50%–95% B (linear gradient).
   - 40–42 min: 95% B (isocratic).
   - 42–43 min: 95%–5% B.
   - 43–50 min: 5% B (re-equilibration).
3. Monitor at 280 nm for flavan-3-ol detection.
4. Set flow rate to 0.7 mL/min.
5. Injection volume: 10 μL.

3.3.3 Identification and Quantification

1. Identify free flavan-3-ols (terminal units): catechin, epicatechin, gallocatechin, epigallocatechin (retention times 12–18 min).
2. Identify benzyl thioether derivatives (extension units): 4β-benzylthio-catechin, 4β-benzylthio-epicatechin, etc. (retention times 20–30 min).
3. Prepare calibration curves using authentic standards.
4. Calculate the following parameters:

I. $$\textit{Mean degree of polymerization } (mDP) = \frac{(\textit{Terminal units} + \textit{Extension units})}{(\textit{Terminal units})}$$

where, Terminal units = concentration of free flavan-3-ols
Extension units = concentration of benzyl thioether derivatives

II. $$\textit{Procyanidin} : \textit{Prodelphinidin ratio}(PC : PD) \frac{(\textit{catechin} + \textit{epicatechin derivatives})}{(\textit{gallocatechin} + \textit{epigallocatechin derivatives})}$$

III. $$\textit{cis} : \textit{trans ratio} = \frac{\textit{epicatechin derivatives}}{\textit{catechin derivatives}}$$

3.4 Phloroglucinolysis Analysis

3.4.1 Phloroglucinolysis Reaction

1. Weigh 2 mg of purified tannin sample into a 2 mL screw-cap vial.
2. Add 1 mL of phloroglucinol reagent (50 g/L phloroglucinol in 0.2 M HCl in methanol).
3. Add 20 μL of ascorbic acid solution (40 mg/mL).
4. Seal the vial tightly and vortex.
5. Incubate at 50°C for 20 min.
6. Stop the reaction by cooling in ice bath and adding 200 μL of sodium acetate solution (200 mM).
7. Centrifuge at 10,000 × *g* for 5 min.
8. Transfer supernatant to an HPLC vial.
9. Analyze immediately or store at −20°C for up to 24 h.

3.4.2 HPLC Analysis of Phloroglucinolysis Products

1. Use the same chromatographic conditions as described in Sect. 3.3.2.
2. Identify peaks:
   - Free flavan-3-ols (terminal units): catechin, epicatechin at retention times 12–18 min.
   - Phloroglucinol adducts (extension units): 4β-phloroglucinol-catechin, 4β-phloroglucinol-epicatechin at retention times 25–35 min.
3. Quantify using external calibration with flavan-3-ol standards.
4. Calculate mDP and composition ratios as described in Sect. 3.3.3.

3.5 MALDI–TOF MS Analysis

3.5.1 Sample Preparation for MALDI-TOF MS

1. Prepare matrix solution: Dissolve 10 mg of 2,5-dihydroxybenzoic acid (DHB) in 1 mL of water or methanol: water (50:50, v/v).
2. Prepare tannin sample solution: Dissolve 1–5 mg of purified tannin extract in 1 mL of water or methanol to obtain 1–5 mg/mL concentration.
3. For improved ionization, add 10 μL of 1 M NaCl solution to 1 mL of sample.
4. Deionize the mixture using Amberlite IRP-64 cation-exchange resin (*see* **Note 14**):
   - Add a small amount of equilibrated resin to the sample.
   - Mix gently for 2 min.
   - Filter to remove resin.
   - Repeat deionization two to three times.

3.5.2 Sample Application and Analysis

1. Mix sample solution with matrix solution at 1:3 (v/v) ratio on a clean surface.
2. Apply 1 μL of the sample-matrix mixture to the MALDI target plate using dried-droplet method (*see* **Note 15**).
3. Allow to air-dry at room temperature (typically 10–15 min).
4. Load the target plate into the MALDI-TOF MS instrument.
5. Set instrument parameters:
   - Mode: Positive reflectron mode.
   - Laser wavelength: 337 nm (nitrogen laser).
   - Accelerating voltage: 20.0 kV.
   - Reflectron voltage: 23.0 kV.
   - Laser power: 10%–25% above ionization threshold.
   - Number of shots: 100–200 per spectrum.
   - Mass range: m/z 200–5000.
6. Calibrate using external peptide standards.
7. Acquire spectra from multiple positions on each sample spot.

3.5.3 Data Interpretation

1. Identify oligomer series based on mass increments:
   - Procyanidins: 288 Da repeat units (catechin/epicatechin).
   - Prodelphinidins: 304 Da repeat units (gallocatechin/epigallocatechin).
   - Mixed oligomers: variable mass increments.
   - Galloylated derivatives: +152 Da per galloyl group.
2. Determine the degree of polymerization by dividing molecular mass by monomer unit mass.
3. Identify sodium adducts [M + Na]+ and potassium adducts [M + K]+.
4. Assess oligomer distribution by measuring relative intensities of each oligomer series.
5. Calculate number-average degree of polymerization from the distribution of oligomers.

3.6 Spectrophotometric Quantification Methods

3.6.1 Total Phenolics by Folin-Ciocalteu Method

1. Dilute tannin extract to an appropriate concentration (typically 100–500 μg/mL).
2. Add 100 μL of diluted extract to a test tube.

3. Add 500 μL of Folin-Ciocalteu reagent (diluted 1:10 with water).
4. Mix and incubate for 5 min at room temperature.
5. Add 400 μL of sodium carbonate solution (7.5%, w/v).
6. Mix thoroughly and incubate in darkness for 30 min.
7. Measure absorbance at 765 nm against a reagent blank.
8. Calculate total phenolic content using a gallic acid standard curve.
9. Express results as mg gallic acid equivalents per g dry weight.

3.6.2 Condensed Tannins by Butanol–HCl Method

1. Prepare butanol–HCl reagent: Mix 950 mL n-butanol with 50 mL of 37% HCl. Prepare fresh.
2. Add 0.5 mL of diluted tannin extract (100–500 μg/mL) to a screw-cap test tube.
3. Add 3 mL of butanol–HCl reagent and 0.1 mL of ferric reagent.
4. Cap tightly and heat at 95°C for 60 min in a heating block.
5. Cool to room temperature.
6. Measure absorbance at 550 nm against a reagent blank.
7. Calculate condensed tannin content using a procyanidin B1 standard curve.
8. Express results as mg procyanidin equivalents per g dry weight.

4 Notes

1. Acetone:water (70:30, v/v) is the most efficient solvent for extracting condensed tannins. Alternatively, methanol:water (50:50, v/v) can be used.
2. Benzyl mercaptan has a strong, unpleasant odor. Always work in a well-ventilated fume hood and store at 4°C in a tightly sealed dark container.
3. Phloroglucinol reagent oxidizes rapidly—prepare fresh before each use. Discard if it turns brown.
4. DHB is the most commonly used MALDI matrix for proanthocyanidins. For hydrolyzable tannins, CHCA or sinapinic acid may provide better results.
5. Immediate freezing in liquid nitrogen is essential to stop polyphenol oxidase activity. If unavailable, freeze-dry immediately or dry at 40°C.

6. Ground samples stored in a desiccator at 4°C remain stable for several months. For best results, extract within 1 month of preparation.
7. Tannin extracts oxidize during storage. Store at −20°C and add ascorbic acid (1 mg/mL) as antioxidant for extended storage.
8. Sephadex LH-20 effectively separates tannins from other phenolics, proteins, and carbohydrates. Monitor fractions by UV spectroscopy at 280 nm.
9. Start with 1 mg/mL sample concentration for HPLC analysis. High concentrations cause column fouling; too dilute samples reduce sensitivity.
10. Negative ionization mode is generally preferred for tannin analysis as phenolics readily form $[M-H]^-$ ions.
11. Use PTFE-lined screw-cap glass vials for thiolysis reactions. Regular plastic caps may dissolve in acidic methanol.
12. Optimal thiolysis conditions are 50°C for 30 min. Higher temperatures may cause epimerization or artifact formation.
13. Analyze thiolysis and phloroglucinolysis products immediately. If storage is necessary, keep at −20°C for no more than 24 h.
14. Deionization using cation-exchange resin removes excess salts that suppress MALDI ionization. Multiple cycles improve results.
15. Apply 0.5–1 μL of sample-matrix mixture to MALDI target plate using dried-droplet method. Allow to air-dry completely before analysis.

References

1. Serrano J, Puupponen-Pimiä R, Dauer A, Aura A-M, Saura-Calixto F (2009) Tannins: current knowledge of food sources, intake, bioavailability and biological effects. Mol Nutr Food Res 53:S310–S329
2. Reed JD, Krueger CG, Vestling MM (2005) MALDI-TOF mass spectrometry of oligomeric food polyphenols. Phytochemistry 66:2248–2263
3. Monagas M, Quintanilla-López JE, Gómez-Cordovés C, Bartolomé B, Lebrón-Aguilar R (2010) MALDI-TOF MS analysis of plant proanthocyanidins. J Pharm Biomed Anal 51:358–372
4. Aron PM, Kennedy JA (2008) Flavan-3-ols: nature, occurrence and biological activity. Mol Nutr Food Res 52:79–104
5. Prior RL, Fan E, Ji H, Howell A, Nio C, Payne MJ, Reed J (2010) Multi-laboratory validation of a standard method for quantifying proanthocyanidins in cranberry powders. J Sci Food Agric 90:1473–1478
6. Karonen M, Loponen J, Ossipov V, Pihlaja K (2004) Analysis of procyanidins in pine bark with reversed-phase and normal-phase high-performance liquid chromatography–electrospray ionization mass spectrometry. Anal Chim Acta 522:105–112

Part III

Molecular Methods for Phytoconstituent Analysis

Part III

Check for updates

Chapter 23

Terpenoids: Genetic Engineering of Terpenoid Pathway

Amit Gamit and Dhruti Amin

Abstract

Terpenoids are the most structurally diverse and abundant group of natural products, synthesized through the mevalonate (MVA) and methylerythritol phosphate (MEP) pathways. Their importance in pharmaceuticals, fragrances, pigments, and biofuels arises from their remarkable biological activity and structural complexity. Genetic engineering of terpenoid biosynthetic pathways enables enhanced production and diversification of structures. This chapter outlines experimental methods for identifying, cloning, and heterologously expressing key terpenoid biosynthetic genes, such as 3-hydroxy-3-methylglutaryl-CoA reductase (HMGR) and isopentenyl diphosphate isomerase (IDI). The described protocols cover genomic DNA isolation from plants or microbes, PCR amplification of target genes, vector construction, and transformation into *Escherichia coli* or *Saccharomyces cerevisiae*. Bioinformatic tools, such as KEGG, MEV, and antiSMASH, are used for pathway annotation and metabolic flux analysis. Overall, this integrated approach supports rational engineering of terpenoid biosynthesis to enhance yields, generate novel analogues, and advance industrial applications.

Keywords Terpenoids, Metabolic engineering, MVA pathway, MEP pathway, Gene cloning

1 Introduction

Isoprene Terpenoids, also known as isoprenoids, represent a wide group of naturally occurring secondary metabolites, the precursors of which are isopentenyl diphosphate (IPP) and dimethylallyl diphosphate (DMAPP), which are 5-carbon compounds. Their biosynthesis follows two convergent and rather independent pathways: the mevalonate (MVA) pathway that is active in both the eukaryote and archaeal domains and the methylerythritol phosphate (MEP) pathway active in bacteria and the plastids of land plants [1]. Terpenoids are categorized as monoterpenes (C10), sesquiterpenes (C15), diterpenes (C20), triterpenes (C30), and tetraterpenes (C40) based on the number of units of isoprene added. They serve the functions of pigments (carotenoids), phytohormones (gibberellins), and defensive agents (phytoalexins). As an industry, terpenoids find application in pharmacological preparations (e.g., artemisinin, paclitaxel), flavoring

Dhruti Amin et al. (eds.), *Phytoconstituent Analysis*, Springer Protocols Handbooks,
https://doi.org/10.1007/978-1-0716-5400-2_23,

agents (e.g., menthol), and as biomers to produce biofuels (e.g., farnesene) [2].

Synthesis of terpenoids is based upon the successive condensation of IPP and DMAPP units, which are catalyzed by the prenyltransferases to yield geranyl diphosphate (GPP), farnesyl diphosphate (FPP), and geranylgeranyl diphosphate (GGPP). These intermediates can be further cyclized and functionalized in different ways by terpene synthases (TPSs) and cytochrome P450 monooxygenases into a broad range of structurally diverse terpenoid scaffolds [3].

The recent developments in synthetic biology and metabolic engineering have allowed systematic re-engineering of terpenoid pathways to achieve better production [Fig. 1] [4]. Among the strategies are overexpression of rate-limiting enzymes in the mevalonate pathway as well as HMG-CoA reductase; introduction of heterologous genes of terpene synthases, and deletion of competitor pathways (i.e., fatty acid biosynthesis).

Microbial systems, such as the case of the plant terpenoid biosynthesis through modular cloning, have shown high-yield biosynthesis of commercially desirable terpenes in microbial systems: *Escherichia coli* and organisms, such as *S. cerevisiae* [5]. As an example, Artemisinin precursor pathway-engineered *E. coli* was able to generate amorpha-11-diene, which serves as a key precursor in the synthesis of antimalarial drugs [2]. KEGG, MEV, and PlantCyc can be used as bioinformatics to map terpenoid pathways, and the analysis of metabolic flux can be used to optimize the supply of precursors and cofactor balancing [6]. With genetic control of enzymes in the

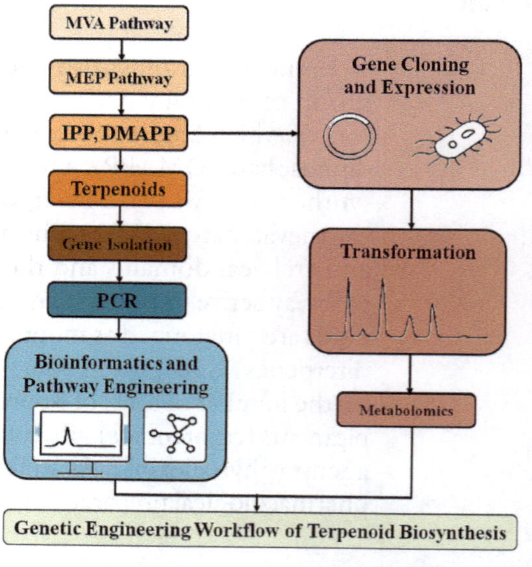

Fig. 1 Workflow for genetic engineering of the terpenoid biosynthesis pathway, including gene cloning and transformation, bioinformatics analysis, and modification of the pathway via genetic engineering

pathway, engineering of promoters and control of cofactor, terpenoid biosynthesis can be rationally engineered to become a production process at industrial scale.

2 Materials

2.1 Isolation of Genomic DNA

1. Fresh plant tissue (e.g., *Artemisia annua*, *Mentha spicata*).
2. Liquid nitrogen and mortar-pestle for tissue homogenization.
3. CTAB extraction buffer (2% CTAB, 1.4 M NaCl, 20 mM EDTA, 100 mM Tris–HCl, pH 8.0).
4. Lysis buffer (50 mM Tris–HCl, 1 mM EDTA, 100 mM SDS).
5. Phenol:chloroform:isoamyl alcohol (25:24:1).
6. Isopropanol and 70% ethanol.
7. TE buffer (10 mM Tris–HCl, 1 mM EDTA).
8. Centrifuge and sterile tubes.
9. Water bath (65°C).
10. Agarose gel electrophoresis setup.
11. SYBR Safe or ethidium bromide stain.
12. Micropipettes and sterile tips.

2.2 Amplification of Terpenoid Pathway Genes

1. Taq DNA polymerase (1 U per reaction).
2. 10× PCR buffer.
3. $MgCl_2$ (2 mM).
4. dNTP mix (200 μM each).
5. Forward and reverse primers for genes.
6. Nuclease-free water.
7. Template DNA (~50 ng).
8. Gel electrophoresis setup for PCR product verification.

2.3 Cloning and Vector Construction

1. Restriction enzymes: EcoRI and XhoI.
2. Plasmid vectors (e.g., pET28a, pYES2).
3. T4 DNA ligase and 1× ligation buffer.
4. Gel extraction kit (e.g., QIAquick, Qiagen).
5. Competent *E. coli* DH5α cells.
6. LB agar plates containing ampicillin (100 μg/mL).
7. Heat block or water bath (42°C).
8. Colony PCR reagents and primers.
9. Sterile microcentrifuge tubes.
10. Sanger sequencing facility.

2.4 Heterologous Expression

1. Expression host strains [*E. coli* BL21(DE3), *S. cerevisiae* INVSc1].
2. Induction agent: IPTG (1 mM).
3. LB or YPD growth medium.
4. Spectrophotometer for OD_{600} measurement.
5. Centrifuge and cell lysis buffer.
6. SDS-PAGE setup and reagents (12% resolving gel).
7. Protein ladder and Coomassie stain.

2.5 Metabolite Extraction and GC–MS Analysis

1. Organic solvents: *n*-hexane and ethyl acetate (1:1, v/v).
2. Nitrogen evaporator or gentle N_2 gas stream setup.
3. GC–MS instrument with capillary column (30 m × 0.25 mm × 0.25 μm).
4. Analytical-grade hexane for sample reconstitution.
5. Microliter syringe (10 μL).
6. Authentic terpenoid standards for comparison.
7. Temperature control and data acquisition software.

2.6 Bioinformatics and Pathway Engineering

1. Computer with high-speed internet.
2. Software/tools:
   a. BLASTx (NCBI).
   b. KEGG Pathway Mapper.
   c. Pfam and InterPro databases.
   d. MEV (Metabolic Explorer Viewer) or COBRA Toolbox.
   e. CellDesigner or Pathway Tools.
3. Gene sequence files (FASTA format).

2.7 Pathway Optimization and Validation

1. Expression plasmids with strong promoters (T7, GAL1).
2. Cytochrome P450 expression constructs and redox partner genes.
3. Co-transformation reagents for *E. coli* or *S. cerevisiae*.
4. GC–MS and LC–MS systems for metabolite profiling.
5. Internal standards for quantification.
6. Data analysis software for yield and metabolomic profiling.

3 Methods

3.1 Isolation of Genomic DNA [1]

1. Use fresh foliage [e.g., *Artemisia annua*, *Mentha spicata*) *Escherichia coli* engineered to express the mevalonate (MVA) or methylerythritol phosphate (MEP) pathway] as sources.
2. Homogenize 100 mg of freshly harvested leaves in liquid nitrogen and extract using 700 μL CTAB buffer (2% CTAB, 1.4 M NaCl, 20 mM EDTA, 100 mM Tris–HCl, pH 8.0).
3. Incubate the suspension at 65°C for 30 min, extract DNA using phenol:chloroform:isoamyl alcohol (25:24:1), and centrifuge at 12,000 rpm for 10 min.
4. Precipitate DNA with isopropanol, wash with 70% ethanol, air-dry the pellet, and dissolve in TE buffer (10 mM Tris–HCl, 1 mM EDTA).
5. Assess the integrity of isolated DNA by 1% agarose gel electrophoresis stained with SYBR Safe, following the method of [1].

3.2 Amplification of Terpenoid Pathway Genes [2, 6, 7]

Amplify genes as in Table 1 that encode key enzymes, such as HMGR, DXS (1-deoxy-D xylulose 5-phosphate synthase), IDI (isopentenyl diphosphate isomerase), GPPS (geranyl diphosphate synthase), and TPS (terpene synthase), using polymerase chain reaction (PCR).

1. Perform PCR amplification using standard 50 μL reaction mixtures which include 1× PCR buffer, 2 mM $MgCl_2$, and 1 U Taq DNA polymerase; 200 μM dNTPs; 0.2 μM forward and reverse primers; and make up volume with PCR-grade nuclease-free water.
2. Add 2 μL of template DNA (≈50 ng) extracted from the protist samples (*see* **Note 1**).
3. PCR settings include: initial denaturation at 95°C for 5 min follow by denaturation at 94°C for 30 s, annealing at 55°C for 45 s, and extension at 72°C for 60 s/kb with final extension at 72°C for 10 min for 35 amplification cycles.

Table 1
List of PCR primers

Gene	Forward primer (5′–3′)	Reverse primer (5′–3′)	Reference
HMGR	ATGGCTTCTGCTGGAGTAC	TCAGTCCAGTTGGTGAGGA	[7]
DXS	GCTGCTATGGCTTCAGTG	TTAGCCCATCCAGACCTC	[6]
TPS	ATGGCAGTGGAACCGTGT	TTAGCGTCCACGTTGATG	[2]

3.3 Cloning and Vector Construction [8] (see Note 2)

1. Digest the PCR-amplified product and plasmid vector (e.g., pET28a or pYES2) with restriction endonucleases *EcoRI* and *XhoI*.
2. Purify the digested DNA fragments using a gel extraction kit.
3. Set up the ligation reaction using T4 DNA ligase (1× reaction mix) and incubate at 16°C overnight.
4. Transform competent *Escherichia coli* DH5α cells by heat shock at 42°C for 45 s.
5. Plate the transformed cells on Luria–Bertani (LB) agar containing ampicillin (100 μg/mL) and incubate overnight at 37°C.
6. Screen putative transformants using colony PCR or restriction digestion analysis, following the method of [8].

3.4 Heterologous Expression [5]

1. Transform the selected expression host, such as *Escherichia coli* BL21(DE3) or *Saccharomyces cerevisiae* INVSc1, using the standard transformation protocol (chemical or electroporation method) depending on the host system.
2. Inoculate a single confirmed colony into LB or YPD medium containing the appropriate antibiotic and incubate at 37°C with shaking until the culture reaches an optical density (OD_{600}) of approximately 0.6.
3. Induce protein expression by adding 1 mM IPTG and continue incubation at 18°C for 16 h to promote proper protein folding and solubility.
4. Harvest the induced cells by centrifugation at 10,000 rpm for 10 min at 4°C. Resuspend the cell pellet in lysis buffer (e.g., 50 mM Tris–HCl, pH 8.0; 150 mM NaCl; 1 mM PMSF) and lyse the cells using sonication or enzymatic lysis.
5. Clarify the lysate by centrifugation and analyze the soluble protein fraction using 12% SDS–polyacrylamide gel electrophoresis (SDS–PAGE). Stain the gel with Coomassie Brilliant Blue to visualize protein bands.
6. Confirm the presence of the expected protein band corresponding to the molecular weight of the target enzyme, as reported by [5], and proceed with purification or further enzymatic activity assays as required.

3.5 Metabolite Extraction and GC–MS Analysis [4]

1. Extract the culture broth containing expressed terpenoid metabolites using a two-solvent system consisting of *n*-hexane and ethyl acetate in a 1:1 volumetric ratio. Mix thoroughly and allow the phases to separate, then collect the organic phase containing the metabolites.

2. Evaporate the organic solvent under a gentle stream of nitrogen gas until dryness. Dissolve the resulting residue in 1 mL of *n*-hexane to obtain a concentrated extract suitable for analysis.
3. Inject 1 μL of the prepared extract into a gas chromatograph–mass spectrometer (GC–MS) equipped with a capillary column (30 m × 0.25 mm × 0.25 μm). Ensure that the injector is operated in splitless mode for optimal detection of low-concentration metabolites.
4. Set the GC temperature program as follows: initial temperature 70°C (hold for 2 min), increase at a rate of 10°C per min to 280°C, and hold at 280°C for 10 min. Operate the MS in electron ionization mode for compound identification.
5. Identify terpenoid compounds based on their retention times and mass fragmentation patterns by comparing spectra with authentic standards and reference libraries [4]. Quantification can be performed using calibration curves derived from known standards.

3.6 Bioinformatics and Pathway Engineering [9, 10]

1. Annotate and verify cloned gene sequences using BLASTx (NCBI) to identify homologous proteins and confirm gene identity. Map the annotated genes onto the KEGG Pathway Mapper [9] to visualize their roles in the terpenoid biosynthetic network.
2. Predict and characterize functional enzyme domains by analyzing the sequences with Pfam and InterPro databases [10]. This step helps identify key catalytic and binding regions within terpenoid biosynthetic enzymes.
3. Perform metabolic flux balance analysis (FBA) using MEV (Metabolic Explorer Viewer) or the COBRA Toolbox to assess the flow of metabolites through the engineered pathway. Use this data to evaluate pathway efficiency and metabolic bottlenecks.
4. Identify rate-limiting steps within the terpenoid biosynthetic pathway and design strategies to optimize these points, such as gene overexpression, promoter modification, or knockout of competing pathways.
5. Construct a comprehensive pathway model using software, such as CellDesigner or Pathway Tools [11], to visualize enzyme interactions, regulatory nodes, and overall metabolic architecture for targeted pathway optimization.

3.7 Pathway Optimization and Validation [6, 11]

Pathway optimization and validation are crucial steps in metabolic engineering to enhance the efficiency of biosynthetic routes and maximize product yield. These processes involve fine-tuning genetic expression, enzyme activity, and cofactor balance to ensure efficient metabolite flow through the desired pathway. The optimized pathway is then validated through analytical and comparative approaches to confirm its functional performance and product diversity.

1. Promoter-Based Overexpression: Pathway optimization is achieved by employing strong and tightly regulated promoters, such as T7 and GAL1, to drive the overexpression of key bottleneck enzymes, including 3-hydroxy-3-methylglutaryl-CoA reductase (HMGR) and 1-deoxy-D-xylulose-5-phosphate synthase (DXS). These promoters enhance transcriptional activity and improve overall metabolic flux through the terpenoid biosynthesis pathway.
2. Gene Co-Transfer and Chromosomal Integration: The target genes are introduced via co-transformation using both plasmid-based expression systems and multiple chromosomal integration strategies to ensure stable and high-level expression. This dual approach enhances genetic stability and supports sustained metabolite production during prolonged cultivation.
3. Co-Expression with Redox Partners: To facilitate efficient oxidative transformations, the respective enzymes are co-expressed with compatible redox partners, such as cytochrome P450 monooxygenases and their electron transfer proteins. This setup ensures proper folding, electron transfer, and catalytic activity of oxygenase-dependent reactions.
4. Yield Estimation and Comparative Analysis: The product yield is quantitatively analyzed using Gas Chromatography–Mass Spectrometry (GC–MS). The experimental strain is compared with a control strain lacking the optimized pathway to assess enhancement in metabolite production and confirm successful pathway modulation.
5. Metabolomic Profiling and Product Validation: Comprehensive metabolomic profiling is conducted using Liquid Chromatography–Mass Spectrometry (LC–MS) to verify the diversity and identity of the synthesized compounds. This analytical validation follows the approach described by [6], confirming pathway efficiency and product variation resulting from genetic optimization.

4 Notes

1. Template Purity: DNA templates with A260/A280 absorbance ratio of 1.8–2.0 are used to obtain optimum efficiency of PCR.
2. Verify the accuracy of the insert sequence through Sanger sequencing before proceeding to expression studies.

References

1. Vranová E, Coman D, Gruissem W (2013) Network analysis of the MVA and MEP pathways for isoprenoid synthesis. Annu Rev Plant Biol 64:665–700. https://doi.org/10.1146/annurev-arplant-050312-120116
2. Martin VJJ, Pitera DJ, Withers ST, Newman JD, Keasling JD (2003) Engineering a mevalonate pathway in *Escherichia coli* for production of terpenoids. Nat Biotechnol 21(7):796–802. https://doi.org/10.1038/nbt833
3. Christianson DW (2017) Structural and chemical biology of terpenoid cyclases. Chem Rev 117(17):11570–11648. https://doi.org/10.1021/acs.chemrev.7b00287
4. Peralta-Yahya PP, Ouellet M, Chan R, Mukhopadhyay A, Keasling JD, Lee TS (2011) Identification and microbial production of a terpene-based advanced biofuel. Nat Commun 2:483. https://doi.org/10.1038/ncomms1494
5. Albertsen L, Chen Y, Bach LS, Rattleff S, Maury J, Brix S, Nielsen J (2011) Diversion of flux toward sesquiterpene production in *Saccharomyces cerevisiae* through metabolic engineering. Appl Environ Microbiol 77(3):1033–1040. https://doi.org/10.1128/AEM.01137-10
6. Vickers CE, Williams TCR, Peng B (2017) Recent advances in synthetic biology for engineering isoprenoid production in microorganisms. Curr Opin Chem Biol 40:47–56. https://doi.org/10.1016/j.cbpa.2017.05.018
7. Kirby J, Keasling JD (2009) Biosynthesis of plant isoprenoids: perspectives for microbial engineering. Annu Rev Plant Biol 60:335–355. https://doi.org/10.1146/annurev.arplant.043008.091955
8. Keasling JD (2010) Manufacturing molecules through metabolic engineering. Science 330(6009):1355–1358. https://doi.org/10.1126/science.1193990
9. Kanehisa M, Sato Y, Kawashima M, Furumichi M, Tanabe M (2023) KEGG as a reference resource for gene and protein annotation. Nucleic Acids Res 51(D1):D587–D592. https://doi.org/10.1093/nar/gkac1072
10. Marchler-Bauer A, Derbyshire MK, Gonzales NR, Lu S, Chitsaz F, Geer LY, Bryant SH (2017) CDD/SPARCLE: functional classification of proteins via subfamily domain architectures. Nucleic Acids Res 45(D1):D200–D203. https://doi.org/10.1093/nar/gkw1129
11. Nielsen J, Keasling JD (2016) Engineering cellular metabolism. Cell 164(6):1185–1197. https://doi.org/10.1016/j.cell.2016.02.004

References

Check for updates

Chapter 24

Glycosides: Gene Expression Studies for Glycosides

Rukhsar Ansari, Priyang Kanojiya, Anjali Desai and Sanket Ray

Abstract

The chapter outlines a comprehensive experimental procedure for high-grade RNA extraction, transcriptome profiling, and gene expression analysis of the biosynthetic gene encoding the glycosides in plants. Glycosides are important secondary metabolites that play a pivotal role in the defense, signaling, and stress-responsive processes in plants. Owing to the biochemical diversity of polysaccharide-rich plant tissues containing polyphenols and secondary metabolites, standard RNA isolation procedures do not successfully provide high-quality RNA for subsequent molecular uses. We present herein several optimized RNA extraction methods, including modified SDS-LiCI, Triton X-100 lysis, SDS-based, CTAB-TRIzol, hot borate, and other tissue-specific adjustments, that should overcome these difficulties and provide reproducible, high-purity RNA from a variety of plant materials. The chapter also outlines the step-by-step workflow in the quality evaluation of RNA, the building up of RNA-Seq libraries, next-generation sequencing, preprocessing of the data, assembly of the transcriptomes, gene expression quantification, and differential expression analysis. The identification and transcriptional profiling of UDP-glycosyltransferases (UGTs) and other key enzymes in the biosynthesis of glycosides are emphasized. The procedures in the article provide a platform for studying the molecular regulation of the metabolism of the glycosides and can also be adapted for functional genomics in model and non-model plant species.

Keywords Glycosides, RNA extraction, UDP-glycosyltransferases, Gene expression, Transcriptome analysis

1 Introduction

1.1 Glycosides as Principal Plant Metabolites

Plants produce glycosides as secondary metabolites, which are, in most instances, important contributors to the plants' pharmacological activities, in signal communication, and in defense mechanisms [1]. Glycosides are molecules with a sugar moiety (glycone) attached to a non-sugar organic compound (aglycone) [2, 3]. The major types of plant glycosides, i.e., flavonoids, saponins, cyanogenic glycosides, and cardiac glycosides, are responsible for a large number of vital biological activities, which are described below per type:

Dhruti Amin et al. (eds.), *Phytoconstituent Analysis*, Springer Protocols Handbooks,
https://doi.org/10.1007/978-1-0716-5400-2_24,

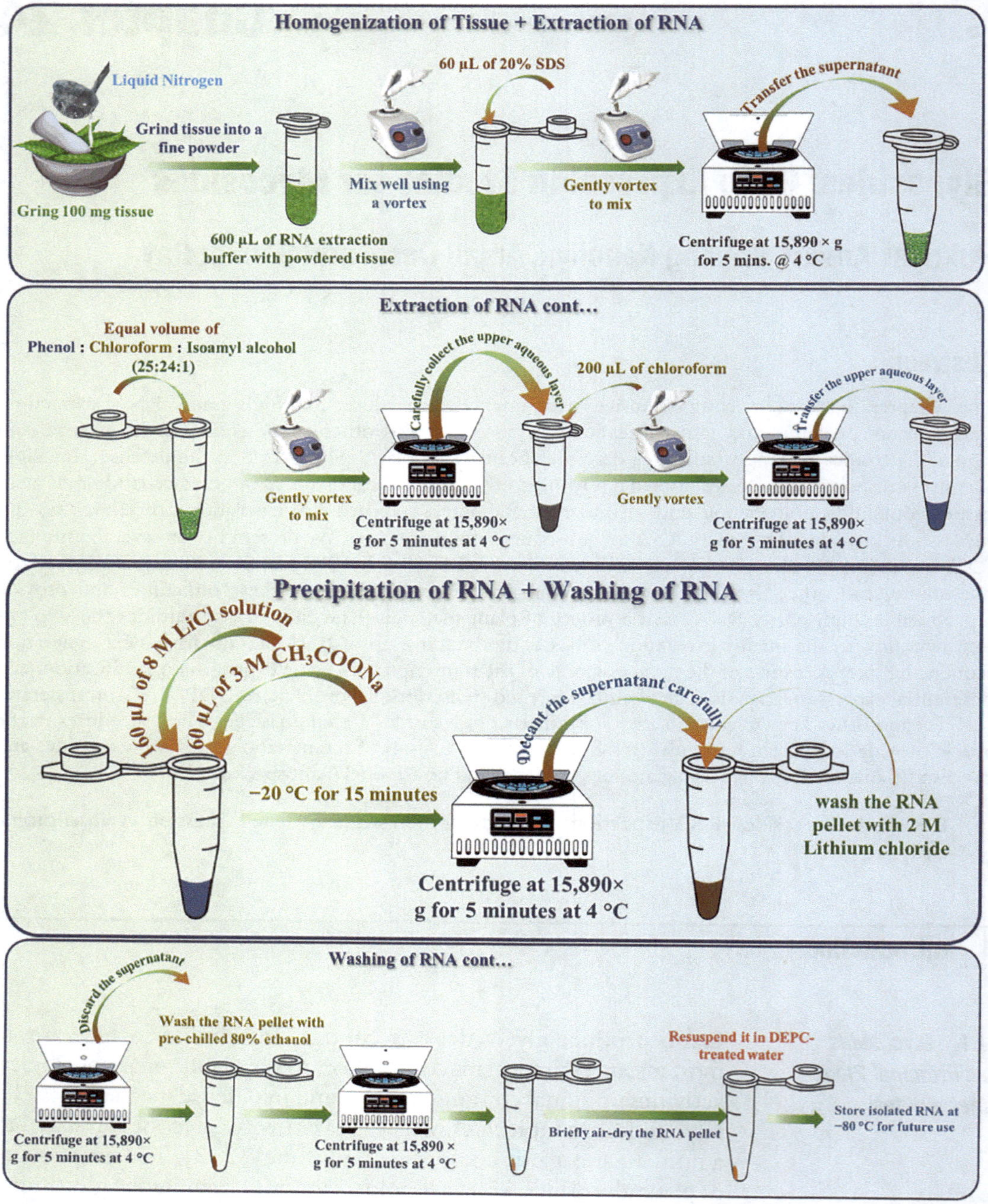

Fig. 1 Schematic diagram of total RNA extraction from starch-rich seeds using the modified SDS-LiCl method

- Flavonoid gycosides: Act as natural sunscreens by absorbing harmful UV radiation [1], function as potent antioxidants [2], and are essential in plant–microbe interactions, such as root nodulation signaling in legumes [3].

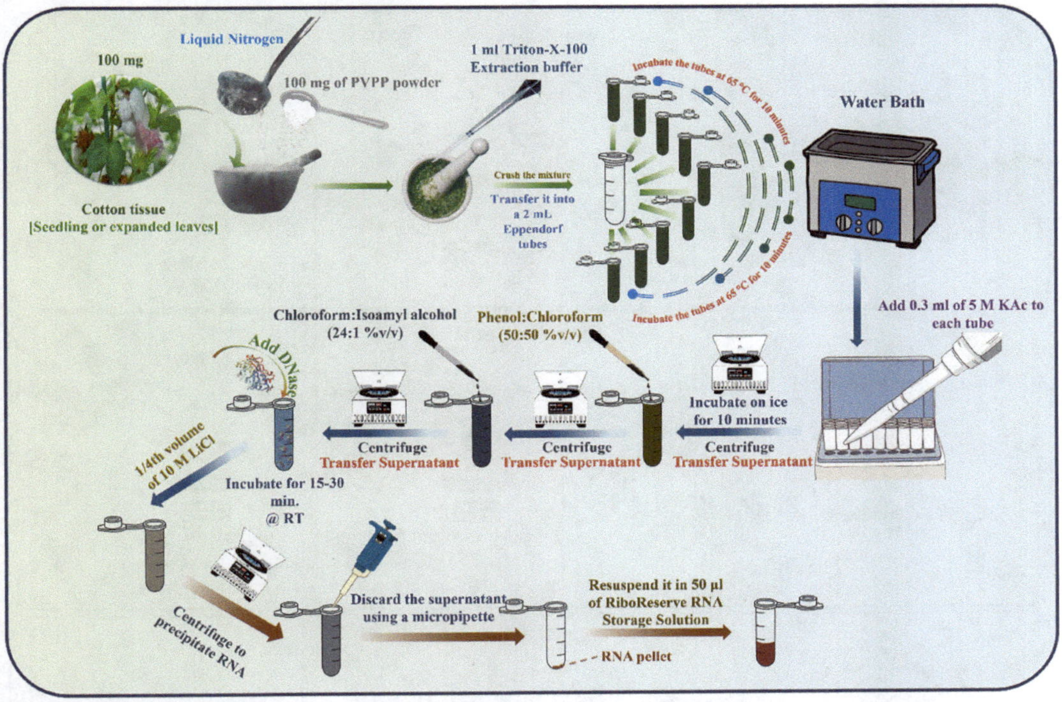

Fig. 2 A flow chart demonstrating the major steps in the Triton X-100 lysis-based RNA extraction protocol for polyphenol-rich plant tissues

- Cyanogenic glycosides: When damaged tissues are enzymatically hydrolyzed, they release highly toxic hydrogen cyanide (HCN), preventing herbivores and pathogens [4]. Besides defense, the chemical also functions as a dynamic store for reduced carbon and nitrogen and can be mobilized and redeployed if the nutrients are in short supply [5].
- Saponins: Detergent-like action (forming stable foams in the aqueous solution) [6]. In biological contexts, they exhibit strong anti-inflammatory [7], anti-fungal activity [8], insecticidal [9], and are capable of altering the permeability in the membranes [10], playing a generalized role in plant defense.
- Cardiac glycosides: Cardiac glycosides are the triterpenoid or steroid glycosides that possess a typical steroid nucleus and are highly specialized, extremely powerful defense molecules, and extremely poisonous to insect and vertebrate herbivores [11]. Cardiac glycosides are also pharmacologically significant because they act on the cardiac muscle (inhibition of the Na^+/K^+-ATPase pump) [12].

The biosynthesis of these structurally complicated molecules is closely controlled at the genetic level. The production of particular genes coding for specific enzymes in their pathways regulates the

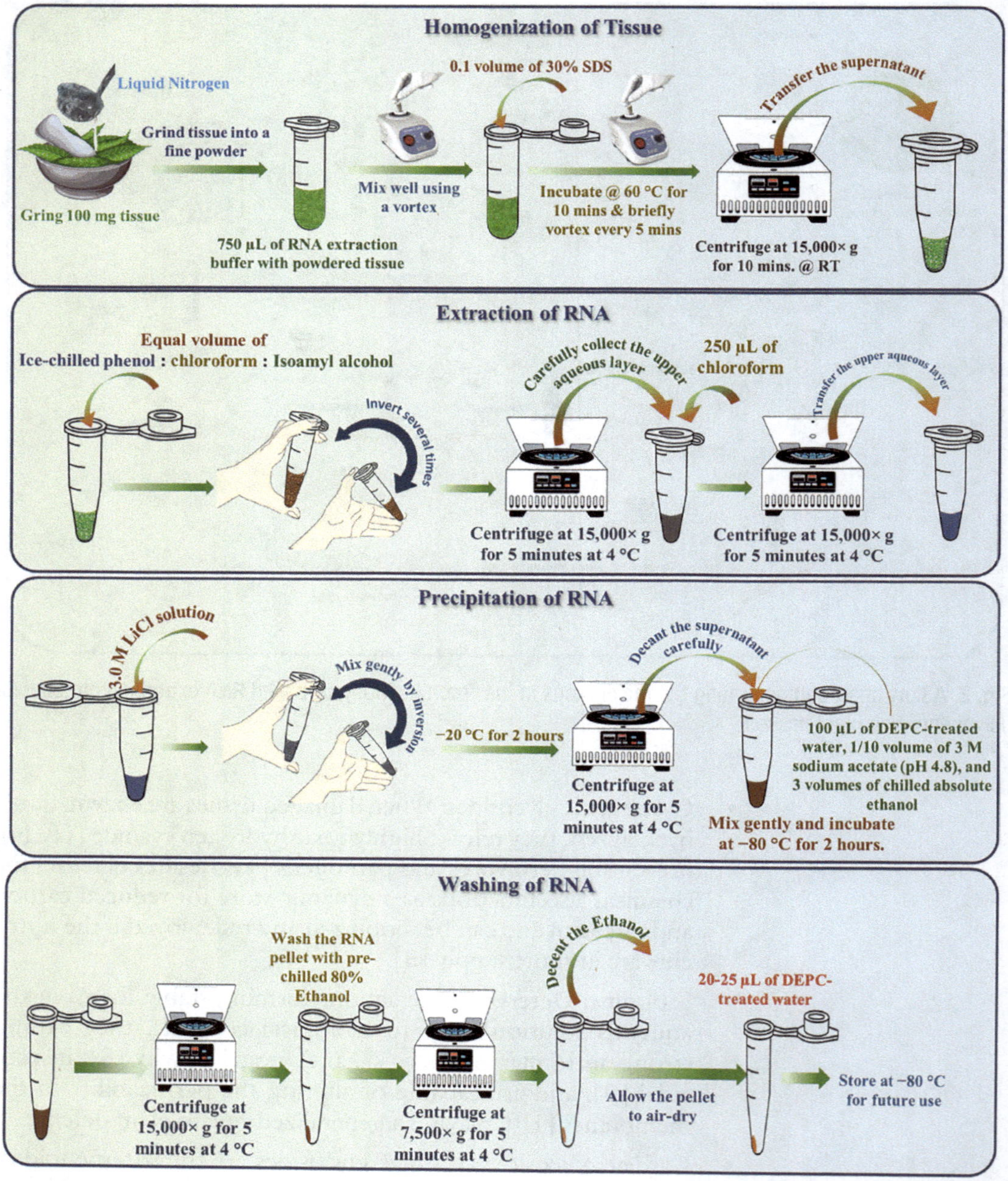

Fig. 3 Stepwise procedure of RNA isolation from complex plant tissues rich in polyphenols and polysaccharides using the modified SDS method

transcriptional output of glycoside synthesis. The most important enzymes within this pathway are glycosyltransferases (GTs), especially the vast family of UDP-glycosyltransferases (UGTs) [13]. Glycosides are produced when aglycones are coupled to sugar groups by UGTs, a process that drastically alters the water solubility, transport, and bioactivity of the compound [14]. Therefore, the transcription of primary UGT genes is the most direct way to see how a plant initiates its chemical defense or signaling pathway in reaction to an altered environment [15].

1.2 Transcript Analysis by Quantitative Real-Time PCR (qPCR)

Alterations in the transcript abundance of these biosynthetic genes are a reflection of the metabolic status of the plant and its continued regulation of glycoside biosynthesis. Quantitative real-time PCR (qPCR) is a good method to quantify such small changes. qPCR is a popular method that is sensitive, powerful, and precise to track the abundance of target gene expression.

The qPCR process essentially includes [16]:

1. High-quality RNA extraction: Purifying pure RNA from plant tissues.
2. Reverse transcription: Converting the RNA template to complementary DNA (cDNA).
3. Real-time amplification: Amplifying the cDNA with gene-specific primers and fluorescent dyes, such as SYBR Green.

Crucially, quantitative data relies heavily on normalization against stably expressed reference (housekeeping) genes. This step accounts for variability in the quality and quantity of the initial RNA starting material and the efficiency of the reverse transcription reaction, thereby ensuring reliable comparative quantification between different samples [17].

1.3 Challenges in Plant RNA Isolation

Isolation of high-purity RNA is necessary for the effective conduct of the subsequent molecular studies in plant species, especially in the event of differential gene expression studies caused by divergent biotic and abiotic stimuli [18]. Intact, high-quality RNA from plant tissues, however, is hard to obtain, primarily due to the presence of secondary metabolites. Most plant species contain high amounts of:

- Polysaccharides and starch are also found to co-precipitate with RNA, so we end up with a viscous RNA pellet, which resuspends with difficulty. It ends up with low yield and low purity [19].
- Polyphenols: Covalently bonded oxidized polyphenols attach to nucleic acids and proteins and lead to the breakdown of RNA and inhibition of downstream enzymatic activities (reverse transcriptase and Taq polymerase) [20, 21].

These challenging metabolites in tissues (in particular, high-starch seeds, roots, and woody tissues with high content of phenolics) tend to produce irregular or low-grade outcomes with commercial kits and prevalent techniques. Therefore, generic RNA extraction procedures tend to have the following deficiencies:

- Flexibility in the need for multiple, labor-intensive extraction buffers or procedures.
- Methods are restricted to the isolation of RNA from distinct tissues only, i.e., young leaves, only in the case of simple examples such as roots or seeds.
- The requirement for expensive TRIZOL (using hazardous organic solvents) and/or commercial column-based kits, which are too costly for large-scale experiments.

1.4 The Demand for a Robust Protocol

The accurate quantification of gene expression depends entirely on the integrity and purity of the isolated nucleic acids. Because RNA yield and quality fluctuate considerably between plant species and tissues, and due to secondary metabolites that stop RNA from being extracted are always present, following standard extraction protocols is not enough. For glycoside gene expression studies to be reliable, the methodology needs to be very well optimized.

2 Materials

2.1 Sample Collection

1. Healthy plant tissues.
2. Sterile scissors, scalpel, or forceps.
3. RNase-free microcentrifuge tubes (1.5–2.0 mL).
4. Liquid nitrogen or dry ice for freezing.
5. Icebox or cooler for transport.
6. Permanent marker and labels.
7. RNase-free gloves.

2.2 Total RNA Extraction

2.2.1 Total RNA Extraction from Starch-Rich Seeds Using the Modified SDS-LiCl Method [22]

1. **Extraction Buffer**
   - 2.5 M NaCl.
   - 2.5% polyvinylpyrrolidone (PVP).
   - 100 mM Tris-HCl (pH 8.0).
   - 25 mM EDTA.
   - 2.5% β-mercaptoethanol.
   - 2% SDS.
2. **Purification Reagents**
   - Acidic phenol (pH 4.3) saturated with 0.1 M citrate buffer.

3. **Precipitation Reagents**

- 2 M lithium chloride (LiCl).
- 3 M sodium acetate (pH 4.8).

2.2.2 Triton X-100 Lysis-Based RNA Extraction Protocol for Tissues Rich in Polyphenols [23]

1. Chloroform.
2. Isoamyl alcohol.
3. Dithiothreitol (DTT).
4. Ethylenediaminetetraacetic acid (EDTA).
5. Potassium acetate (KAc).
6. Lithium chloride (LiCl).
7. Sodium chloride (NaCl).
8. Sodium hydroxide (NaOH).
9. Phenol.
10. Polyvinylpyrrolidone, cross-linked, insoluble (PVPP).
11. Sodium dodecyl sulfate (SDS).
12. RiboReserve RNA Storage Solution.
13. DNase.
14. Triton X-100 extraction buffer.
15. 5 M potassium acetate (pH 6.5).
16. Phenol:chloroform (PC, 50:50, v/v)
17. Chloroform:isoamyl alcohol (CI, 24:1, v/v).
18. TE buffer (10 mM Tris, pH 8.0; 1 mM EDTA, pH 8.0).
19. 10 M lithium chloride.

2.2.3 Modified SDS-Based RNA Extraction from Polyphenol- and Polysaccharide-Rich Tissues [24]

1. Tris base, molecular biology grade.
2. Ethylenediaminetetraacetic acid disodium salt (EDTA Na_2).
3. Polyvinylpyrrolidone (PVP K30, M.W. 50,000).
4. Sodium chloride (NaCl).
5. β-Mercaptoethanol.
6. Sodium acetate, anhydrous.
7. Chloroform.
8. DEPC nuclease-free water.
9. Absolute ethanol, molecular biology grade.
10. Ethidium bromide solution.
11. Acetic acid, glacial.
12. Isoamyl alcohol.
13. Phenol.
14. Lithium chloride solution.

15. RNase Away Surface Decontamination Solution.
16. Sodium dodecyl sulfate (SDS).
17. Agarose MB grade.
18. Orange G loading dye.
19. EDTA 0.5 M, pH 8.0.
20. TAE electrophoresis buffer 50×.
21. Tris-HCl 1 M (pH 8.0).
22. 30% SDS.
23. 0.1 M citrate buffer (pH 4.2).
24. 3 M sodium acetate (pH 4.8).
25. Saturated phenol with citrate buffer (pH 4.2).
26. Phenol/chloroform/isoamyl alcohol (25:24:1, v/v/v).
27. RNA extraction buffer.

2.2.4 RNA Extraction Method with Modified Extraction Buffer [25]

1. Extraction buffer (EB): contains 0.25 M NaCl, 0.05 M Tris-HCl (pH 7.5), 20 mM EDTA, and 1% (w/v) SDS.
2. Polyvinylpyrrolidone (PVP) (M.W. 40,000).
3. β-Mercaptoethanol.
4. Chloroform:isoamyl alcohol (CI): 24:1 (v/v).
5. Phenol:chloroform:isoamyl alcohol (PCI): 1:1 (v/v).
6. 3 M sodium acetate: pH 5.2.
7. Absolute ethanol: 100%.
8. 10 M lithium chloride (LiCl).
9. DEPC-treated water.

2.2.5 Modified CTAB-TRIzol Method [26]

1. CTAB extraction buffer.
2. 5% (v/v) β-mercaptoethanol (β-ME).
3. TRIzol.
4. Chloroform:isoamyl alcohol (CIA) (24:1).
5. 10 M LiCl.
6. Ethanol (70%).
7. RNase-free water.

2.2.6 Modified Hot Borate RNA Extraction Method [27]

1. Borate extraction buffer.
2. 1 M KCl (potassium chloride).
3. 5 M LiCl (lithium chloride).
4. 2 M LiCl (lithium chloride).
5. 10 mM Tris-HCl (pH 7.5).
6. 5 M KAc (potassium acetate, pH 6.5).

7. 1 M DTT (dithiothreitol).
8. Proteinase K solution (20 mg/mL).
9. Absolute ethanol (200 proof).
10. 70% ethanol.
11. RNase-free water (nuclease-free water).
12. DNase I solution (2 units/μL).
13. RNeasy Mini Kit (Qiagen).

2.3 RNA Quality and Quantification

2.3.1 RNA Concentration and Purity assessment Using the Nanodrop Spectrophotometer [28]

1. Purified RNA samples.
2. Thermo Scientific NanoDrop Spectrophotometer.

2.3.2 RNA Quality Assessment Using 1% Agarose Gel Electrophoresis (Mini Gel Apparatus)

1. Agarose, molecular biology grade.
2. 1× TAE buffer.
3. Ethidium bromide solution (10 mg/mL).
4. Distilled water.
5. Gel loading buffer (10×).
6. RNA sample (150–200 ng).
7. 200-mL flask.
8. Mini gel electrophoresis apparatus.
9. Power supply.
10. Gel documentation system.

2.3.3 RNA Integrity Analysis Using Agilent Bioanalyzer 2100

Kit Components (Agilent RNA 6000 Nano Kit)

- RNA 6000 Nano Chips.
- Agilent RNA 6000 Nano Gel Matrix (red cap).
- RNA Nano Dye Concentrate (blue cap).
- Agilent RNA 6000 Nano Marker (green cap).
- Agilent RNA 6000 Ladder (yellow cap).
- Spin filters.
- Electrode cleaners.
- Syringe and Safe-Lock Eppendorf tubes.

Required Equipment

- Agilent 2100 Bioanalyzer.
- Chip Priming Station.
- IKA vortex mixer.

- Microcentrifuge (≥1500 × *g*).
- Heating block or thermal cycler for denaturation.
- Pipettes and RNase-free tips.

2.4 RNA-Seq Library Preparation

The core reagents are supplied with the TruSeq Stranded mRNA Library Prep kit. Key components include:

- RNA Purification Beads (RPB).
- Bead Binding Buffer (BBB), Bead Washing Buffer (BWB), and Elution Buffer (ELB).
- Fragment, Prime, Finish Mix (FPF).
- First Strand Synthesis Act D Mix (FSA).
- Second Strand Marking Master Mix (SMM).
- A-Tailing Mix (ATL).
- Ligation Mix (LIG).
- PCR Master Mix (PMM) and PCR Primer Cocktail (PPC).
- Resuspension Buffer (RSB).

The following consumables must be sourced separately:

- Agencourt AMPure XP beads.
- SuperScript II Reverse Transcriptase.
- Ethanol (200 proof) for preparing fresh 80% EtOH.
- Nuclease-free water, pipette tips, and 96-well PCR plates.
- Sequencing index adapters (purchased separately).

2.5 Quantitative and Qualitative Analysis of Library

Qubit Fluorometer:

- Qubit dsDNA HS buffer and Qubit dsDNA HS Reagent.
- Qubit dsDNA HS Standards.
- Qubit fluorometer
- **Agilent Bioanalyzer 2100 :**
  - DNA dye concentrate and DNA gel matrix.
  - Chip priming station with syringe.
  - DNA assay chip.
  - DNA marker and ladder.
  - Vortex mixer.
  - Agilent Bioanalyzer 2100 system.
- **qPCR Assay:**
  - Library samples (diluted 1:10,000 and 1:20,000).
  - PCR-grade water.
  - Kapa Primer Premix (10×).
  - ROX Low (50×) (or ROX High).
  - KAPA SYBR FAST qPCR Master Mix (2×).

- Six DNA standards with known concentrations.
- qPCR plate and system.

- **Library Sample Pooling:**
  - Library samples at 10 or 20 nM.
  - Library dilution buffer (10 mM Tris-HCl (pH 8.5) with 0.1% Tween 20).
  - Microcentrifuge tubes.

2.6 High-Throughput Sequencing

- RNA-Seq libraries (quantified and validated).
- Illumina sequencing platform (e.g., NovaSeq 6000 or HiSeq 4000).
- Flow cell and sequencing reagent kit (as per manufacturer's protocol).
- Pipettes and sterile RNase-free tips.
- Ice bucket for reagent handling.
- Computer with sequencing control and data acquisition software.

2.7 Quality Assessment and Preprocessing of Raw Reads

- Raw sequencing data files (FASTQ format) obtained from Sect. 2.6.
- Computer workstation or server with sufficient storage capacity.
- Quality control software such as FastQC or MultiQC.
- Trimming software such as Trimmomatic or Cutadapt.
- Metadata file containing sample identifiers and sequencing parameters.
- Backup drive or external hard disk for data storage.

2.8 Transcriptome Assembly and Alignment

- Quality-filtered RNA-Seq reads from Sect. 2.7 (clean FASTQ files).
- Reference genome sequence (FASTA format) and annotation file (GTF/GFF), if available.
- Bioinformatics software for alignment (e.g., HISAT2, STAR, or TopHat2).
- Software for de novo assembly (e.g., Trinity).
- High-performance computing server with large memory capacity (≥64 GB RAM recommended).
- Data storage device or backup drive.

2.9 Gene Expression Quantification

- Aligned RNA-Seq files (BAM/SAM) from Sect. 2.8 or assembled transcriptome (FASTA).
- Gene annotation files (GTF/GFF), if using a reference genome.

- Expression quantification software: HTSeq, featureCounts, or Salmon/Kallisto.
- Computer workstation or server with sufficient memory and storage.

2.10 Differential Gene Expression Analysis

- Normalized gene expression data from Sect. 2.9.
- Biological replicates for each condition.
- Differential expression analysis software (e.g., DESeq2 or edgeR).
- Computer with R or equivalent statistical environment.

2.11 Functional Annotation of Genes

- FASTA sequences of transcripts or differentially expressed genes.
- Gene annotation databases: NCBI, UniProt, GO, KEGG.
- Annotation tools: BLAST, InterProScan, or Blast2GO.
- Computer with internet access and sufficient memory.

2.12 Pathway Analysis

- List of differentially expressed genes (from Sect. 2.10).
- Functional annotations and pathway mapping (from Sect. 2.11).
- Pathway analysis software: KEGG Mapper, PlantCyc, or Pathview.

3 Methods

3.1 Sample Collection

1. **Select Plant Material**
   1. Choose healthy plants at a similar growth stage.
   2. Decide which tissue (leaf, root, stem, etc.) is needed for glycoside studies.
2. **Collect Tissue**
   1. Cut the tissue carefully using sterilized scissors or a scalpel.
   2. Place each sample immediately into a prelabeled, RNase-free tube.
3. **Freeze Immediately**
   1. Snap-freeze the tissue in liquid nitrogen or place it on dry ice.
   2. Keep samples frozen during transport to the lab.
4. **Store Samples**
   1. Store tissues at −80°C until RNA extraction.
   2. Avoid thawing and refreezing the samples.

5. **Record Details**
    1. Note sample ID, tissue type, collection date, plant stage, and conditions.
    2. Optionally, take photos for reference.

3.2 Isolation of Plant Total RNA

3.2.1 Optimized SDS-LiCl Method for Total RNA Isolation from Starch-Rich Seeds [22] (Fig. 1)

This section explains how to get high-quality total RNA from plant tissues, with a focus on seeds that are high in starch, such as wheat, rice, and maize. Traditional RNA extraction methods often have trouble with these samples because they have a lot of starch in them. This can cause the samples to solidify, give low yields, and lower the purity of the RNA. The methods described here are a quick way to solve these problems.

Homogenization of Tissue:

1. Grind 100 mg of fresh plant tissue to a fine powder using a mortar and pestle with liquid nitrogen.
2. Add 600 μL of RNA extraction buffer to the powdered tissue.

Extraction of RNA:

3. Add 60 μL of 20% SDS (final concentration of 2%) to the mixture.
4. Gently vortex to mix and then centrifuge at 15,890 × *g* for 5 min at 4°C.
5. Carefully collect the upper aqueous phase and transfer it to a new tube.
6. Add an equal volume of Phenol:chloroform:isoamyl alcohol (25:24:1). Gently vortex and centrifuge at 15,890 × *g* for 5 min at 4°C.
7. Transfer the upper aqueous phase to a new tube and add 200 μ L of chloroform. Gently vortex and centrifuge at 15,890 × *g* for 5 min at 4°C.

Precipitation of RNA:

8. Transfer the upper aqueous phase to a clean tube.
9. Add 160 μL of 8 M lithium chloride (final concentration of 2 M) and 60 μL of 3 M sodium acetate.
10. Incubate the tube at −20°C for 15 min to precipitate the RNA.

Washing of RNA:

11. Centrifuge the mixture at 15,890 × *g* for 5 min at 4°C to pellet the RNA.

12. Discard the supernatant and wash the RNA pellet with 2 M lithium chloride. Centrifuge again at 15,890 × *g* for 5 min at 4°C.
13. Discard the supernatant and wash the pellet with pre-chilled 80% ethanol. Centrifuge at 15,890 × *g* for 5 min at 4°C.
14. Briefly air-dry the RNA pellet and resuspend it in DEPC-treated water.
15. Store the isolated RNA at −80°C.

3.2.2 Triton X-100 Lysis-Based RNA Extraction Protocol for Tissues Rich in Polyphenols [23] (Fig. 2)

Homogenization of Tissue

1. Take 100 mg of cotton tissue (seedlings or expanded leaves) and grind it to a fine powder using a mortar and pestle with liquid nitrogen.
2. Add 100 mg of polyvinylpyrrolidone (PVPP) powder and mix thoroughly.

Extraction and Lysis

3. Add 1 mL of freshly prepared and pre-heated Triton X-100 extraction buffer to the ground material.
4. Crush the mixture and transfer it into a 2 mL Eppendorf tube.
5. Incubate the tubes in a 65°C water bath for 10 min.

Contaminant Removal and Phase Separation

6. Add 0.3 mL of 5 M potassium acetate to each tube. Incubate on ice for 10 min.
7. Centrifuge the mixture. Transfer the supernatant to a fresh tube.
8. Add an equal volume of phenol:chloroform (PC, 50:50 v/v). Mix by inverting the tube and then centrifuge.
9. Transfer the supernatant to a new tube.
10. Add an equal volume of chloroform:isoamyl alcohol (CI, 24:1 v/v). Mix by inverting the tube and then centrifuge.
11. Transfer the supernatant to a fresh tube.

DNase Treatment and RNA Precipitation

12. Add DNase to the supernatant and incubate for 15–30 min at room temperature.
13. Add 1/4th volume of 10 M lithium chloride to the supernatant. Mix by inverting the tube.
14. Centrifuge to precipitate the RNA pellet.

RNA Pellet Recovery and Storage

15. Discard the supernatant using a micropipette.
16. Air-dry the RNA pellet.

17. Resuspend the pellet in 50 µL of RiboReserve RNA Storage Solution.
18. Store the RNA for long-term use.

3.2.3 Modified SDS-Based RNA Extraction from Polyphenol and Polysaccharide Rich Tissues [24] (Fig. 3)

Homogenization of Tissue

1. Immediately transfer samples from the −80°C freezer to liquid nitrogen to prevent thawing.
2. Grind 100 mg of tissue into a fine powder using a pre-chilled mortar and pestle with liquid nitrogen.

RNA Extraction

3. Add 750 µL of RNA extraction Buffer to the powdered tissue in an RNase-free 2 mL microcentrifuge tube and mix vigorously with a vortex (Fig. 4).
4. Add 0.1 volume of 30% SDS to the homogenate and incubate at 60°C for 10 min, briefly vortexing every 5 min. Allow the sample to cool before proceeding.
5. Centrifuge the homogenate at 15,000 × *g* for 10 min at room temperature.
6. Carefully transfer the supernatant to a new 1.5 mL RNase-free microcentrifuge tube.
7. Add an equal volume of ice-chilled phenol/chloroform/isoamyl alcohol to the supernatant and mix by inverting the tube several times.
8. Centrifuge the mixture at 15,000 × *g* for 5 min at 4°C.
9. Carefully collect the upper aqueous layer, avoiding the middle whitish layer, and transfer it to a new 1.5 mL microcentrifuge tube.
10. Add 250 µL of chloroform to the aqueous layer, shake vigorously, and centrifuge at 15,000 × *g* for 5 min at 4°C.
11. Transfer the final upper aqueous layer to a new microcentrifuge tube.

RNA Precipitation

12. Add 3.0 M lithium chloride (LiCl) solution to the sample and mix gently by inversion.
13. Incubate the sample at −20°C for 2 h.
14. Centrifuge the solution at 15,000 × *g* for 5 min at 4°C to pellet the RNA, and then carefully decant the supernatant.
15. Add 100 µL of DEPC-treated water, 1/10 volume of 3 M sodium acetate (pH 4.8), and 3 volumes of chilled absolute ethanol. Mix gently and incubate at −80°C for 2 h.

Washing of RNA

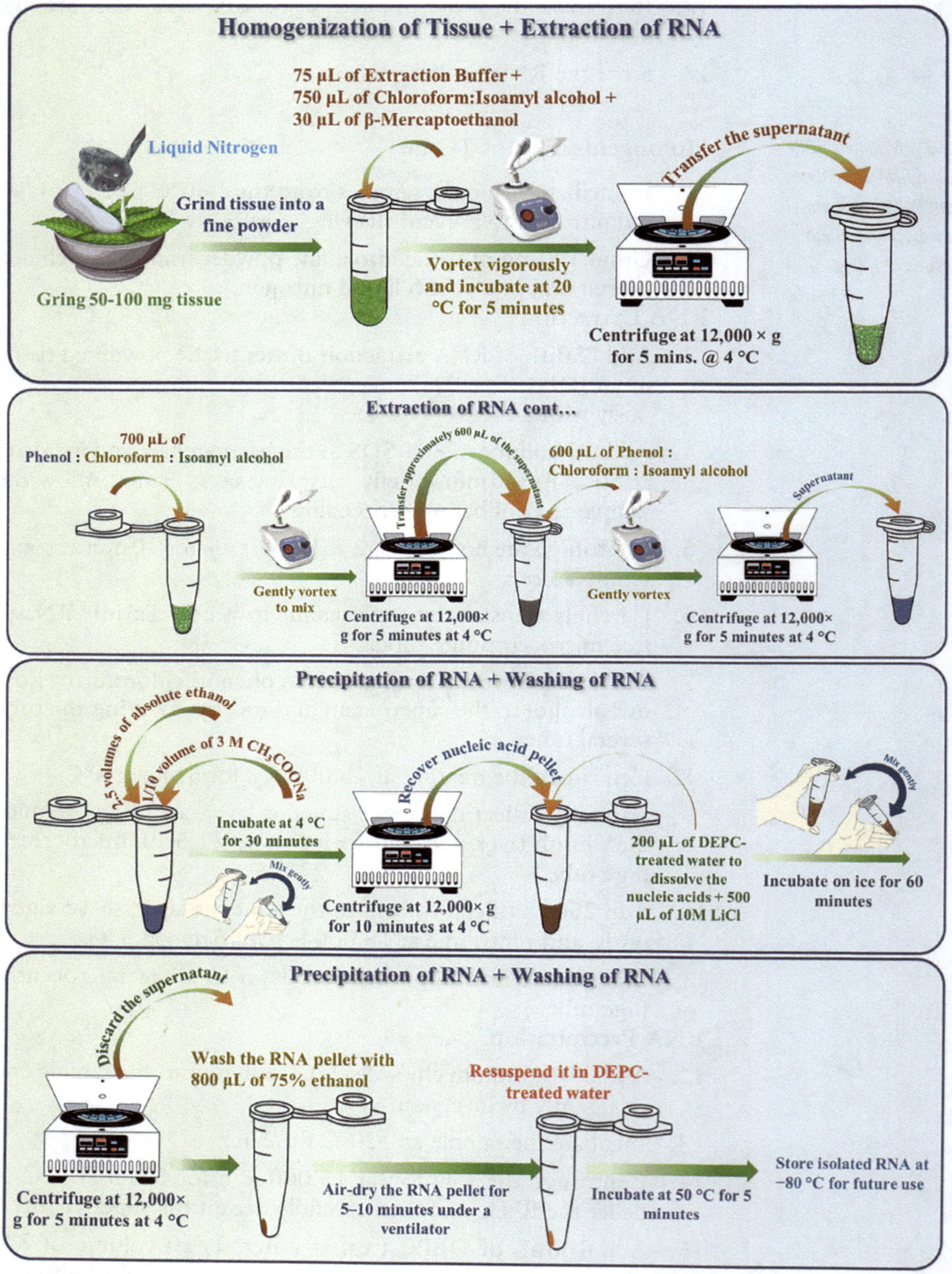

Fig. 4 Diagrammatic overview of total RNA extraction using the modified buffer formulation

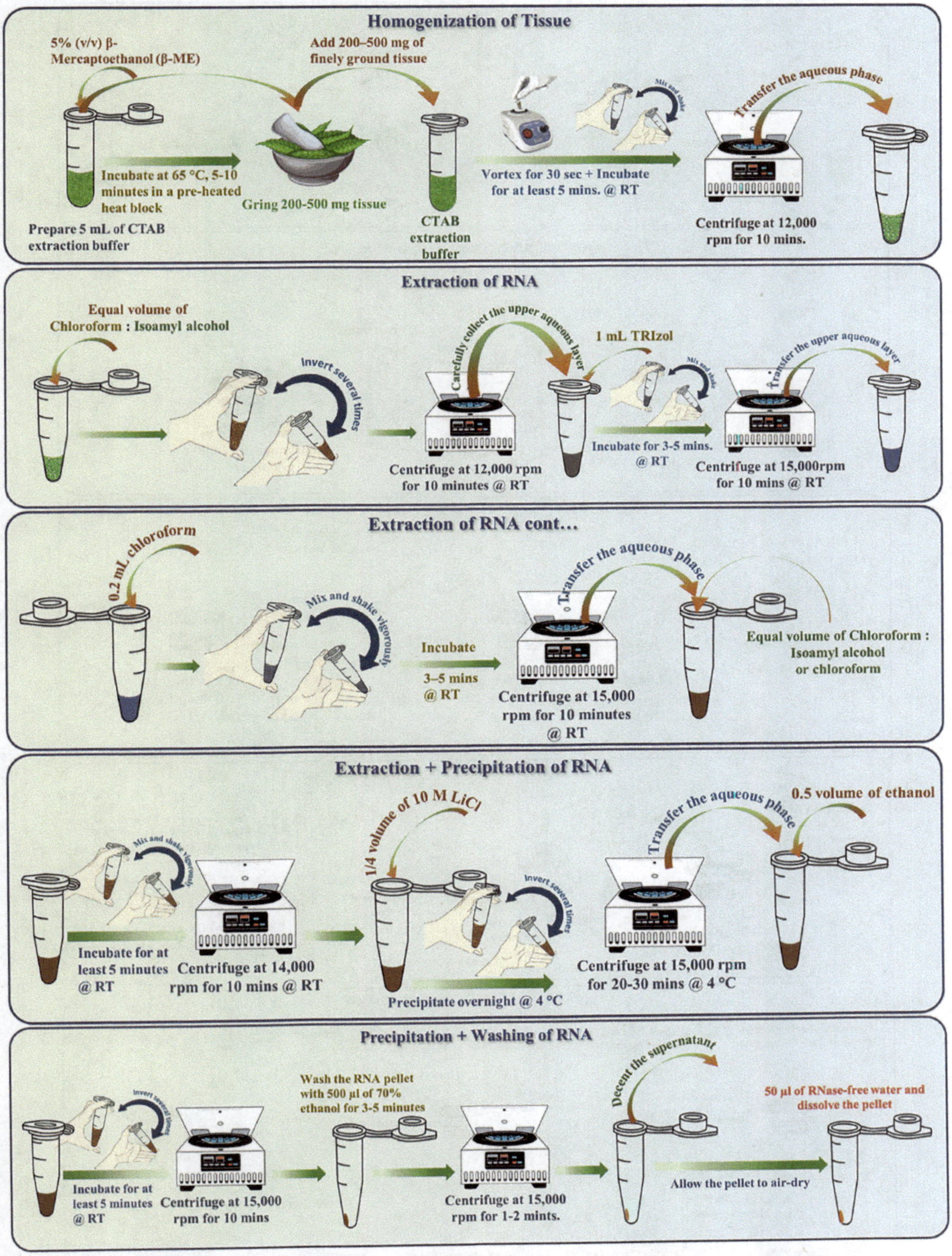

Fig. 5 Process diagram depicting the modified CTAB-TRIzol-based RNA extraction method

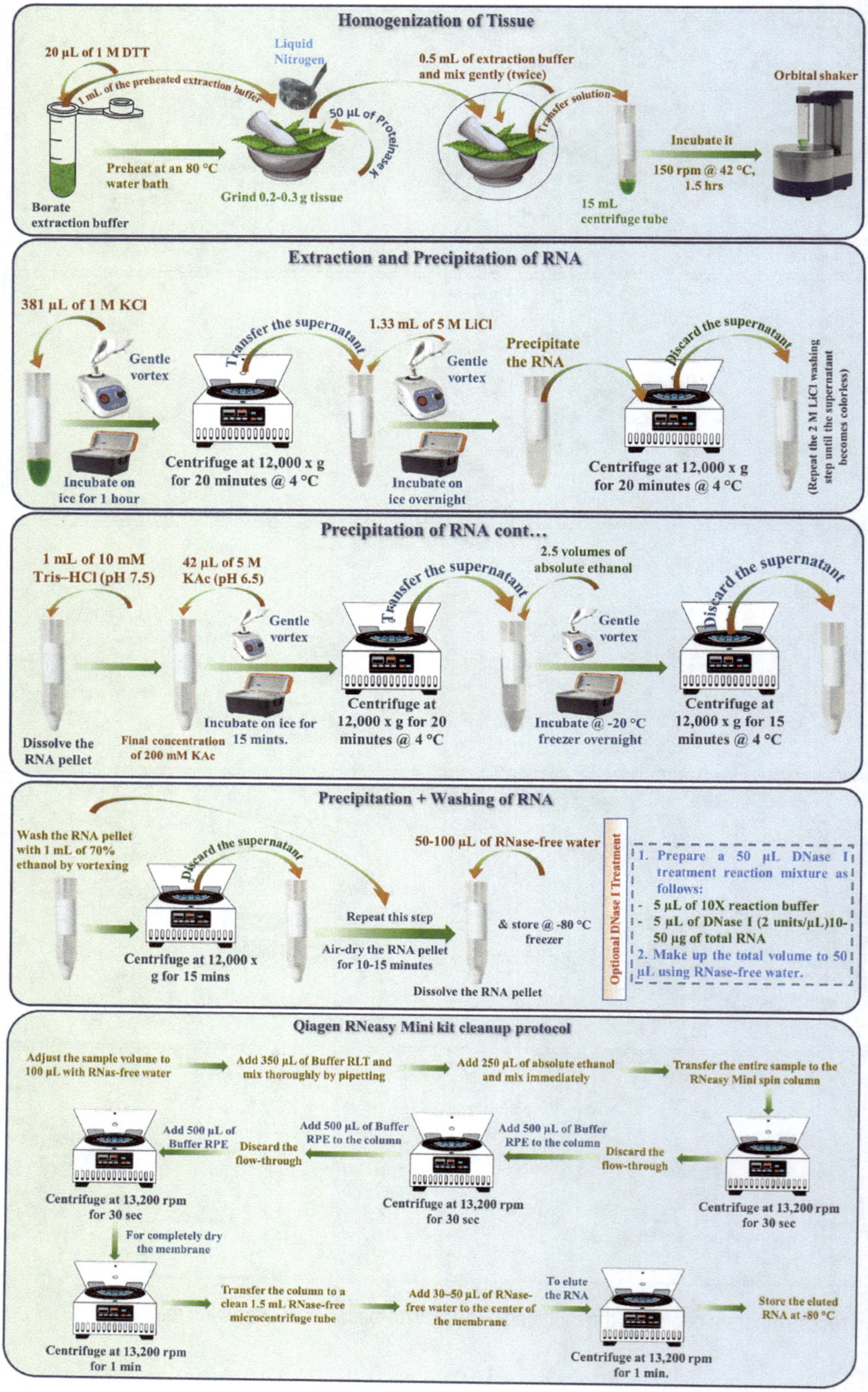

Fig. 6 Workflow of RNA isolation from plant tissues using the modified hot borate extraction protocol

16. Centrifuge at 12,500 × g for 5 min at 4°C and decant the supernatant.
17. Wash the RNA pellet with pre-chilled 80% ethanol by centrifuging at 7500 × *g* for 5 min at 4°C.
18. Decant the ethanol and allow the pellet to air-dry, taking care not to over-dry it, which can prevent it from dissolving.
19. Dissolve the RNA pellet in 20–25 μL of DEPC-treated water and store it immediately at −80°C for future use.

3.2.4 RNA Extraction Method with Modified Extraction Buffer [25]

1. **Homogenization of Tissue**
   1. Grind 50–100 mg of plant tissue (stem, leaf, or flower) in liquid nitrogen to create a fine powder.
2. **Extraction of RNA**
   2. In a 2.0 mL RNase-free microcentrifuge tube, mix 75 μL of extraction buffer (EB), 750 μL of chloroform:isoamyl alcohol (CI), and 30 μL of β-mercaptoethanol.
   3. Add the tissue powder to the tube, vortex vigorously, and incubate at 20°C for 5 min.
   4. Centrifuge the mixture at 4°C for 5 min at 12,000 × *g*.
   5. Transfer approximately 700 μL of the supernatant to a new 2.0 mL RNase-free microcentrifuge tube. Add 700 μL of phenol:chloroform:isoamyl alcohol (PCI), mix gently, and centrifuge at 4°C for 5 min at 12,000 × *g*.
   6. Transfer approximately 600 μL of the supernatant to a new 2.0 mL RNase-free microcentrifuge tube. Add 600 μL of PCI, mix, and centrifuge at 4°C for 5 min at 12,000 × g.
3. **Precipitation of RNA**
   7. Add 1/10 volume of 3 M sodium acetate (pH 5.2) and 2.5 volumes of absolute ethanol to the supernatant. Mix gently and incubate at 4°C for 30 min.
   8. Centrifuge at 4°C for 10 min at 12,000 × *g* to recover the nucleic acid pellet.
   9. Add 200 μL of DEPC-treated water to dissolve the nucleic acids, then add 500 μL of 10 M LiCl. Mix gently and incubate on ice for 60 min.
   10. Centrifuge at 4°C for 10 min at 12,000 × *g*, discard the supernatant, and retain the pellet.
4. **Washing of RNA**
   11. Wash the pellet with 800 μL of 75% ethanol.
   12. Air-dry the pellet for 5–10 min under a ventilator.

13. Add 50 μL of DEPC-treated water to dissolve the RNA and incubate at 50°C for 5 min.
14. The RNA samples can be used immediately for downstream applications or stored at −80°C for future use.

3.2.5 Modified CTAB-TRIzol method [26] (Fig. 5)

1. **Homogenization of Tissue**
    1. Prepare 5 mL of CTAB extraction buffer by adding 5% (v/v) β-mercaptoethanol (β-ME) to it.
    2. Incubate the buffer at 65°C for 5–10 min in a preheated heat block.
    3. Add 200–500 mg of finely ground tissue to the prepared CTAB extraction buffer.
    4. Vortex for 30 s to mix and shake thoroughly.
    5. Incubate for at least 5 min at room temperature (RT) on a nutating mixer or similar shaker, then centrifuge at 12,000 rpm for 10 min.
2. **Extraction of RNA**
    6. Transfer the aqueous phase to a new tube. Add an equal volume of chloroform:isoamyl alcohol (CIA) (24:1). Mix by shaking thoroughly, incubate at RT for at least 5 min on a shaker, and centrifuge at 12,000 rpm for 10 min.
    7. Repeat the previous step.
    8. Transfer the aqueous phase to a new tube.
    9. Dissolve the RNA pellet in 1 mL of TRIzol (Ambion).
    10. Vortex and shake vigorously, then incubate for 3–5 min at RT.
    11. Centrifuge the tubes at 15,000 rpm for 10 min.
    12. Transfer the aqueous phase to a 1.5 mL microcentrifuge tube.
    13. Add 0.2 mL of chloroform to the aqueous phase, mix well, and shake vigorously.
    14. Incubate for 3–5 min at RT and centrifuge for 10 min at 15,000 rpm (>20,000 × *g*).
    15. Transfer the aqueous phase to a 1.5 mL microcentrifuge tube.
    16. Add an equal volume of CIA or chloroform.
    17. Mix well by vigorous shaking, incubate for at least 5 min on a nutating mixer at RT, and centrifuge for 10 min at 14,000 rpm.
    18. Note: This step can be repeated for challenging tissue types (e.g., fruit, tuber) to completely eliminate proteins.

3. **Precipitation of RNA**

19. Add 1/4 volume of 10 M LiCl.
20. Mix by inversion and precipitate overnight at 4°C (or 4–6 h at −20°C).
21. Centrifuge at 15,000 rpm (>20,000 × *g*) for 20–30 min at 4°C.
22. Carefully decant or pipette out all the supernatant.
23. Transfer the aqueous phase to a 1.5 mL microcentrifuge tube and add 0.5 volume of ethanol (if using isopropanol, add 1 volume).
24. Mix well by inversion.
25. Centrifuge at 15,000 rpm for 10 min.
26. Carefully decant or pipette out the supernatant.

4. **Washing of RNA**

27. Wash the RNA pellet with 500 μL of 70% ethanol for 3–5 min.
28. Centrifuge at 15,000 rpm for 1–2 min.
29. Carefully decant or pipette out the supernatant.
30. Repeat the washing step and air-dry the pellet.
31. Add 50 μL of RNase-free water and dissolve the pellet completely.

3.2.6 Modified Hot Borate RNA Extraction Method [27] (Fig. 6)

1. **Homogenization of Tissue**

1. Add 20 μL of 1 M DTT into 2 mL of borate extraction buffer and preheat at an 80°C water bath.
2. Grind 0.2–0.3 g of plant tissue with a liquid nitrogen-cooled mortar and pestle.
3. Add 1 mL of the preheated extraction buffer to the mortar and grind further.
4. Add 50 μL of Proteinase K (20 mg/mL) into the mortar and mix well.
5. Add 0.5 mL of extraction buffer and mix gently, then add another 0.5 mL of extraction buffer and mix gently.
6. Transfer the solution into a 15 mL centrifuge tube and incubate in an orbital shaker at 150 rpm at 42°C for 1.5 h.

2. **Extraction of RNA**

7. Add 381 μL of 1 M KCl to the tube and mix by gentle vortex. Incubate on ice for 1 h
8. Centrifuge at 4°C for 20 min at 12,000 *g*.

9. Transfer the supernatant (about 2 mL) to another 15 mL centrifuge tube.

3. **Precipitation of RNA**

10. Add 1.33 mL of 5 M LiCl
11. (final concentration of 2 M LiCl) and mix with a gentle vortex. Incubate on ice overnight to precipitate the RNA.
12. Centrifuge at 4°C for 20 min at 12,000*g* and discard the supernatant.
13. Wash the RNA pellet with 1 mL of cold 2 M LiCl by vortexing or pipetting.
14. Centrifuge at 4°C for 20 min at 12,000*g* and discard the supernatant.
15. Repeat the 2 M LiCl washing step until the supernatant becomes colorless.
16. Dissolve the RNA pellet in 1 mL of 10 mM Tris-HCl (pH 7.5).
17. Add 42 μL of 5 M KAc (pH 6.5) to a final concentration of 200 mM KAc and mix by gentle vortex.
18. Incubate on ice for 15 min, then centrifuge at 4°C for 20 min at 12,000*g*.
19. Transfer the supernatant to another 15 mL centrifuge tube. Add 2.5 volumes of absolute ethanol and mix well by gentle vortex.
20. Incubate at a −20°C freezer overnight.
21. Centrifuge at 4°C for 15 min at 12,000*g* and discard the supernatant.

4. **Washing of RNA**

22. Wash the RNA pellet with 1 mL of 70% ethanol by vortexing.
23. Centrifuge at 4°C for 15 min at 12,000*g* and discard the supernatant.
24. Repeat the 70% ethanol washing step once.
25. Air-dry the RNA pellet for 10–15 min.
26. Dissolve the RNA pellet with 50–100 μL of RNase-free water and store at a −80°C freezer.
27. (Optional DNase I Treatment) Set up a 50-μL DNase I treatment reaction mixture containing 5 μL of 10× reaction buffer, 5 μL of DNase I (2 units/μL), and 10–50 μg of total RNA. Make up the volume to 50 μL with RNase-free water.

28. Incubate the reaction mixture at 37°C for 20 min, then follow the Qiagen RNeasy Mini kit cleanup protocol as below:

- Adjust the sample volume to 100 μL with RNase-free water.
- Add 350 μL of Buffer RLT and mix thoroughly by pipetting.
- Add 250 μL of absolute ethanol and mix immediately.
- Transfer the entire sample to the RNeasy Mini spin column.
- Centrifuge for 30 s at 13,200 rpm (16,000 ×*g*) and discard the flow-through.
- Add 500 μL of Buffer RPE to the column and centrifuge for 30 s at 13,200 rpm.
- Discard the flow-through, add another 500 μL of Buffer RPE, and centrifuge for 30 s at 13,200 rpm.
- Centrifuge for 1 min at 13,200 rpm to completely dry the membrane.
- Transfer the column to a clean 1.5 mL RNase-free microcentrifuge tube.
- Add 30–50 μL of RNase-free water to the center of the membrane.
- Centrifuge for 1 min at 13,200 rpm to elute the RNA.
- Store the eluted RNA at −80°C.

3.3 RNA Quality and Quantification

3.3.1 RNA Concentration and Purity Assessment Using the Nanodrop Spectrophotometer [28]

Purify nucleic acid samples before measurement to ensure accurate results.

1. Use a NanoDrop Spectrophotometer to find out the quantity of total RNA remains in the sample.
2. The absorbance at 260 nm (A260) can be utilized to determine out the RNA concentration. (An absorbance of 1.0 means that there is about 40 μg/mL of RNA in the sample.)
3. Use the A260/A280 ratio and the A260/A230 ratio to check the purity the sample RNA.
4. To get good results from RNA-Seq library preparation, it is best to use RNA samples with an A260/A280 ratio of more than 1.9 and an A260/A230 ratio of more than 1.6 (Thermo Fisher Scientific 2009).

3.3.2 RNA Quality Assessment Using 1% Agarose Gel Electrophoresis [29]

1. Put 0.3 g of agarose and 30 mL of 1× TAE buffer in a flask and mix them.
2. Take note of the initial weight. Put the solution on a hot plate and stir it gently until the agarose is completely dissolved. Do not let it boil over. Use distilled water to bring the weight back to its original level.
3. After about 5 min, add 3 μL of ethidium bromide (10 mg/mL stock) and mix gently.
4. Pour the solution into a gel casting tray with a comb, avoiding bubbles. Allow it to solidify for 10–15 min.
5. Take off the comb. Put the gel tray in the electrophoresis tank so that the wells are facing the negative electrode. Submerge the gel in 1× TAE buffer.
6. Combine 150–200 ng of RNA (in less than 9 μL) with 1 μL of 10× gel loading buffer. Carefully load the mixture into the wells.
7. Plug it in and run the gel for 60–90 min at 70 V.
8. Take the gel out of the tank and use a gel documentation system to observe the RNA bands.

3.3.3 RNA Integrity Analysis Using Agilent Bioanalyzer 2100

The integrity and concentration of total RNA are to be determined using an Agilent Bioanalyzer 2100. All procedures must be performed with the RNA 6000 Nano Kit according to the official manufacturer's guidelines [30].

Preparation of Gel–Dye Mix

1. Allow the RNA gel matrix and RNA dye concentrate to equilibrate to room temperature for 30 min before use. Protect the dye from light.
2. Pour 550 μL of RNA gel matrix into a spin filter and centrifuge at 1500 × *g* for 10 min at room temperature.
3. Aliquot 65 μL of the filtered gel into RNase-free microcentrifuge tubes. These aliquots can be stored at 4°C for up to 1 month.
4. Vortex the prewarmed RNA dye concentrate for 10 s and spin down. Add 1 μL of the dye to a 65 μL aliquot of filtered gel.
5. Vortex the tube thoroughly to mix the gel and dye. Centrifuge the mixture at 13,000 × *g* for 10 min at room temperature. Use the prepared gel–dye mix within 1 day.

Sample Denaturation and Chip Loading

6. Heat denatures the RNA ladder and all RNA samples at 70°C for 2 min using a thermal cycler or water bath to minimize secondary structure. Immediately cool the vials on ice.
7. Place a new RNA Nano chip on the chip priming station. Load 9.0 μL of the gel–dye mix into the well marked with a dark circle.

8. Prime the chip for 30 s using the chip priming station.
9. After priming, pipette 9.0 µL of the gel–dye mix into the other two wells marked "G."
10. Pipette 5 µL of RNA 6000 Nano marker (green cap) into the ladder well and all 12 sample wells.
11. Pipette 1 µL of the denatured RNA ladder into the ladder well and 1 µL of each denatured RNA sample into its respective sample well.
12. If any sample wells are unused, fill them with 5 µL of marker and 1 µL of RNase-free water.

Assay Run and Analysis

13. Place the loaded chip horizontally in the vortex mixer adapter and vortex at 2400 rpm for 60 s.
14. Insert the chip into the Agilent 2100 Bioanalyzer and start the run within 5 min of loading to prevent evaporation.
15. Use the 2100 Expert Software to start the assay, enter sample information, and analyze the results. The software will generate an electropherogram, a ribosomal ratio, and an RNA Integrity Number (RIN) for each sample.

Interpretation of Results (RNA Integrity Number, RIN)

The RNA Integrity Number (RIN) is a score from 1 (highly degraded) to 10 (fully intact) that the 2100 Expert software calculates to provide a standard, objective assessment of RNA quality.

- **RIN 8–10: Very Good Quality.** The RNA is unchanged and has only a little amount of contamination. This is perfect for RNA-Seq, microarrays, and qPCR, which are all sensitive downstream applications.
- **RIN 6–7: Good Quality.** The RNA has lost some of its quality, but it can still be used for a lot of things, such as qPCR and library preparation. Higher quality RNA is better, though.
- **RIN < 6: Low Quality.** The RNA has been damaged a lot. It is undesirable to use this RNA for downstream applications because it can give results that are not reliable or fair, especially in quantitative experiments.

3.4 RNA-Seq Library Preparation (Provided in: TruSeq Stranded mRNA Reference Guide (1000000040498 v00))

Commercially available kits can be used to efficiently and reliably prepare RNA sequencing (RNA-Seq) libraries. The manufacturer's protocol is closely adhered to in the comprehensive process for creating strand-specific mRNA libraries, which is described below.

The TruSeq Stranded mRNA Sample Preparation Guide (Part No. 15031047 Rev. E) from Illumina, Inc., which describes the necessary steps from mRNA purification to the final amplification of the sequencing-ready library, serves as the basis for this procedure.

Isolating and Breaking Down Messenger RNAs

1. Allow the RNA Purification Beads (RPB) to drop to room temperature for approximately half an hour to prepare them.
2. Make 3.0 μg of RNA in total; the final volume of the liquid must be 50 μL.
3. Add 50 μL of RPB to the RNA sample and mix it well by pipetting.
4. Heat the mixture to 65°C for 5 min, then quickly move it to 4°C to cool down.
5. Let the mRNA and beads stick together by leaving them at room temperature for 5 min.
6. Put the tube on a magnetic stand for 5 min, then discard the supernatant. For a quick rinse to get rid of unbound RNA, add 200 μL of Bead Washing Buffer.
7. Add 50 μL of Elution Buffer to the beads and incubate them at 80°C for 2 min to get the mRNA out.
8. For a quick rinse to get rid of unbound RNA, add 200 μL of Bead Washing Buffer.
9. Add 50 μL of Elution Buffer to the beads and incubate them at 80°C for 2 min to get the mRNA out.
10. Add 50 μL of Bead Binding Buffer to the mRNA and let it rest at room temperature for 5 min to re-bind.
11. Put the tube back on the magnetic stand and throw discard the supernatant again.
12. To get rid of any leftover junk, add 200 μL of Bead Washing Buffer.
13. Put 19.5 μL of Fragment, Prime, Finish Mix right on the beads.
14. Incubate the mixture at 80°C for 2 min to break up the RNA. Then, collect 17 μL of the supernatant and move it to a new tube.

First-Strand Complementary DNAs Synthesis

15. To the primed mRNA sample, add 8 μL of a 1:9 mixture of the SuperScript II Reverse Transcriptase and First Strand Synthesis Act D Mix.
16. Put in a thermocycler and follow this schedule: First, hold it at 25°C for 10 min, then at 42°C for 15 min, and finally at 70°C for 15 min.
17. The first-strand cDNA can be kept at −20°C or used instantly.

Second-Strand Complementary DNAs Synthesis and Cleanup

18. Mix 20 μL of Second Strand Marking Master Mix with the first strand cDNA and keep it at 16°C for 1 h.
19. Add 90 μL of AMPure XP beads that have been warmed up to the cDNA and let it rest for 15 min.
20. Place the tube on a magnetic stand, throw discard the supernatant, and then rinse the beads twice with 200 μL of freshly made 80% ethanol each time.
21. Let the beads dry in the air for a short time, then put them back in 20 μL of Resuspension Buffer.
22. Take the 17.5 μL of the supernatant with the double-stranded cDNA to a different tube.
23. The double-stranded cDNA can be kept for up to 7 days at −20°C.

Adenylate 3′-Ends (Tailing Reaction)

24. Mix 12.5 μL of A-Tailing Mix with 17.5 μL of blunt-ended double-stranded cDNA.
25. Use a thermocycler to incubate the sample for 30 min at 37°C, and then raise the temperature to 70°C for 5 min.
26. Move directly to the step of adapter ligation.

Ligate Adapters and Cleanup

27. Put together the adapter, Ligation Mix, 30 μL of A-tailed cDNA, and Resuspension Buffer.
28. Incubate the sample for 10 min at 30°C.
29. After adding 5 μL of Stop Buffer, perform two cleanup procedures with 80% ethanol washes and AMPure XP beads.
30. Use 22.5 μL of Resuspension Buffer to elute the final adapter-ligated product.

Enrich DNA Fragments with PCR and Cleanup

31. Prepare a PCR reaction using 20 μL of the adapter-ligated product, 5 μL of PCR Primer Cocktail, 25 μL of PCR Master Mix.
32. Run the PCR under the following conditions:
    - Initial denaturation at 98°C for 30 s
    - 15 cycles of:
      - 98°C for 10 s
      - 60°C for 30 s
      - 72°C for 30 s
33. Use AMPure XP beads and 80% ethanol washes to clean up the amplified product.
34. Use 32.5 μL of Resuspension Buffer to elute the final DNA library and keep it at −20°C until further use.

3.5 Quantitative and Qualitative Analysis of the Library

3.5.1 Quantify Library Samples with Qubit Fluorometer [31]

Purpose: Measures total double-stranded DNA concentration in the library.

1. Prepare the working solution by mixing 1.99 mL of Qubit dsDNA HS buffer and 10 µL of Qubit dsDNA HS Reagent.
2. In separate tubes, prepare two standards and the library sample. For each standard, combine 190 µL of the working solution with 10 µL of the standard. For the library sample, combine 199 µL of the working solution with 1 µL of the sample.
3. Vortex the tubes for 2–3 s and incubate at room temperature for 2 min.
4. Read the tubes using the Qubit fluorometer.

3.5.2 Analyze the Size of Library Samples (Provided by: High Sensitivity DNA Kit for 2100 Bioanalyzer Systems Kit Guide)

Purpose: Checks the fragment size distribution and detects any adapter dimers or degradation.

Prepping the Reagents

1. Take the High Sensitivity DNA Dye Concentrate (blue cap) and the High Sensitivity DNA Gel Matrix (red cap) out of the fridge. Let them sit at room temperature for exactly 30 min to equilibrate.
2. Mix the Gel–Dye: Vortex the blue-capped Dye Concentrate for 10 s to ensure the DMSO is thawed and mixed.
3. Pipette 15 µL of the Dye Concentrate into the Gel Matrix vial.
4. Cap the vial, vortex for 10 s to mix thoroughly.
5. Transfer the mix to a spin filter and centrifuge it for 15 min (at 2240 *g*, or approx. 6000 rpm). This filtering step is crucial for removing particulates.
6. Protect the final Gel–Dye Mix from light and store at 2–8°C when not in use.

Priming and Loading the Chip

7. Prime the Chip: Get a fresh High Sensitivity DNA chip and place it securely on the Chip Priming Station. Follow the priming station procedure exactly:
   - Pipette 9.0 µL of the prepared (and room temperature) Gel–Dye Mix into the central well.
   - Close the lid, set the plunger to 1 mL, and push it down until the clip holds it.
   - Wait for exactly 60 s, release the clip, wait 5 s, and then slowly pull the plunger back up to 1 mL.

- Open the lid and add another 9.0 μL of the Gel–Dye Mix into the two remaining G wells.

8. Add Markers: Add 5 μL of the green-capped High Sensitivity DNA Marker to all 12 wells (the ladder well and the 11 sample wells).
9. Add Ladder and Samples: Add 1 μL of the yellow-capped High Sensitivity DNA Ladder into the well marked with the ladder symbol.
10. In the remaining 11 sample wells, add 1 μL of your prepared library sample.

Run the Assay

11. Vortex the Chip: Place the loaded chip horizontally on the IKA vortex mixer adapter and vortex for 60 s (1 min) at 2400 rpm.
12. Transfer the chip immediately to the Bioanalyzer instrument and start the run within 5 min of vortexing.
13. Once the 45-min run is complete, immediately remove the chip.
14. Evaluate the electropherogram for the expected library smear pattern. Look for:
    - Proper peak size (your target region)
    - A flat baseline
    - Absence of small peaks (e.g., adapter dimers) near the lower marker
15. Perform a quick electrode cleaning using the electrode cleaner chip and 350 μL of deionized water.

3.5.3 Quantify the Concentration of Library Samples with qPCR Assay

Purpose: Measures amplifiable library molecules

1. Dilute each library sample to a 1:10,000 and 1:20,000 dilution.
2. Prepare a qPCR master mix with PCR-grade water, Kapa Primer Premix, ROX, and KAPA SYBR FAST qPCR Master Mix.
3. Add 8 μL of the master mix and 2 μL of the appropriate standards, samples, or water to the wells of a qPCR plate. It is recommended to run each sample in triplicate.
4. Run the qPCR program with initial activation at 95°C for 5 min, followed by 35 cycles of denaturation, annealing, and extension.
5. Analyze the data to calculate the concentration of the library samples based on a standard curve.

3.5.4 Library Sample Pooling

Purpose: Combines multiple libraries at equal molar concentrations for multiplexed sequencing

Library sample pooling can be performed following the procedure suggested by Illumina, as summarized below (refer to Library Quantification and Quality Control Quick Reference Guide, Illumina).

1. Validate the sequence compatibility of the indexes used in the libraries.
2. Dilute each library sample to a concentration of 20 nM (or 10 nM) using a library dilution buffer [32].
3. Transfer an equal volume from each diluted library sample into a new tube and mix well.
4. Store the prepared library pool at −20°C, ready for sequencing.

3.6 High-Throughput Sequencing [33]

The goal of this step is to use a high-throughput sequencing platform to get high-quality sequence reads from the RNA-Seq libraries that have been prepared. The goal of sequencing is to get detailed transcriptome data that can be used to find and measure the genes that are involved in the production of glycosides in plants.

Cluster Generation on the Flow Cell

1. Follow the instructions for the sequencing platform to load the pooled library mixture onto the Illumina flow cell. (Inside the flow cell, each cDNA fragment binds to complementary oligonucleotides fixed on the surface and undergoes bridge amplification, forming dense clusters of identical DNA fragments.)
2. Proceed with sequencing once cluster generation is complete.

Sequencing by Synthesis

3. Start the sequencing run using the Illumina Control Software (ICS) once the cluster generation is completed.
4. Choose the paired-end sequencing configuration (e.g. 2 × 150 bp) to make sure that the alignment is more accurate and the assembly coverage is better.
5. Fluorescently labeled nucleotides are incorporated one base at a time during the sequencing-by-synthesis process.
6. The camera system recognizes each incorporated base and records the fluorescent signal that originates from it for base calling.
7. Until all template fragments have been read in accordance with the chosen run configuration, keep going through the sequencing cycle.

Data Acquisition and Output

8. The fluorescent signals produced during base incorporation will be recorded by the sequencing software.
9. These signals are automatically saved as FASTQ files after being transformed into nucleotide sequences.
10. The read sequences and matching quality scores are included in every FASTQ file.
11. To allocate each read to its corresponding sample, the data are demultiplexed using barcode indices.
12. Following generation, the raw data files are moved to a safe storage system for additional analysis, such as expression quantification, genome assembly, and quality evaluation.

3.7 Quality Assessment and Preprocessing of Raw Reads [34]

Validating and improving the quality of the raw sequencing reads derived from high-throughput sequencing is the aim of this step. This ensures that only high-quality reads are used to assemble the transcriptome to investigate how the genes involved in glycoside biosynthesis are expressed in plants.

1. **Retrieval of Raw Data**

   Firstly, get the FASTQ-format raw sequencing data files that the Illumina sequencing system created. Check that all of the files are named correctly, with the names of the samples, the conditions, and the number of times they were run. To keep everything organized, make a special folder structure with names such as "Sample_1," "Sample_2," and so on.

2. **Initial Quality Assessment**

   Use tools such as FastQC or MultiQC to check the quality of the data before executing anything else. Examine thoroughly the reports that have been given for the following parameters:

   - The quality of the per-base sequence: The mean Phred quality score for most bases should be at least 30.
   - GC content distribution: It should be in line with what is expected for the genome of the plant species.
   - Adapter contamination: This means that there are still sequencing adapters that need to be removed.
   - Levels of sequence duplication: A lot of duplication could mean that the library was over-amplified during preparation.

3. **Data Cleaning and Trimming**

   Remove unwanted sequences and bases based on the quality reports. Trimming is usually done to get rid of:

- Adapter sequences that were added during library preparation
- Bases that are not of good quality (especially at the ends of reads)

- Reads that are shorter than the acceptable limit, which is usually less than 50 bp.

After trimming, reads that are unclear and contain undefined bases ("N") are filtered out so that only high-confidence reads are kept.

4. **Reassessment of Quality**
   After trimming, use the same software to do a second quality check. To make sure that the overall read quality has improved a lot, compare the results after trimming to the original reports.
5. **Data Documentation**

Keep the quality reports for all samples and save them. Keep track of important numbers, including the total number of reads before and after trimming, the average read length, and the mean quality score. The documentation is important to keeping the analysis clear and reproducible of the analysis.

Expected Output

After preprocessing, "clean reads" represent those that are free of adapter contamination, low-quality bases, and sequencing artifacts. The reads that come out will have high Phred quality scores (usually >30) and a consistent GC content. The next step is to either assemble the transcriptome or align the references with these clean reads.

3.8 Transcriptome Assembly and Alignment

The objective of this section is to assemble and align high-quality RNA-Seq reads to a reference genome (if available) or construct the transcriptome de novo. This step aims to generate a comprehensive set of transcripts that represent the expressed genes in plant tissues related to glycoside biosynthesis [35].

1. **Selection of Assembly Strategy**
   Begin by determining whether a suitable reference genome is available for your plant species.
   - If a reference genome exists, proceed with reference-based alignment [36].
   - If the species lacks a reference genome, use de novo transcriptome assembly [37].

2. **Reference-Based Alignment**
   - Obtain the genome sequence and annotation files from databases such as NCBI or Ensembl Plants [38].
   - Prepare the genome index using alignment software (e.g., HISAT2) [39].
   - Align the high-quality reads to the reference genome [39].

- Convert the alignment results into sorted and indexed files for downstream analysis (using tools such as SAMtools) [40].
- Evaluate alignment statistics such as the percentage of mapped reads, coverage depth, and alignment accuracy.

3. **De Novo Transcriptome Assembly (for Non-Model Plants)** [37]
   - For plants that do not have a reference genome, use the Trinity software or a similar assembler.
   - Enter paired-end clean reads and set the best memory and CPU settings for assembly.
   - When the assembly is done, the output file (usually called "Trinity.fasta") will have the transcript sequences that were put back together.
   - Check the quality of the assembly by looking at things such as the N50 value, the total number of transcripts, and the average length of the transcripts.
4. **Assembly Validation and Redundancy Removal**
   - Use tools such as BUSCO (Benchmarking Universal Single-Copy Orthologs) to check for completeness [41].
   - Filter out short or duplicate contigs to make downstream expression analysis more accurate.
5. **Data Organization and Storage**
   - Keep all the assembled transcript files (FASTA) and alignment files (BAM/SAM) in organized folders.
   - Keep a metadata file that lists the software versions, assembly parameters, and quality metrics so that the process can be repeated.

Expected Output

Following this step, one of the following will be obtained:

Either a de novo assembled transcriptome (FASTA file) with non-redundant contigs signifying distinct transcripts, or a reference-aligned dataset (BAM files) displaying the location and abundance of each transcript.

These aligned or assembled sequences serve as the basis for glycoside biosynthesis gene annotation, differential expression analysis, and gene quantification.

3.9 Gene Expression Quantification

This step aims to analyze the aligned or assembled RNA-Seq data to quantify the abundance of transcripts in each sample. Consequently, possible to identify the genes involved in the biosynthesis of plant glycosides by providing a measure of gene expression levels [42].

1. **Preparation of Input Files** [43]

- For reference-based quantification: ensure BAM files can be sorted and indexed.
- For de novo quantification: use the assembled transcripts (FASTA) as a reference.
- Confirm that annotation files (GTF/GFF) match the reference genome.

2. **Mapping Reads to Transcripts or Genes**
   - Employ an appropriate quantification instrument:
     - For gene-level counting based on references, use FeatureCounts [44] or HTSeq-count [45].
     - For estimating transcript-level abundance, use Kallisto [46] or Salmon [47].
   - Provide the software with the reference/transcriptome and clean reads.
   - If applicable, specify paired-end reads. Select the appropriate alignment quality, read counting, and strandedness parameters.

3. **Normalization of Counts**
   - Normalize raw counts to account for sequencing depth and transcript length:
     - FPKM (Fragments Per Kilobase per Million mapped reads) [43]
     - TPM (Transcripts Per Million) [48]
   - Gene expression can be accurately compared across samples or experimental conditions by using normalized values.

4. **Output Generation**
   - For every gene or transcript, the program creates tables with read counts or normalized expression values.
   - Make sure every file is labeled with the sample ID and condition, and save the results in well-organized folders.

Expected Output

- Raw read counts and/or normalized expression values (FPKM/TPM) for every sample are included in a gene expression matrix.
- This matrix forms the foundation for determining which genes related to glycoside biosynthesis are highly expressed, lowly expressed, and differentially expressed under particular conditions.

3.10 Differential Gene Expression Analysis

To determine which genes, particularly those involved in glycoside biosynthesis, exhibit significantly different expression levels in various tissues, treatments, or conditions.

1. Preparation of Expression Matrix
   - Combine normalized expression values (raw counts are typically recommended as input for DGE packages into a single matrix [49].
   - Include all biological replicates and ensure correct sample labels.
2. Designing the Statistical Model
   - Define experimental conditions (e.g., control vs. treated, leaf vs. root).
   - Specify the model formula according to the study design in DESeq2 [49] or edgeR [50].
3. Performing Differential Expression Analysis
   - Import the expression matrix into the software.
   - Fit the statistical model to count data.
   - Estimate fold changes and significance levels (*p*-values, adjusted for multiple testing using FDR) [51].
4. Filtering Significant Genes
   - Apply thresholds such as |log2 fold change| ≥1 and FDR < 0.05 [34].
   - Extract genes meeting the criteria for upregulation or downregulation.
5. Visualization [34]
   - To visualize DGE results, create plots such as heatmaps, volcano plots, or MA plots.
   - Locate gene clusters linked to glycoside biosynthesis.

Expected Output

- A table containing fold change, *p*-values, and adjusted *p*-values for genes with differential expression.
- Plots that illustrate which genes are upregulated and downregulated.
- Gene candidates for functional analysis that may be involved in glycoside biosynthesis.

3.11 Functional Annotation of Genes

To predict the roles of genes, especially those involved in glycoside biosynthesis, by using databases and sequence similarity to give them biological meaning.

1. Sequence Similarity Search
   - Use BLAST to compare sequences to known databases (NCBI NR or Swiss-Prot).
   - Set an e-value threshold (e.g., 1e-5) to identify significant matches.
2. Gene Ontology Annotation
   - Assign GO terms [52] based on BLAST results or InterProScan domain predictions [53].
   - Categorize genes into biological process, molecular function, and cellular component.
3. Pathway Mapping
   - Assign annotated genes to KEGG pathways with an emphasis on the biosynthesis of secondary metabolites.
   - Emphasize the processes involved in the synthesis and modification of glycosides.
4. Data Organization
   - Generate tables summarizing gene IDs, predicted functions, GO terms, and pathway annotations.

Expected Output

- Annotated gene list with predicted functions and GO terms.
- KEGG pathway maps showing candidate glycoside biosynthesis genes.
- Summary tables linking expression levels to functional roles.

3.12 Pathway Analysis

To find metabolic pathways that are enriched with genes that are expressed differently, emphasizing pathways that are involved in the plant biosynthesis of glycosides.

1. Input Preparation
   - Compile a list of DEGs with associated functional annotations.
   - Include fold changes and significance values for weighting during enrichment analysis [34].
2. Pathway Mapping
   - Use KEGG Mapper or similar tools to place DEGs into known metabolic pathways.
   - Highlight pathways associated with secondary metabolites, specifically glycosides.
3. Enrichment Analysis

- Use statistical tests (such as the hypergeometric test or Fisher's exact test) to find pathways that are significantly enriched with DEGs.
- Optimize for multiple testing using Bonferroni [54] or FDR correction.

4. Visualization

- Create pathway diagrams that display genes that are upregulated and downregulated.
- Emphasize important enzymes (such as glycosyltransferases) that are involved in the biosynthesis of glycosides.

4 Notes

4.1 Sample Collection

4.1.1 Modified SDS-LiCl Method [22]

Principle

By overcoming the difficulties caused by the high starch and secondary metabolite content in plant seeds, the fundamental idea behind this modified SDS-LiCl method is to effectively isolate high-quality RNA. It uses a multistep procedure to accomplish this:

1. Lysis and Denaturation: The high SDS concentration functions as a potent detergent, dissolving cellular membranes and denaturing proteins, such as ribonucleases. By doing this, the RNA is released from the cells and shielded from deterioration.
2. Contaminant Removal: The high concentration of NaCl stabilizes the RNA and reduces polysaccharide contamination, while polyvinylpyrrolidone binds to and precipitates phenolic compounds.
3. Phase Separation: RNA is separated from residual polysaccharides and co-precipitating proteins using acidic phenol. While most proteins and DNA denature and separate into the organic phase or the interface, RNA stays in the aqueous phase.
4. Selective Precipitation: Transfer RNAs, 5S ribosomal RNAs, and DNA stay in solution, while large RNA molecules are selectively precipitated by lithium chloride. By increasing the ionic strength, sodium acetate increases precipitation efficiency.

4.1.2 Importance and Advantages

This approach works especially well for gene expression research in difficult plant tissues, such as the starch-rich seeds of maize, rice, and wheat, where conventional methods frequently do not work. The key advantages are:

- The optimized concentrations of key chemicals and the inclusion of a purification step with acidic phenol result in a higher yield of cleaner RNA.
- This protocol is designed to prevent the sample solidification and co-precipitation of starch that commonly occurs with other methods.
- While optimized for seeds, this method is also effective for other difficult plant tissues, including roots, and can be scaled for both small- and large-scale extractions.

4.1.3 Preparation of Reagents

- Extraction Buffer: To prepare the extraction buffer, dissolve 2.5 M NaCl, 2.5% polyvinylpyrrolidone, 100 mM Tris-HCl (pH 8.0), 25 mM EDTA, and 2% SDS in RNase-free water. Add 2.5% β-mercaptoethanol just before use in a fume hood due to its volatile nature and pungent odor.
- Acidic Phenol: Commercial acidic phenol solutions are available. Ensure the pH is maintained at 4.3. Phenol is highly corrosive and should be handled with appropriate personal protective equipment in a fume hood.
- Lithium Chloride: Prepare a 2 M stock solution of lithium chloride in RNase-free water.
- Sodium Acetate: Prepare a 3 M stock solution of sodium acetate (pH 4.8) in RNase-free water. Acetic acid can be used to adjust the pH.

4.2 Triton X-100 Lysis-Based RNA Extraction Protocol for Tissues Rich in Polyphenols

4.2.1 Principle of the Method [23]

To ensure that the final product is pure and undamaged, this procedure for extracting RNA from plant tissue that is high in polyphenols depends on a number of steps. Because it immediately freezes the tissue, mechanical grinding in liquid nitrogen is essential for stopping RNA degradation and RNase activity. Several ingredients are included in the Triton X-100 Extraction Buffer to ensure efficient lysis and contamination removal:

- SDS and Triton X-100 are detergents that denature proteins, including RNases, and damage cell membranes.

- In order to stop polyphenolic compounds from interfering with the RNA, PVPP is added specifically to bind and precipitate them.
- The following phenol-based organic solvent extractions: Chloroform and chloroform:isoamyl alcohol are used to separate denatured proteins and lipids (in the organic phase) from RNA (in the aqueous phase).
- The final DNase treatment step is crucial to remove any residual genomic DNA contamination, which can obstruct subsequent applications such as quantitative PCR.
- Lithium chloride is used for selective precipitation, which precipitates large RNA molecules while leaving smaller contaminants in solution.

4.2.2 Preparation of Reagents

- Triton X-100 Extraction Buffer: Mix 5 mL of 0.6 M Tris-base (pH 9.1), 4 mL of 5 M NaCl, 20 mL of 0.5 M EDTA/NaOH (pH 9.0), 20 mL of 10% SDS, and 0.5 mL of Triton X-100 to make 100 mL. Before using, add 0.31 g of fresh dithiothreitol (DTT) and use double-distilled water to bring the volume up to 100 mL.
- 5 M Potassium Acetate (KAc): Mix 80 mL of distilled water with 49.07 g of potassium acetate to dissolve it. Use glacial acetic acid to bring the pH down to 6.5 and increase the final volume to 100 mL.
- Phenol:Chloroform (PC, 50:50, v/v): Make this by combining equal parts of chloroform and phenol. The proper personal protective equipment (PPE) must be worn when handling this mixture in a fume hood.
- Chloroform:Isoamyl Alcohol (CI, 24:1, v/v): Make this by combining one part isoamyl alcohol with 24 parts chloroform.
- Lithium chloride (10 M): To get 100 mL, dissolve 42.39 g of lithium chloride in 80 mL of distilled water.

4.2.3 RiboReserve RNA Storage Solution

A commercially available product named RiboReserve RNA Storage Solution is used to protect RNA from RNase degradation. It offers a constant condition for keeping RNA samples for extended periods of time at room temperature. For researchers who need to transport samples or store them for later use without a freezer, this can be especially helpful.

Storing the RNA pellet in TE buffer or RNase-free water at −80°C for extended periods of time is a popular substitute for RiboReserve.

4.2.4 Limitations of the Protocol

This protocol successfully tackles the primary difficulties associated with handling polyphenolic-rich plant tissues by employing PVPP to eliminate phenolic compounds and utilizing LiCl precipitation to selectively extract RNA from polysaccharides. But a major drawback with this method is that it needs a separate DNase treatment step. This is necessary to get rid of any remaining DNA contamination, particularly if the RNA is going to be used for sensitive downstream applications such as quantitative reverse transcription PCR (RT-qPCR) or transcriptome analysis.

4.3 Modified SDS Based RNA Extraction from Polyphenol and Polysaccharide Rich Tissues [24]

4.3.1 Principle of the Method

The difficulties of RNA extraction from tissues high in polyphenols and polysaccharides, such as those of *Musa* species, are addressed by this modified SDS-based protocol. Sodium dodecyl sulfate (SDS), an anionic detergent, is used to first lyse the cells. In addition to inactivating RNases that would otherwise break down the RNA, the high concentration of SDS (30%) and a heat incubation step at 60°C effectively rupture cell membranes and separate proteins from the RNA. After lysis, the RNA is separated from other cellular components using a low-pH phenol-based extraction method. While most proteins and DNA are denatured and separate into the organic phenol layer or the interface at this acidic pH, RNA stays in the aqueous phase. LiCl precipitation, the last purification stage, is essential for RNA isolation. LiCl reduces contamination by selectively precipitating large RNA molecules while leaving the majority of polysaccharides and DNA fragments in solution. The RNA yield is further increased by the subsequent addition of ethanol and sodium acetate.

4.3.2 Preparation of Reagents

- Tris-HCl 1 M (pH 8.0): To make this, dissolve 6.055 g of Tris base in 30 mL of DEPC-treated water, then add concentrated HCl to raise the pH to 8.0. Finally, add DEPC-treated water to make the total volume 50 mL.
- 30% SDS: Mix 30 g of SDS with DEPC-treated water. Stir on a heated stir plate at 60°C until everything is dissolved, then add water to make 100 mL.
- 0.1 M Citrate Buffer (pH 4.2): This is prepared by mixing two stock solutions. Solution A: 0.1 M citric acid monohydrate (4.202 g in 200 mL DEPC-treated water). Solution B: 0.1 M trisodium citrate dihydrate (5.88 g in 200 mL DEPC-treated water). Combine 54 mL of Solution A with 46 mL of Solution B.
- 3 M Sodium Acetate (pH 4.8): Dissolve 4.92 g of sodium acetate in 12 mL of DEPC-treated water, adjust the pH to 4.8 with glacial acetic acid, and bring the final volume to 20 mL.

- Saturated Phenol with Citrate Buffer (pH 4.2): Melt phenol crystals at 65°C. Add an equal volume of the 0.1 M citrate buffer, mix for 15 min, and allow the phases to separate. Store at 4°C overnight before use.
- Phenol/Chloroform/Isoamyl Alcohol (25:24:1): Combine 50 mL of saturated phenol solution with 48 mL of chloroform and 2 mL of isoamyl alcohol. Before using, keep it in the fridge for at least 24 h.
- RNA Extraction Buffer: Mix 15 mL of DEPC-treated water, 5 mL of 100 mM Tris-HCl (pH 8), 2.5 mL of 0.5 M EDTA-Na2, 1.25 g of PVP K-30, and 7.305 g of NaCl. Before using, stir at 40°C, autoclave, and add 15 μL of β-mercaptoethanol to every 750 μL of buffer.
- Other Solutions: Prepare 0.5 M EDTA, 50× TAE electrophoresis buffer, and other standard solutions as needed.

4.4 RNA Extraction Method with Modified Extraction Buffer [25]

4.4.1 Principle of the Method

This RNA extraction method is based on a modified CTAB–phenol–chloroform method that works best on plant tissues that have a lot of polyphenols and polysaccharides. Adding SDS to the Extraction Buffer (EB) makes it easier to break down cells and proteins, which helps release nucleic acids from plant tissue. Polyvinylpyrrolidone (PVP) is necessary to bind and remove phenolic compounds that would otherwise oxidize and break down RNA. β-Mercaptoethanol is a strong reducing agent that stops oxidative damage and stops RNases from working. After lysis, proteins, lipids, and polysaccharides are removed in steps using chloroform:isoamyl alcohol (CI) and phenol:chloroform:isoamyl alcohol (PCI). RNA remains in the aqueous phase. Next, sodium acetate and ethanol are used to make the total nucleic acids fall out of solution. The next step, LiCl precipitation, selectively separates RNA by precipitating high-molecular-weight RNA and leaving DNA and polysaccharide contaminants in solution. The last steps of washing and dissolving make sure that the RNA is very pure and intact, which makes it good for downstream uses like RT-PCR or sequencing.

4.4.2 Preparation of Reagents

- 1 M Tris-HCl (pH 8.0): Weigh 12.1 g of Tris base and mix it with 80 mL of DEPC-treated water. Use concentrated HCl to bring the pH up to 8.0. Make the total volume 100 mL. Autoclave and keep at room temperature.
- 0.5 M EDTA (pH 8.0): To make this, mix 9.31 g of disodium EDTA dihydrate (Na2EDTA·2 H2O) with about 40 mL of DEPC-treated water. Use NaOH pellets to bring the pH to 8.0 (EDTA only dissolves when the pH is 8). Make the final volume 50 mL. Store at room temperature after autoclaving.

- 30% SDS (w/v): Dissolve 30 g of SDS in 80 mL of DEPC-treated water with gentle heating (55–60°C) and stirring until fully dissolved. Bring the volume to 100 mL and store at room temperature.
- 3 M Sodium Acetate (pH 5.2): Dissolve 24.6 g of sodium acetate trihydrate in 60 mL of DEPC-treated water. Adjust pH to 5.2 with glacial acetic acid and bring the final volume to 100 mL. Autoclave and store at 4°C.
- 10 M Lithium Chloride (LiCl): Dissolve 42.4 g of LiCl in 70 mL of DEPC-treated water. Bring the final volume to 100 mL. Filter-sterilize through a 0.22 μm membrane and store at 4°C.
- Chloroform:Isoamyl Alcohol (24:1): Mix 24 volumes of chloroform with 1 volume of isoamyl alcohol (e.g., 96 mL chloroform + 4 mL isoamyl alcohol). Store in an amber glass bottle at 4°C.
- Phenol:Chloroform:Isoamyl Alcohol (25:24:1): Combine 25 mL of Tris-saturated phenol (pH 8.0), 24 mL chloroform, and 1 mL isoamyl alcohol. Mix thoroughly and store at 4°C in the dark.
- Extraction Buffer (EB):

To prepare 100 mL of EB, mix the following:

  - 0.25 M NaCl: 1.46 g
  - 0.05 M Tris-HCl (pH 7.5): 0.61 g Tris base
  - 20 mM EDTA (pH 8.0): 4 mL of 0.5 M stock
  - 1% SDS (w/v): 1 g
  - 4% PVP (w/v): 4 g PVP-40

Dissolve all components in DEPC-treated water, adjust pH to 7.5, and bring the volume to 100 mL. Store at room temperature. Add β-mercaptoethanol (2%–3% v/v) freshly to the buffer immediately before use.

- 75% Ethanol (v/v): Mix 75 mL of absolute ethanol with 25 mL of DEPC-treated water. Chill before use.
- DEPC-Treated Water (0.1%): Add 1 mL of DEPC to 1 L of water, mix overnight, and autoclave the next day to deactivate residual DEPC. Store at room temperature.

4.5 Modified CTAB-TRIzol method [26]

4.5.1 Principle of the Method

This protocol is a modified method of the one described by [19]. It was designed to quickly get RNA out of challenging plant tissues that have a lot of polysaccharides and phenolic compounds. The first step in the method is to use a CTAB buffer to homogenize the tissue and break up the cells. After that, there is an important phase separation step with chloroform and TRIzol to get rid of proteins and other cell parts, leaving the RNA in the water phase. After that, LiCl and ethanol are used to make the RNA solid. Finally, the RNA pellet is washed to get rid of any extra RNA and make sure it is ready for use in other steps.

4.5.2 Reagent Preparation

- CTAB Extraction Buffer: Prior to use, the protocol recommends adding 5% (v/v) β-mercaptoethanol (β-ME) to the CTAB extraction buffer. The recipe for the base CTAB buffer itself is not included in the text that is provided. The concentrations of components such as CTAB, Tris, EDTA, and NaCl, which are frequently included in standard CTAB protocols, can differ based on the type of plant.
- TRIzol Reagent: This is a ready-to-use, commercially available monophasic solution of guanidine thiocyanate and phenol. Its purpose is to inhibit RNase activity and lyse cells.
- Chloroform: Isoamyl alcohol (CIA, 24:1 v/v): 24 parts chloroform and 1 part isoamyl alcohol are mixed to make this solution. By separating the organic and aqueous phases, this mixture increases the extraction efficiency of nucleic acids.
- A 10 M lithium chloride solution is made by dissolving 424 g of anhydrous LiCl in 1 L of distilled water.

4.6 Modified Hot Borate RNA Extraction Method [27]

4.6.1 Principle of the Method

The purpose of this modified hot borate RNA extraction protocol is to separate high-quality RNA from plant tissues, which frequently contain contaminants such as polyphenols, RNases, and other substances that can break down RNA and obstruct subsequent uses. The method relies on several key principles:

- Hot Borate Extraction: The borate extraction buffer, heated to 80°C, quickly inactivates RNases, preventing RNA degradation.
- Detergents and Chelating Agents: The buffer contains SDS and NP-40 for cell lysis and protein denaturation, while EGTA chelates divalent cations to further inhibit RNases.
- Protein and Contaminant Removal: Proteinase K is added to digest proteins. PVP-40 and sodium deoxycholate help to remove polyphenols and other secondary metabolites.
- Several Precipitation and Washing Steps: The method removes impurities such as DNA, proteins, and polysaccharides while selectively isolating RNA through a sequence of precipitations with LiCl, potassium acetate, and ethanol. The removal of any remaining contaminants is guaranteed by the numerous washing stages, especially the 2 M LiCl wash.
- To guarantee the highest purity and integrity of the RNA, which makes it appropriate for sensitive applications such as quantitative PCR, the DNase I treatment and the RNeasy Mini Kit-based cleanup that follows are optional but advised.

4.6.2 Preparation of Reagents

- Borate Extraction Buffer: Weigh 7.63 g of sodium tetraborate decahydrate, 1.14 g of EGTA, 1.0 g of SDS, 1.0 g of sodium deoxycholate, and 2.0 g of PVP-40 to make 100 mL. Add approximately 75–80 mL of Millipore ultrapure water and 0.5 mL of NP-40. Adjust the volume to 100 mL after stirring until dissolved. Store at room temperature after autoclaving.
- 1 M KCl: Dissolve 7.456 g of KCl in about 70 mL of Millipore ultrapure water. Stir until dissolved, adjust the volume to 100 mL, autoclave, and store at room temperature.
- 5 M LiCl: Dissolve 21.195 g of LiCl in about 70 mL of Millipore ultrapure water. Stir until dissolved, adjust the volume to 100 mL, autoclave, and store at room temperature.
- 2 M LiCl: Dissolve 8.478 g of LiCl in about 70 mL of Millipore ultrapure water. Stir until dissolved, adjust the volume to 100 mL, autoclave, and store at 4°C.
- 10 mM Tris-HCl (pH 7.5): Dissolve 0.121 g of Tris base in about 50 mL of Millipore ultrapure water. Adjust the pH to 7.5 with 5 N HCl, adjust the volume to 100 mL, autoclave, and store at room temperature.
- 5 M KAc (pH 6.5): Dissolve 49.07 g of KAc in about 50 mL of Millipore ultrapure water. Adjust the pH to 6.5 with glacial acetic acid, adjust the volume to 100 mL, autoclave, and store at room temperature.
- 1 M DTT: Dissolve 154.24 mg of DTT in 1 mL of Millipore ultrapure water, divide into five 200-μL aliquots, and store at −20°C.
- 70% ethanol: Mix 70 mL of absolute ethanol with 30 mL of Millipore ultrapure water.
- Other Reagents: Proteinase K solution, absolute ethanol, RNase-free water, DNase I solution, and the RNeasy Mini Kit are commercially available and should be stored as instructed.

4.7 RNA Quality and Quantification

4.7.1 RNA Concentration and Purity assessment using the Nanodrop spectrophotometer

Importance and Interpretation of Ratios

Spectrophotometer measurements include the absorbance of all molecules in the sample that absorb at the wavelength of interest.

- A260 /A280 Ratio: This ratio is used to assess the purity of DNA and RNA. A ratio of ~2.0 is generally accepted as "pure" for RNA. A ratio lower than this may indicate contamination by protein, phenol, or other substances that absorb near 280 nm.
- A260 /A230 Ratio: This ratio serves as a secondary measure of nucleic acid purity. For pure nucleic acids, the expected values

are typically in the range of 2.0–2.2 and are often higher than the A260 /A280 values. A ratio that is significantly lower than expected may indicate the presence of contaminants that absorb at 230 nm, such as organic compounds or salts.

4.7.2 RNA Quality Assessment using 1% Agarose Gel Electrophoresis

Importance and Interpretation

A critical step in evaluating the isolated RNA's visual quality is gel electrophoresis. The presence of distinct, sharp bands that correspond to the 26S and 18S ribosomal RNAs shows that the RNA is high quality and intact. On the other hand, if there is a faint or smeared pattern, it indicates that the RNA was damaged during handling or extraction. Reliability in downstream applications such as RNA-Seq depends on the integrity of the RNA, which is confirmed by little to no smearing. An additional general indicator of RNA quality is the ratio of the 26S to 18S rRNA band intensity.

4.7.3 RNA Integrity Analysis Using Agilent Bioanalyzer 2100

Importance and Interpretation

The Agilent Bioanalyzer 2100 is an important tool for checking the integrity of RNA, which is necessary for RNA-Seq and other downstream applications to work. It shows RNA fragments in a graph and figures out the RNA Integrity Number (RIN).

- RIN Values: The RIN is a number between 1 and 10 that shows how well the RNA is working. A high RIN value (usually between 8 and 10) means that the RNA is of high quality and has not broken down much, which is perfect for gene expression studies.
- Understanding the Results: The Bioanalyzer's electropherogram will show clear peaks for the ribosomal RNA subunits. The 26S and 18S rRNA bands are for plants. A good RNA sample will have clear, sharp peaks, while a bad sample will have a smeared pattern. The Bioanalyzer uses an algorithm to figure out the RIN from the entire profile. This gives an objective measure of RNA quality.

4.8 RNA-Seq Library Preparation

Several of the reagents listed, such as Bead Washing Buffer, Resuspension Buffer, and different master mixes, are usually parts of a commercial library preparation kit that are only available to the company that makes them. The kit usually comes with instructions for how to use these reagents, so the user should follow the manufacturer's specific instructions for how to prepare and store them. Some reagents, for instance, need to be kept at −20°C after being used once. To ensure the best binding efficiency, it is a good idea to warm up the beads to room temperature before using them.

4.9 Quantitative and Qualitative Analysis of the Library

Importance and Interpretation

- Qubit Fluorometer: The Qubit fluorometer is a very accurate and sensitive way for quantifying DNA. It uses fluorescent dyes that only stick to the target nucleic acid, which makes the concentration measurement more accurate because it does not pick up on contaminants that might absorb at 260 nm.

- Agilent Bioanalyzer: This machine does a full quality check of the library by making an electropherogram that shows the size distribution of DNA fragments. A good library preparation will show a clean, sharp peak in the expected size range (e.g.,, 200–500 bp). This means that the fragments are the right size and there are not too many primer dimers or adapter dimers.
- qPCR Assay: This is the best way to find out how much of a sequencing library there is. It only counts the pieces that can be amplified by PCR, which are the ones that will be sequenced. This makes sure that the library is at the right concentration for clustering on the flow cell, which stops it from clustering too much or too little.

4.10 High-Throughput Sequencing

Principle

Millions of DNA or cDNA fragments can be sequenced simultaneously by using high-throughput sequencing (HTS), also referred to as next-generation sequencing (NGS). To capture the entire gene expression profile of a plant tissue, transcriptome studies transform RNA into complementary DNA (cDNA), which is subsequently sequenced.

Illumina sequencing technology is used in this protocol because it produces paired-end reads with high depth and accuracy, which are essential for complex plant transcriptomes. Before sequencing by synthesis, each library is subjected to amplification and cluster formation on a flow cell. The instrument's software transforms the resulting fluorescent signals into nucleotide sequences, producing raw data in FASTQ format [33, 34].

Notes:

- To guarantee data quality, make sure libraries have an RNA Integrity Number (RIN) higher than 7.0 before sequencing.
- Because paired-end sequencing increases annotation accuracy and facilitates the assembly of overlapping transcripts, it is advised for plants with complex genomes.
- To avoid RNA degradation, handle all materials in an RNase-free environment.
- The statistical robustness of subsequent gene expression analyses is enhanced by adequate biological replication, which calls for at least three biological replicates per condition.

4.11 Quality Assessment and Preprocessing of Raw Reads

Principle

Following sequencing, technical errors like low-quality bases, adapter sequences, or ambiguous reads are frequently present in the data produced (in FASTQ format). Errors in gene quantification, assembly, and mapping may result from these problems. Consequently, to detect and eliminate problematic sequences, a quality assessment step is required [34].

Quality evaluation is the first step, during which visualization tools like FastQC are used to check parameters like base quality score, GC content, sequence duplication, and adapter contamination. After that, low-quality regions and adapters are eliminated through preprocessing or trimming using programs like Trimmomatic or Cutadapt. This procedure guarantees precise downstream analysis and increases the dependability of transcript quantification [55].

Notes:

- Keep a copy of the original raw data in a different folder at all times. This lets you check things out if data is lost or there are mistakes in processing.
- Do not trim too much, as this can cut down on the amount of data and affect the coverage of the transcript.
- Make sure to use paired-end reads when trimming to keep the read pairing correct.
- To make sure the samples can be compared, the same parameters should be used for all of them.
- All data processing should be done on a computer with enough memory and backup support to keep the data safe.

4.12 Transcriptome Assembly and Alignment

Principle

Transcriptome assembly is the process of putting together full-length transcripts from short sequencing reads taken from RNA-Seq data. There are two main ways to do it:

1. Reference-based assembly, which means that reads are lined up with a known genome sequence from the same or a closely related plant species.
2. De novo assembly, which is often used for non-model plants, is when transcripts are put back together without a reference genome [37].

Reference-based alignment tools such as HISAT2 or STAR map sequencing reads to the reference genome with great accuracy. This helps count known genes and find splice variants. De novo assembly (e.g., using Trinity), on the other hand, groups and joins overlapping reads based only on how similar their sequences are. This creates contiguous sequences (contigs) that represent transcripts [56].

The identification of potential genes involved in glycoside biosynthesis pathways is based on this step.

Notes

- To capture low-abundance transcripts for de novo assembly, make sure the sequencing depth is adequate (ideally ≥30 million reads per sample).
- Paired-end sequencing is especially useful for resolving complex gene structures in plants with large or polyploid genomes.
- Use BUSCO scores to assess the completeness of assembled transcriptomes; a completeness >85% denotes a high-quality assembly.
- To prevent misassembled contigs, do not mix reads from various experimental conditions during assembly.
- For reproducibility and transparency, always keep track of software versions, parameters, and input file names.

4.13 Gene Expression Quantification

Principle

Quantifying gene expression involves determining the number of sequencing reads that correspond to each gene or transcript. Reads are mapped to reference genome genes that have been annotated, and expression levels are quantified using raw counts or normalized values such as TPM (Transcripts Per Million) or FPKM (Fragments Per Kilobase of transcript per Million mapped reads).

Similar normalization techniques are applied for de novo assemblies, which map reads back to the assembled transcripts. Precise measurement enables cross-sample comparison and the detection of genes with differential expression in biosynthetic pathways, including cytochrome P450 enzymes that form glycosides and UDP-glycosyltransferases [57, 58].

4.14 Differential Gene Expression Analysis

Principle

In order to identify statistically significant differences, differential gene expression (DGE) analysis compares normalized expression values (such as TPM, FPKM, or raw counts) across samples or conditions. Biological variability and sequencing depth are taken into account by statistical models. DGE assists in identifying potential genes that may control the synthesis of glycosides, such as cytochrome P450s or glycosyltransferases [49].

DESeq2, edgeR, and limma-voom are popular tools. In order to determine which genes are upregulated or downregulated, these tools model count data using negative binomial distributions or other suitable statistical frameworks.

4.15 Functional Annotation of Genes

Principle

By comparing sequences to carefully curated databases, functional annotation determines the roles of genes. Typical methods consist of:

- BLAST looks for homologs against UniProt or NCBI.
- Gene Ontology (GO) annotation for cellular constituents, molecular activities, and biological processes

- To link genes to metabolic pathways, use KEGG pathway mapping.

This step connects biological processes and pathways related to glycoside synthesis to genes that are differentially expressed [52, 59].

4.16 Pathway Analysis

Principle

Using databases such as KEGG or PlantCyc, pathway analysis links genes with differential expression to metabolic pathways. This makes it easier to understand how variations in gene expression impact biochemical pathways and the synthesis of secondary metabolites. Enrichment analysis establishes whether genes linked to specific pathways are overrepresented in DEGs [59].

Acknowledgments

The authors acknowledge the unwavering support and inspiration of Naran Lala College of Professional and Applied Science in shaping their academic journey and contributing to the realization of this chapter.

References

1. Li L et al (2023) Natural products and extracts from plants as natural UV filters for sunscreens: a review. Plants 6(3):183–195
2. Heim KE, Tagliaferro AR, Bobilya DJ (2002) Flavonoid antioxidants: chemistry, metabolism and structure–activity relationships. J Nutr Biochem 13(10):572–584
3. Redmond JW et al (1986) Flavones induce expression of nodulation genes in Rhizobium. Nature 323(6089):632–635
4. Boter M, Díaz I (2023) Cyanogenesis, a plant defence strategy against herbivores. Int J Mol Sci 24(8):6982
5. Gleadow RM, Møller BL (2014) Cyanogenic glycosides: synthesis, physiology, and phenotypic plasticity. Annu Rev Plant Biol 65(1):155–185
6. Singh B, Kaur A (2018) Control of insect pests in crop plants and stored food grains using plant saponins: a review. LWT 87:93–101
7. Yang-Hua Y (1992) A triterpenoid saponin from Phytolacca esculenta. Phytochemistry 31(7):2552–2554
8. Ohtani K, Mavi S, Hostettmann K (1993) Molluscicidal and antifungal triterpenoid saponins from Rapanea melanophloeos leaves. Phytochemistry 33(1):83–86.
9. Dolma SK et al (2018) Insecticidal activities of tea saponin against diamondback moth, Plutella xylostella and aphid, Aphis craccivora. Phytoparasitica 37(1):52–55
10. Berlowska J et al (2015) Cell lysis induced by membrane-damaging detergent saponins from Quillaja saponaria. Biochim Biophys Acta 75:44–48
11. Boeckler GA, Gershenzon J, Unsicker SB (2011) Phenolic glycosides of the Salicaceae and their role as anti-herbivore defenses. Phytochemistry 72(13):1497–1509
12. Staub H (1959) Pharmacology of cardiac glycosides. Am J Cardiol 3(6):776–793
13. Ren C et al (2025) Flavonoid UDP-glycosyltransferase in plants: functional identification, substrate recognition mechanism, and biotechnology application. Plant J 24(5):4451–4474
14. Vogt T, Jones P (2000) Glycosyltransferases in plant natural product synthesis: characterization of a supergene family. Trends Plant Sci 5(9):380–386
15. Howe GA, Jander G (2008) Plant immunity to insect herbivores. Annu Rev Plant Biol 59:41–66
16. Arya M et al (2005) Basic principles of real-time quantitative PCR. Expert Rev Mol Diagn 5(2):209–219
17. Gutierrez L et al (2008) The lack of a systematic validation of reference genes: a serious

pitfall undervalued in reverse transcription–polymerase chain reaction (RT-PCR) analysis in plants. Plant Methods 6(6):609–618
18. Floris M et al (2009) Post-transcriptional regulation of gene expression in plants during abiotic stress. Plant Physiol 10(7):3168–3185
19. Chang S, Puryear J, Cairney J (1993) A simple and efficient method for isolating RNA from pine trees. Plant Mol Biol Rep 11(2):113–116
20. Salzman R et al (1999) An improved RNA isolation method for plant tissues containing high levels of phenolic compounds or carbohydrates. Plant Mol Biol Rep 17(1):11–17
21. Gasic K, Hernandez A, Korban SS (2004) RNA extraction from different apple tissues rich in polyphenols and polysaccharides for cDNA library construction. Plant Mol Biol Rep 22(4):437–438
22. Vennapusa AR et al (2020) A universal method for high-quality RNA extraction from plant tissues rich in starch, proteins and fiber. Sci Rep 10(1):16887
23. Mandaliya VB, Bhatt P, Thaker VS (2024) High quality RNA extraction protocol for polyphenolics-rich cotton tissue. Appl Biochem Biotechnol 694:115604
24. Saha K et al (2025) Optimized SDS-based protocol for high-quality RNA extraction from Musa spp. Plants 8(1):21
25. Liu L et al (2018) A method for extracting high-quality total RNA from plants rich in polysaccharides and polyphenols using Dendrobium huoshanense. PLoS One 13(5):e0196592
26. Sasi S et al (2023) DNA-free high-quality RNA extraction from 39 difficult-to-extract plant species and its validation for downstream molecular applications. Plant Direct 19(1):84
27. Wan CY, Wilkins TA (1994) A modified hot borate method significantly enhances the yield of high-quality RNA from cotton (Gossypium hirsutum L.). Anal Biochem 223(1):7–12
28. Thermo Fisher Scientific (2009) USA T042 Technical Bulletin. NanoDrop Spectrophotometers. Thermo Fisher Scientific Inc
29. Arunachalam K, Sreeja PS (2025) Agarose gel electrophoresis for RNA integrity assessment. In: Advanced cell and molecular techniques: protocols for in vitro and in vivo studies. Springer, pp 121–124
30. Kuschel M (2000) Analysis of total RNA using the Agilent 2100 Bioanalyzer and the RNA 6000 LabChip® kit. Agilent Technologies Application Note
31. Naidoo A (2024) Towards molecular autopsies: internal validation of the Qubit 1X dsDNA HS Assay Kit. BioTechniques
32. Smollett K et al (2022) Library pooling and quality control for Illumina sequencing. Protocols.io
33. Metzker ML (2010) Sequencing technologies—the next generation. Nat Rev Genet 11(1):31–46
34. Conesa A et al (2016) A survey of best practices for RNA-seq data analysis. Genome Biol 17(1):13
35. Wang Z, Gerstein M, Snyder M (2009) RNA-Seq: a revolutionary tool for transcriptomics. Nat Rev Genet 10(1):57–63
36. Langmead B, Salzberg SL (2012) Fast gapped-read alignment with Bowtie 2. Nat Methods 9(4):357–359
37. Grabherr MG et al (2011) Full-length transcriptome assembly from RNA-Seq data without a reference genome. Nat Biotechnol 29(7):644–652
38. Howe KL et al (2021) Ensembl 2021. Nucleic Acids Res 49(D1):D884–D891.
39. Kim D et al (2019) Graph-based genome alignment and genotyping with HISAT2 and HISAT-genotype. Nat Biotechnol 37(8):907–915
40. Li H et al (2009) The sequence alignment/map format and SAMtools. Bioinformatics 25(16):2078–2079
41. Simão FA et al (2015) BUSCO: assessing genome assembly and annotation completeness with single-copy orthologs. Bioinformatics 31(19):3210–3212
42. Mortazavi A et al (2008) Mapping and quantifying mammalian transcriptomes by RNA-Seq. Nat Methods 5(7):621–628
43. Li H (2009) The sequence alignment/map (SAM) format and SAMtools 1000 genome project data processing subgroup. Bioinformatics 25:1
44. Liao Y, Smyth GK, Shi W (2014) featureCounts: an efficient general-purpose program for assigning sequence reads to genomic features. Bioinformatics 30(7):923–930
45. Anders S, Pyl PT, Huber W (2015) HTSeq—a Python framework to work with high-throughput sequencing data. Bioinformatics 31(2):166–169
46. Bray NL et al (2016) Near-optimal probabilistic RNA-seq quantification. Nat Biotechnol 34(5):525–527
47. Patro R et al (2017) Salmon provides fast and bias-aware quantification of transcript expression. Nat Methods 14(4):417–419
48. Wagner GP, Kin K, Lynch VJ (2012) Measurement of mRNA abundance using RNA-seq data: RPKM measure is inconsistent among samples. Trends Biotechnol 131(4):281–285
49. Love MI, Huber W, Anders S (2014) Moderated estimation of fold change and dispersion for RNA-seq data with DESeq2. Genome Biol 15(12):550
50. Robinson MD, McCarthy DJ, Smyth GK (2010) edgeR: a Bioconductor package for differential expression analysis of digital gene expression data. Bioinformatics 26(1):139–140
51. Benjamini Y, Hochberg Y (1995) Controlling the false discovery rate: a practical and powerful

approach to multiple testing. J R Stat Soc Ser B 57(1):289–300

52. Ashburner M et al (2000) Gene ontology: tool for the unification of biology. Nat Genet 25(1):25–29
53. Jones P et al (2014) InterProScan 5: genome-scale protein function classification. Bioinformatics 30(9):1236–1240
54. Storey JD, Tibshirani R (2003) Statistical significance for genomewide studies. Proc Natl Acad Sci USA 100(16):9440–9445
55. Bolger AM, Lohse M, Usadel B (2014) Trimmomatic: a flexible trimmer for Illumina sequence data. Bioinformatics 30(15):2114–2120
56. Haas BJ et al (2013) De novo transcript sequence reconstruction from RNA-seq using the Trinity platform for reference generation and analysis. Nat Protoc 8(8):1494–1512
57. Trapnell C et al (2012) Differential gene and transcript expression analysis of RNA-seq experiments with TopHat and Cufflinks. Nat Protoc 7(3):562–578
58. Li B, Dewey CN (2011) RSEM: accurate transcript quantification from RNA-Seq data with or without a reference genome. BMC Bioinformatics 12(1):323
59. Kanehisa M et al (2021) KEGG: integrating viruses and cellular organisms. Nucleic Acids Res 49(D1):D545–D551.

Check for updates

Chapter 25

Polyketides: Molecular Approaches to Polyketide Biosynthesis

Mandar Kulkarni, Harshvadan Patel and Ketankumar Panchal

Abstract

Polyketides form a diverse class of biologically active secondary metabolites synthesized by microorganisms, plants, and fungi. Their structural variety and potent pharmacological properties have made them pivotal leads in drug discovery and biotechnology. The molecular basis of polyketide biosynthesis lies in the action of polyketide synthases (PKSs), which adopt modular, iterative, or simple architectures (Types I, II, and III) to build complex carbon scaffolds via sequential condensation of acyl-CoA precursors. Contemporary molecular methods now enable identification of biosynthetic gene clusters, domain-level characterization of PKS modules, heterologous expression of clusters, genome mining, pathway engineering, and synthetic biology-driven refactoring of PKS pathways. In this chapter, we focus on the molecular-level strategies for exploring, manipulating, and optimizing polyketide biosynthesis. Three detailed approaches are elaborated: the first for in silico mining and annotation of PKS gene clusters from genomes, the second for wet-lab PCR amplification and domain-specific analysis of a selected PKS module, and the third for expression of PKS genes using recombinant technology. These approaches offer researchers in natural-product science and molecular microbiology practical tools for accessing the rich polyketide repertoire and for engineering novel derivatives.

Keywords Domain architecture, Heterologous expression, Polyketide synthase, PKS gene cluster mining, Secondary metabolites, Synthetic biology

1 Introduction

Polyketides form a very important group of natural compounds that are produced by microorganisms, plants, and some fungi. Many well-known medicines that we use today, such as erythromycin, tetracycline, doxorubicin, and lovastatin, belong to this group and demonstrate their significant role in modern healthcare [1–3]. Their structures vary widely, and this diversity mainly comes from the way polyketide synthases (PKSs) assemble simple acyl-CoA building blocks into long and modified carbon chains. PKSs are generally classified into Type I, Type II, and Type III systems based on their organization and catalytic behavior [1, 4, 5].

Dhruti Amin et al. (eds.), *Phytoconstituent Analysis*, Springer Protocols Handbooks,
https://doi.org/10.1007/978-1-0716-5400-2_25,

With progressive advancement of biosciences, researchers have been able to understand PKS systems in a much deeper. Each PKS is made up of several catalytic parts, and every domain whether it is KS, AT, ACP, or the reducing domains such as KR, DH, or ER collectively performs a specific reaction that shapes the polyketide chain step by step [4, 6]. Even a small change in one domain often alters the final molecule quite noticeably, and this is one reason why polyketides show so much natural variation. The entire set of genes required for a particular polyketide biosynthesis is arranged together in a compact gene cluster. Because these clusters are usually well-organized, genome-mining tools such as antiSMASH, PRISM, and NaPDoS have become very helpful for quickly identifying PKS regions in newly sequenced genomes [7–9]. These tools not only point out the cluster but also give an idea of the domain arrangement and possible product type, which makes the initial analysis much easier for researchers.

In recent years, interest has shifted from only identifying polyketides to modifying their pathways for improved production or to generate new structural analogs. Heterologous expression approaches using engineered *Escherichia coli* and *Streptomyces* hosts have played an important role by allowing *PKS* genes or entire clusters to be expressed outside the native organism [10, 11]. Gene-editing strategies, including CRISPR-based methods, are now also used to alter PKS modules, inactivate selected domains, or reorganize entire clusters to produce novel or higher-yielding metabolites [11].

The present chapter discusses molecular approaches that are commonly used in the study of polyketide biosynthesis. To maintain practical value, three protocols are included. The first protocol explains how to identify and annotate PKS biosynthetic gene clusters through genome mining. The second protocol describes PCR-based amplification and domain-level characterization of *PKS* genes. The third protocol describes the expression of these clusters in an expression system. Together, these approaches help in linking gene sequences with their biosynthetic functions, which is essential for understanding and engineering polyketide pathways for research and biotechnological applications.

2 Materials

2.1 Genome Mining and Annotation of PKS Biosynthetic Gene Clusters Using antiSMASH [1, 6–8, 11, 12]

1. antiSMASH web server.
2. NCBI Genome/WGS database.
3. BLASTp (optional).
4. NaPDoS or NaPDoS2 (optional).
5. PRISM web server (optional).

6. Computer system.
7. Stable internet connection.
8. Genome FASTA file.
9. GenBank (.gbk) output files.
10. HTML and JSON result files.

2.2 PCR Amplification and Domain-Level Analysis of PKS Genes (KS, AT, ACP) [3, 12–15]

1. Genomic DNA of the selected organism.
2. Nuclease-free water.
3. PCR buffer (10×) with $MgCl_2$.
4. dNTP mix (10 mM each).
5. Forward and reverse primers specific to PKS domains (KS, AT, ACP).
6. *Taq* DNA polymerase or high-fidelity polymerase.
7. PCR mastermix (2×) (optional).
8. Agarose powder.
9. Ethidium bromide or similar safe DNA stain.
10. 1× TAE or 0.5× TBE buffer.
11. Molecular weight DNA ladder.
12. Gel loading dye.
13. Equipment: thermal cycler, microcentrifuge, gel electrophoresis unit, and gel documentation system

2.3 Heterologous Expression of a Polyketide Synthase (PKS) Gene in a Suitable Microbial Host [9, 16–18]

1. Genomic DNA or cloned PKS gene from donor organism.
2. High-fidelity DNA polymerase.
3. Restriction enzymes (compatible with vector MCS).
4. Ligation mix or Gibson Assembly reagents.
5. Expression vector (e.g., pSET152, pUZ8002, pET28b, pETDuet).
6. Competent *E. coli* cells (DH5α for cloning; BL21(DE3) for expression).
7. *Streptomyces* spores or competent protoplasts (if using actinomycete hosts).
8. LB, ISP-2, or TSB media.
9. Antibiotics (kanamycin, apramycin, chloramphenicol, depending on vector).
10. Inducer (IPTG for *E. coli*; thiostrepton for *Streptomyces*).
11. Plasmid isolation kit.
12. DNA gel purification kit.
13. Nuclease-free water.

14. Equipment: PCR thermal cycler, electroporator or heat-shock setup, shaking incubator, centrifuge, spectrophotometer, SDS-PAGE apparatus (optional), and HPLC or LC-MS instrument (for metabolite detection).

3 Methods

3.1 Genome Mining and Annotation of PKS Biosynthetic Gene Clusters Using antiSMASH [1, 6–8, 11]

1. Download the whole-genome sequence of the selected organism from the NCBI database in FASTA format (complete or high-quality draft assemblies). Remove any extra text or annotation lines that do not belong to FASTA formatting. Rename the file with a short, easy-to-identify name for convenience during analysis.
2. Open the antiSMASH web server in a browser and select "Submit new job" from the homepage.
3. Upload the prepared FASTA file in the input section and choose the correct organism category (Bacteria/Fungi).
4. Enable analysis options such as domain prediction, active-site finder, and cluster border detection. Essential analysis options to enable are KnownClusterBlast, ClusterBlast, and SubClusterBlast, optionally ActiveSiteFinder and TIGRFam. Submit the job and wait for antiSMASH to complete the genome analysis.
5. After the run finishes, download all output files from the results page. Save the HTML report, GenBank cluster files, and JSON summary into a dedicated folder. Open the HTML report and examine the list of predicted biosynthetic gene clusters.
6. Identify clusters labeled as Type I PKS, Type II PKS, or Type III PKS. Note the cluster number, region length, and the number of genes in each cluster. Record the predicted product class shown by antiSMASH.
7. Open each PKS cluster GenBank (.gbk) file in a genome viewer such as Artemis (or SnapGene, Geneious). Examine the arrangement of core PKS domains such as KS, AT, ACP, KR, DH, and ER.
8. Verify whether the domain order follows expected PKS architecture (modular/iterative). Check for the presence of accessory or tailoring enzymes such as cyclases, oxidoreductases, or methyltransferases.
9. Copy the amino-acid sequences of KS and AT domains if additional validation is needed. Run BLASTp for these domains

to identify close homologs in the NCBI protein database (optional).

10. Compare the BLAST results to known PKS pathways to understand expected metabolite types.
11. Upload KS or AT domain sequences to NaPDoS/NaPDoS2 for phylogenetic classification (optional). Save NaPDoS classification tables for documentation and comparison.
12. Review antiSMASH-predicted core structures or scaffold sketches when available. Compare predicted scaffolds with related known polyketide structures.
13. Organize the findings into a table summarizing cluster number, PKS type, domain order, cluster size, and predicted product. Export individual cluster sequence in FASTA and GenBank formats for downstream PCR primer design and heterologous expression experiment.
14. Store all downloaded and generated files in a structured project folder. Proceed to downstream validation or wet-lab analysis based on project objectives.

3.2 PCR Amplification and Domain-Level Analysis of PKS Genes (KS, AT, ACP) [3, 12–15]

1. Extract genomic DNA from the selected microorganism using respective standard DNA isolation procedure (kit or manual) followed by DNA quantification.
2. Design primers targeting conserved motifs within KS, AT, or ACP domains using targeted reference PKS genes or alignments. Verify primer specificity using BLAST against the organism's genome or public PKS databases.
3. Prepare the PCR reaction mixture by adding buffer, $MgCl_2$, dNTPs, primers, polymerase, and genomic DNA. Alternately, PCR master mix to final concentration 1× can be used in place of individual components. Include a no-template control (NTC) for contamination check.
4. Mix gently and briefly spin down to ensure all components settle at the bottom of the tube.
5. Place PCR tubes into the thermal cycler and set an initial denaturation step at 95°C for 3–5 min. Program 30–35 amplification cycles depending on DNA template quality. Use an annealing temperature based on primer Tm; perform gradient PCR from Tm − 5°C to Tm + 5°C to optimize annealing temperature. Set extension time based on expected amplicon size (approximately 1 kb/min). Finish the PCR with a final extension at 68–72°C for 5–10 min.
6. Prepare a 1%–1.5% agarose gel using TAE or TBE buffer and add ethidium bromide or similar safe DNA stain after the gel cools slightly.

7. Load an appropriate DNA ladder in the first lane to determine amplicon size later. Mix PCR products with loading dye and load them into separate wells. Run electrophoresis at 80–120 V depending on gel size and thickness.
8. Visualize amplified fragments under UV or blue-light transillumination. Identify KS, AT, or ACP domain amplicons based on expected product size. Excise clear bands from the gel if downstream sequencing is required.
9. Purify the DNA using any standard gel extraction kit. Submit one part of purified products for Sanger sequencing to confirm domain identity and another part for cloning and expression purpose (if planned).
10. Align the obtained sequences (e.g., MEGA, MUSCLE, Clustal Omega) with known PKS domains to validate amplification. Translate nucleotide sequences to amino acids and inspect for signature motifs. Compare the domain type with antiSMASH-predicted PKS clusters for consistency.
11. Store the sequences and gel images in the project archive for future annotation and publication use.

3.3 Heterologous Expression of a Polyketide Synthase (PKS) Gene in a Suitable Microbial Host [9, 16–18]

1. Amplify the PKS gene or module from genomic DNA using high-fidelity polymerase to ensure sequence accuracy. Confirm amplicon size by running the PCR product on an agarose gel.
2. Purify the amplified gene using a gel extraction kit to remove polymerase, salts, and unused primers.
3. Digest both the purified PKS amplicon and the chosen expression vector using compatible restriction enzymes. Clean the digested products to remove enzymes and buffers prior to ligation.
4. Set up a ligation reaction (or Gibson Assembly) using the digested vector and PKS insert at an appropriate molar ratio.
5. Transform the ligation mixture into chemically competent *E. coli* DH5α cells using heat shock or electroporation. Plate the transformants on antibiotic-containing agar plates to select colonies carrying the recombinant plasmid.
6. Screen several colonies using colony PCR or restriction digestion to verify the presence of the PKS insert. Isolate the confirmed recombinant plasmid using a plasmid miniprep kit.
7. Sequence the cloned PKS gene to confirm correctness, particularly the catalytic motifs and reading frame.
8. Transform the verified plasmid into an expression host such as *E. coli*) expression using IPTG (for *E. coli*) or thiostrepton (for *Streptomyces*) at the recommended concentration.

9. Continue incubation under efficient expression conditions (*E. coli* 16–18°C for 16–20 h and *Streptomyces* 28–30°C for 3–7 days).
10. Harvest cells at appropriate time points to monitor PKS expression or metabolite production. If protein verification is required, prepare crude lysates and analyze by SDS-PAGE to check for PKS overexpression.
11. Extract metabolites from the culture broth using ethyl acetate or methanol depending on the target polyketide. Concentrate the extracts using rotary evaporation or nitrogen blow-down.
12. Analyze metabolite extracts using HPLC or LC-MS to detect polyketide production. Compare chromatograms/mass profiles with known standards or predicted masses to confirm product formation.
13. Preserve positive expression strains as glycerol stocks for long-term storage. Store plasmid, sequencing data, and chromatographic profiles in the project archive for downstream publications.

4 Notes

1. Draft or highly fragmented genome assemblies may split PKS clusters across contigs; such incomplete assemblies often lead to missing domains or artificially truncated modules. PKS cluster boundaries predicted by antiSMASH sometimes merge adjacent clusters, especially in actinomycetes; always confirm borders manually using the .gbk file. antiSMASH occasionally mis-assigns hybrid PKS–NRPS clusters; inspect KS and AT arrangements to confirm true modular architecture.
2. Tailoring enzymes such as cyclases, oxidoreductases, or methyltransferases may be placed outside predicted boundaries; check flanking regions for associated genes.
3. KS and AT domains with borderline similarity scores should be re-validated using BLASTp and NaPDoS/NaPDoS2 to avoid false classification. When BLASTp hits do not align with expected PKS subtypes, check the domain truncation status in the viewer, as partial ORFs can generate misleading outputs.
4. For large and complex clusters, JSON output is more informative for programmatic parsing; use it to verify antiSMASH HTML summaries.
5. If the cluster appears incomplete, inspect secondary contigs for orphan KS, AT, or ACP genes that may belong to the same

BGC. antiSMASH does not always differentiate between cis- and trans-AT PKSs; manual alignment of AT sequences may be required for accurate classification. When predicted clusters show similarity to known BGCs, confirm by examining synteny rather than relying solely on antiSMASH similarity scores. Protocol-2: PCR amplification and domain-level analysis of PKS genes (KS, AT, ACP).

6. Use default primer design guidelines and ensure primers target conserved catalytic motifs such as HGTGT (KS) or GHSQG (AT) to avoid non-specific amplification. PKS genes are often large and GC-rich; using a high-fidelity polymerase improves amplification accuracy.
7. Cross-reactivity between similar PKS families is common; sequence verification is essential for domain confirmation. For organisms with multiple PKS pathways, primer binding sites may occur in several clusters; distinguish them using sequencing.
8. Failed amplification of AT or ACP domains may indicate atypical or trans-AT architectures; manual sequence inspection is recommended.
9. When aligning domain sequences, use curated PKS reference datasets for classification rather than raw BLAST hits. Conserved motifs may vary slightly across taxa; avoid discarding sequences solely based on minor motif variations.
10. All procedures were performed using appropriate personal protective equipment (lab coat, gloves, and eye protection) to ensure safe handling of biological and chemical materials.
11. Ethidium bromide and all EtBr-contaminated materials were collected in designated hazardous-waste containers and disposed of according to institutional chemical safety guidelines.
12. PKS genes often exceed 6–10 kb; long PCR requires high-fidelity polymerases with strong processivity.
13. Non-canonical start codons in native PKS genes should be adjusted to standard ATG if expression problems occur in *E. coli*.
14. Codon bias between donor and host organisms can suppress expression; codon-optimized constructs may improve yields.
15. Large plasmids (>10 kb) may cause instability in *E. coli*; use low-copy or integrative vectors for stable maintenance.
16. PKS expression sometimes requires co-expression of tailoring enzymes such as cyclases, oxidoreductases, or phosphopantetheinyl transferase for ACP activation.

References

1. Wang J, Zhang R, Chen X, Sun X, Yan Y, Shen X et al (2020) Biosynthesis of aromatic polyketides in microorganisms using type II polyketide synthases. Microb Cell Fact 19(1):110
2. Staunton J, Weissman KJ (2001) Polyketide biosynthesis: a millennium review. Nat Prod Rep 18(4):380–416
3. Risdian C, Mozef T, Wink J (2019) Biosynthesis of polyketides in Streptomyces. Microorganisms 7(5):124
4. Hertweck C, Luzhetskyy A, Rebets Y, Bechthold A (2007) Type II polyketide synthases: gaining a deeper insight into enzymatic teamwork. Nat Prod Rep 24(1):162–190
5. Hemmerling F, Hahn F et al (2016) Biosynthesis of oxygen and nitrogen-containing heterocycles: polyketides and related pathways. Beilstein J Org Chem 12:1512–1549
6. Blin K, Shaw S, Kloosterman AM, Charlop-Powers Z, van Wezel GP, Medema MH, Weber T (2021) antiSMASH 6.0: improving cluster detection and comparison capabilities. Nucleic Acids Res 49(W1):W29–W35
7. Skinnider MA, Merwin NJ, Johnston CW, Magarvey NA (2017) PRISM 3: expanded prediction of natural product chemical structures from microbial genomes. Nucleic Acids Res 45(W1):W49–W59
8. Ziemert N, Podell S, Penn K, Badger JH, Allen E, Jensen PR (2012) The Natural Product Domain Seeker NaPDoS: a phylogeny-based bioinformatic tool to classify secondary metabolite gene diversity. PLoS One 7(3):e34064
9. Pfeifer BA, Admiraal SJ, Gramajo H, Cane DE, Khosla C (2001) Biosynthesis of complex polyketides in a metabolically engineered strain of E. coli. Science 291(5509):1790–1792
10. Baltz RH (2010) Streptomyces and Saccharopolyspora hosts for heterologous expression of secondary metabolite gene clusters. J Ind Microbiol Biotechnol 37(8):759–772
11. Tong Y, Charusanti P, Zhang L, Weber T, Lee SY (2015) CRISPR-Cas9 based engineering of actinomycetal genomes. ACS Synth Biol 4(9):1020–1029
12. Kieser T, Bibb MJ, Buttner MJ, Chater KF, Hopwood DA (2000) Practical Streptomyces genetics. John Innes Foundation, Norwich
13. Ayuso-Sacido A, Genilloud O (2005) New PCR primers for the screening of type I polyketide synthase genes in actinomycetes. Appl Environ Microbiol 71(11):7001–7007
14. Metsä-Ketelä M, Halo L, Munukka E, Hakala J, Mäntsälä P, Ylihonko K (2002) Molecular evolution of aromatic polyketides and comparative sequence analysis of polyketide ketosynthase and 16S ribosomal DNA genes from various Streptomyces species. Appl Environ Microbiol 68(9):4472–4479
15. Weber T, Kim HU (2016) The secondary metabolite bioinformatics portal: computational tools to facilitate natural product discovery. Nat Prod Rep 33(8):988–94
16. Rodríguez E, Banchio C, Diacovich L, Bibb MJ, Gramajo H (2001) Role of an essential acyl coenzyme A carboxylase in the primary and secondary metabolism of Streptomyces coelicolor A3(2). Appl Environ Microbiol 67(9):4166–4176
17. Komatsu M, Uchiyama T, Omura S, Cane DE, Ikeda H (2010) Genome-minimized Streptomyces host libraries for improved heterologous expression. Proc Natl Acad Sci USA 107(6):2646–2651
18. Yuzawa S, Eng CH, Langley E et al (2017) Estimating active-site distances and designing linker sequences for PKS engineering. ACS Synth Biol 6(3):405–414

Check for updates

Chapter 26

Non-Ribosomal Peptides—Genetic Analysis And Pathways

Amit Gamit and Dhruti Amin

Abstract

Non-ribosomal peptides (NRPs) are structurally diverse secondary metabolites synthesized by large multi-modular enzyme complexes known as non-ribosomal peptide synthetases (NRPSs), which, unlike ribosomal translation, utilize both proteinogenic and non-proteinogenic amino acids for their biosynthesis. Therapeutic NRPs, such as surfactin, vancomycin, and cyclosporin, exhibit antimicrobial, cytotoxic, and immunosuppressive activities. Genetic interrogation of the NRPS pathways presents scientists with the ability to investigate biosynthetic diversity, understand domain organization, and predict their possible peptide products. The current chapter outlines a detailed experimental guideline for the genetic identification and pathway characterization of the NRPS gene clusters of the soil-derived *Streptomyces sp.* isolates. The process involves culturing, genomic DNA extraction, and amplification using degenerate primers targeting conserved NRPS domains, followed by sequencing and bioinformatic analysis (antiSMASH, NaPDoS, NRPSPredictor2) to determine domain organization, substrate specificity, and biosynthetic pathways. This combined molecular-computation workflow thus allows the identification of new pharmacologically active natural products.

Keywords Nonribosomal peptides (NRPs), Nonribosomal peptide synthetases (NRPS), Biosynthetic pathways, Domain architecture, Metabolic engineering

1 Introduction

The non-ribosomal peptides (NRPs) represent an extraordinarily heterogeneous group of bioactive secondary metabolites that bacteria and fungi produce with a unique set of enzymes that do not rely on the ribosome [1]. This is biosynthesized by large multi-modular enzymes, known as non-ribosomal peptide synthetases (NRPSs), which are mediators of the process and are linear assembly lines; each module adds a particular monomer to the growing peptide chain [2].

An average NRPS module consists of three major domains: an adenylation (A) domain, which binds and activates an amino acid; peptidyl carrier protein (PCP, also denoted as T) domain, which transports the activated residue; and condensation (C) domain, which mediates a peptide bond formation [3]. The discharge of the

Dhruti Amin et al. (eds.), *Phytoconstituent Analysis*, Springer Protocols Handbooks,
https://doi.org/10.1007/978-1-0716-5400-2_26,

mature peptide happens via the terminal thioesterase (TE) domain through hydrolysis or cyclization [4]. Incorporating both standard and non-standard amino acids, this modular system enables the vast structural and functional diversity seen in non-ribosomal peptides (NRPs) (Fig. 1) [5]. Here, NRPs have irreplaceable ecological and biological functions, i.e., as antibiotics (e.g., vancomycin, surfactin), cytotoxins (e.g., bleomycin), siderophores (e.g., pyoverdine), and immunosuppressants (e.g., cyclosporin) [6]. Considering the current rise in the severity of antimicrobial resistance as a medical condition, the study of NRPS biosynthetic pathways has become the foremost goal of drug discovery in natural products and synthetic biology. An understanding of the organization of NRPS genes has enabled metabolic engineering aimed at producing new bioactive molecules with improved pharmacological properties [7].

NRPS systems are characterized by genetics, requiring the identification and characterization of biosynthetic gene clusters (BGCs) encoding NRPS enzymes. These clusters are usually operonic, with each operon containing genes that encode various NRPS modules alongside other tailoring enzymes required for product maturation [8]. These clusters also preserve adenylation domain sequences best, which permits the amplification of these sequences using degenerate primers that recognize consensus motifs, such as A3 (STGxDxLxVxR) and A10 (core lysine motif) [9].

Recent improvements in genomic sequencing technology and computational biology have had a significant impact on the study of NRPS. AntiSMASH [10], NaPDoS [11], and NRPSPredictor2 [12] are automated tools that identify BGCs, annotate domain architecture, and predict substrate specificities. Additionally, phylogenetic analyses of adenylation and condensation domains help clarify lineage connections among microbial species and identify potential cryptic or

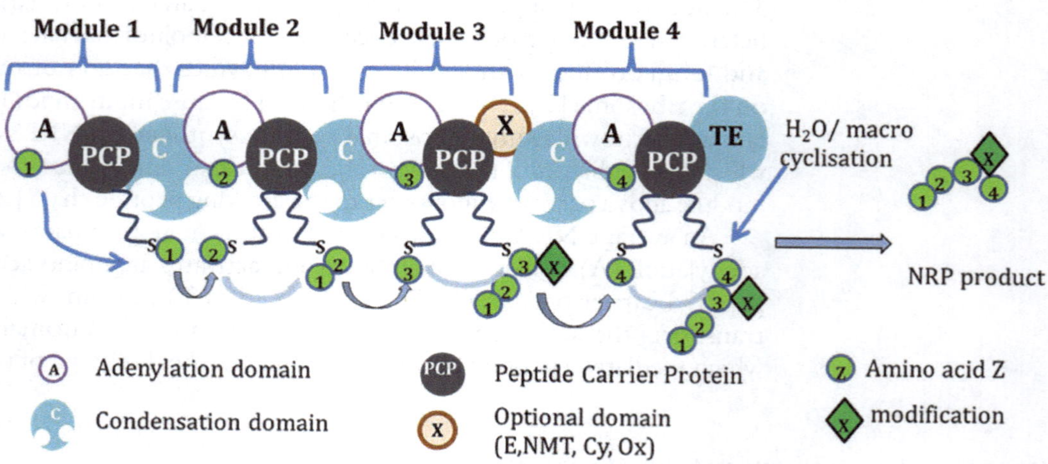

Fig. 1 NRPs biosynthetic pathway

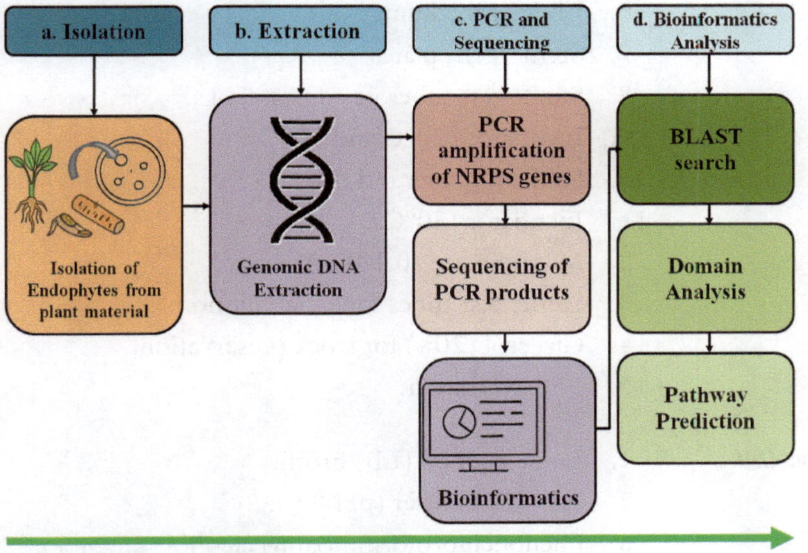

Fig. 2 Workflow of NRPs pathway genetic analysis showing (**a**) isolation of endophytes from plant sample, (**b**) Extraction of DNA, (**c**) PCR amplification of *NRP* gene and sequencing, and (**d**) Bioinformatics analysis of NRP synthesis pathway

silent clusters [13]. For example, the *NRPS* gene cluster of *Streptomyces coelicolor* contains the canonical order C A T C A T TE, which predicts a cyclic lipopeptide structure [14]. A similar modular structure is seen in the srfA cluster of the biosurfactant surfactin in *Bacillus subtilis* [15]. Not only can gene clusters be genetically dissected to elucidate biosynthetic mechanisms, but syntactic hybrids can also be readily designed through rational design. In short, molecular genetics, combined with PCR-mediated domain amplification and bioinformatics-overseen annotation of these amplifications, allows researchers to discover new *NRPS* gene clusters and determine the chemical structures of their products (Fig. 2). The materials needed and the stepwise procedure required to detect, amplify, and analyze *NRPS* genes are found in the subsequent sections.

2 Materials

2.1 Collection and Isolation

1. Healthy seeds of *Hordeum vulgare* (Barley).
2. Ethanol (70%, v/v).
3. Sodium hypochlorite solution (5%).
4. Sterile distilled water.
5. YPDA medium (peptone, yeast extract, dextrose, agar).

6. LB agar medium.
7. Sterile Petri plates.
8. Sterile test tubes or conical flasks.
9. Incubator (28°C and 30°C).
10. Sterile pipettes and tips
11. Incubator (30°C).
12. Micropipette (0.1–1 mL range).
13. Sterile test tubes for serial dilution.
14. Glycerol (20%) for stock preservation.
15. −80 °C freezer.

2.2 Genomic DNA Extraction

1. Luria–Bertani (LB) broth.
2. Tris–HCl buffer (pH 8.0).
3. Phenol:chloroform:isoamyl alcohol (25:24:1).
4. Isopropanol and 70% ethanol.
5. TE buffer (10 mM Tris–HCl, 1 mM EDTA).
6. Centrifuge and sterile centrifuge tubes.
7. Shaker incubator (30°C, 200 rpm).
8. Agarose gel electrophoresis setup.
9. Ethidium bromide or SYBR Safe stain.

2.3 PCR Amplification of NRPS Genes

1. PCR tubes and thermal cycler.
2. Taq DNA polymerase (1 U/reaction).
3. 10× PCR buffer.
4. $MgCl_2$ (2 mM).
5. dNTP mix (100 μM each).
6. Degenerate primers (Table 1).
7. Nuclease-free water.
8. Template genomic DNA (~50 ng).

2.4 Gel Electrophoresis

1. Agarose powder (0.5%–1.2%).
2. 1× TAE buffer.
3. SYBR Safe or ethidium bromide.
4. Gel casting tray and combs.
5. DNA ladder (1 kb).
6. UV/blue-light transilluminator.
7. Power supply and gel electrophoresis unit.

Table 1
Common degenerate primers for NRPS screening

Domain	NRPS synthetases genes	primer name	sequence(5′–3′)	References
Adenylation domain and Thiolation domain	Plipastatin	Ap1-F	AGMCAGCKSGCMASATCMCC	[17]
		Tp1-R	GCKATWWTGAARRCCGGCGG	
	Surfactin	As1-F	CGCGGMTACCGVATYGAGC	[18]
		Ts1-R	ATBCCTTTBTWDGAATGTCCGCC	
	Kurstakin	Akrs-F	TCHACWGGRAATCCAAAGGG	[19]
		Tks-R	CCACCDKTCAAAKAARKWAC	

2.5 PCR Product Purification and DNA Sequencing

1. Sterile scalpel or gel cutter.
2. Gel extraction kit (e.g., QIAquick Gel Extraction Kit, Qiagen).
3. NanoDrop or spectrophotometer.
4. Sterile microcentrifuge tubes.
5. Access to Sanger sequencing facility.

2.6 Bioinformatics and Sequence Analysis

1. Computer with internet access.
2. Software/tools:
   a. NCBI BLASTn / BLASTx.
   b. antiSMASH v6.0.
   c. NRPSPredictor2.
   d. MIBiG database.
   e. ClustalW (for multiple sequence alignment).
   f. MEGA X (for phylogenetic analysis).
3. Sequence files (FASTA or chromatogram format).

2.7 Pathway Reconstruction and Interpretation

1. Geneious Prime or Pathway Tools software.
2. Reference NRPS gene clusters (e.g., *srfA* of *Bacillus subtilis*).
3. Domain annotation data (C, A, T, TE modules).
4. Computational tools for peptide sequence prediction and visualization.

3 Methods

3.1 Collection and Isolation [16]

1. Select Approximately 20 healthy seeds and wash thoroughly under running tap water to remove adhering dust and debris.
2. Surface sterilize the seeds using 70% ethanol for 1 min, followed by treatment with 5% sodium hypochlorite for 5 min, and then rinse three times with sterile distilled water.
3. Soak the sterilized seeds overnight in 25 mL of sterile distilled water to soften the tissues and facilitate easy mashing.
4. Inoculate the seed suspension onto YPDA medium medium (pH 6.5) and incubate at 28°C until the appearance of microbial colonies.
5. Sub culture the Individual colonies on LB agar plates and incubate at 30°C overnight.
6. Select the morphologically different colonies purify them by repeating streaking and keep the isolates in 20% glycerol stocks at −80°C.

3.2 Genomic DNA Extraction

1. Grow purified bacterial isolate in 50 mL of LB broth medium at 37°C and 200 rpm for 24 h.
2. Harvest the cells by centrifuging the culture at 10,000 rpm for 5 min in autoclaved tubes.
3. Resuspend the cell pellets in 500 μL of Tris–HCl buffer (pH 8).
4. Add phenol:chloroform:isoamyl alcohol (25:24:1) solution and gently mix for 5 min.
5. Centrifuge the mixture at 12,000 rpm for 10 min and transfer the supernatant to a clean tube.
6. Precipitate the DNA using isopropanol, then wash the pellet with 70% ethanol.
7. Air-dry the pellet and dissolve it in TE buffer (10 mM Tris-HCl, 1 mM EDTA).
8. Assess DNA integrity using agarose gel electrophoresis containing ethidium bromide.

3.3 PCR Amplification of NRPS

Amplify the conserved NRPS domains with degenerate primers to adenylation (A) and condensation (C) domains. Carry out PCR amplification using any of the below-mentioned primer set/pairs (*see* **Note 1**).

1. Perform PCR amplification using standard 50 μL reaction mixtures which include 1× PCR buffer, 2 mM $MgCl_2$, and 1 U Taq DNA polymerase; 100 μM deoxynucleoside triphosphates;

0.2 μM forward and reverse primers; and make up volume with PCR grade nuclease free water.

2. Add 2 μL of template DNA (≈50 ng) extracted from the protist samples (*see* **Note 2**).
3. PCR settings includes: initial denaturation at 94°C for 2 min follow by 35 cycles of three steps; denaturation for 30 s at 94°C; annealing step for 30 s at 43°C with (As1-F/Ts2-R), at 58°C with (Ap1-F/Tp1-R), and at 44.4°C with (Aks-F/Tks-R); and an extension step of 45 s at 72°C except with Ap1-F/Tp1-R primers (75 s at 72°C) and (Aks-F/Tks-R) with final extension for 5 min at 72°C. (*see* **Note 3**).

3.4 Gel Electrophoresis [20]

1. Load 1.2% agarose gel into 1× TAE buffer with SYBR Safe dye (*see* **Note 4**).
2. Various samples that have been subjected to PCR are loaded into the proper wells with 5 μl of PCR product and 1 kb ladder.
3. Run the gel at 80–100 V for 45 min.
4. Observe the bands with the help of UV or blue-light transillumination and make a photo.
5. It is assumed that the size of the product can range between 700 and 800 bp, and that is a conserved NRPS domain.

3.5 Purification and DNA Sequencing

1. The expected band in the PCR is cut off with a sterile scalpel.
2. Microfilter the fragment removed using a commercial gel isolation kit (e.g., QIAquick, Qiagen).
3. Measure the concentration of the DNA using a Nano drop spectrophotometer.
4. Send the genomic DNA to bidirectional Sanger sequencing.

Sequencing results are obtained as chromatograms or FASTA files for subsequent analysis. Bidirectional sequencing ensures high confidence in base calling [2].

3.6 Bioinformatics and Sequence Analysis (see Note 5)

1. Analysis BL21x/BLASTn or BLASTx BL21x (NCBI) to identify homologous nonribosomal peptide synthetase (NRPS) genes [21].
2. Predict biosynthetic gene clusters (BGCs) and annotate constituent domains with antiSMASH version 6.0 [10].
3. Record the functional grouping of each NRPS module, i.e., condensation (C), adenylation (A), thiolation (T), and thioesterase (TE) domains.
4. Amino Acid specificity of the adenylation domains of proteins: infer using NRPSPredictor2 [12].
5. Check the biogenetic pathway against entries in MIBiG0 repository [13].

6. To examine phylogenetic relationships, query the sequences of the adenylation domains with ClustalW, and a neighbor-joining tree was created with MEGA X [22]. (*see* **Note 6**).

3.7 Pathway Reconstruction and Interpretation (see Note 7)

1. Identify the genes of the non-ribosomal synthetase to peptide (NRPS) open reading frames (ORFs).
2. Use Geneius Prime or Pathway Tools to have detailed visualization of genomes, and systematic definition of the arrangement of modules.
3. Extend the biosynthetic logic to be reconstituted systematically and in a defined manner by laying out initiation, elongation, as well as termination modules.
4. Determine the sequence of the peptide using domain order and the specificity of substrates according to [4].
5. Detect structural novelty by comparing the predicted products to the characterized non-ribosomal peptides (e.g., surfactin, bacitracin) according to [7].

4 Notes

1. Primer design: Primer sets based on conserved motifs identified in families of well-characterized *NRPS* genes found in the GenBank database are designed.
2. Template Purity: DNA templates with A260/A280 absorbance ratio of 1.8–2.0 are used to obtain optimum efficiency of PCR.
3. PCR Optimization: Vary the concentration of $MgCl_2$ between 1.5 and 3.0 Mm with regard to the needs of the template to increase the fidelity of amplification.
4. Safety- propose the mutagenic property of ethidium bromide and use less toxic nucleic-acid staining, e.g., SYBR Safe, to reduce the risk.
5. Bioinformatics Tests: Independently cross-test domain architecture against an assortment of sequence-based prediction sites, such as Pfam, InterPro, and SMART, as advised by [23].
6. AS example: The *srfA* cluster of *Bacillus subtilis* encodes surfactin synthetase modules that correspond to lipopeptide biosynthesis [14].
7. Alternate Tools: Complementary computation tools that are used in the process of validating the delineation of the NRPS domain boundaries can be used; these include PRISM, NaPDoS, and SeMPI tools.

References

1. Marahiel MA, Stachelhaus T, Mootz HD (1997) Modular peptide synthetases involved in non-ribosomal peptide synthesis. Chem Rev 97(7):2651–2674.
2. Finking R, Marahiel MA (2004) Biosynthesis of non-ribosomal peptides. Annu Rev Microbiol 58:453–488.
3. Sieber SA, Marahiel MA (2005) Molecular mechanisms underlying nonribosomal peptide synthesis: approaches to new antibiotics. Chem. Rev 105(2):715–738
4. Stachelhaus T, Mootz HD, Marahiel MA (1999) The specificity-conferring code of adenylation domains in nonribosomal peptide synthetases. Chem Biol 6(8):493–505.
5. Kohli RM, Trauger JW, Schwarzer D, Marahiel MA, Walsh CT (2001) Generality of peptide cyclization catalyzed by isolated thioesterase domains of nonribosomal peptide synthetases. Biochemistry 40(23):7099–7108.
6. Süssmuth RD, Mainz A (2017) Nonribosomal peptide synthesis—principles and prospects. Angew Chem Int Ed Engl 56(14):3770–3821.
7. Walsh CT (2016) Insights into the chemical logic and enzymatic machinery of NRPS assembly lines. Chem Rev 116(5):2436–2486.
8. Challis GL, Naismith JH (2004) Structural aspects of non-ribosomal peptide biosynthesis. Curr Opin Struct Biol 14(6):748–756.
9. Ayuso-Sacido A, Genilloud O (2005) New PCR primers for the screening of NRPS and PKS-I systems in actinomycetes: detection and distribution of these biosynthetic gene sequences in major taxonomic groups. Microbial Ecol 49(1):10–24.
10. Blin K, Shaw S, Kloosterman AM, Charlop-Powers Z, van Wezel GP, Medema MH, Weber T (2021) antiSMASH 6.0: Improving cluster detection and comparison capabilities. Nucleic Acids Res 49(W1):W29–W35.
11. Ziemert N, Podell S, Penn K, Badger JH, Allen EE (2012) The natural product domain seeker NaPDoS: a phylogeny-based bioinformatic tool to classify secondary metabolite gene diversity. PLoS ONE 7(3):e34064.
12. Röttig M, Medema MH, Blin K, Weber T, Rausch C, Kohlbacher O (2011) NRPSPredictor2—a web server for predicting NRPS adenylation domain specificity. Nucleic Acids Res 39(2):W362–W367.
13. Kautsar SA, Blin K, Shaw S, Weber T, Medema MH (2020) MIBiG 2.0: a repository for biosynthetic gene clusters. Nucleic Acids Res 48(D1):D454–D458.
14. Weber T, Marahiel MA (2001) Exploring the domain structure of modular non-ribosomal peptide synthetases. FEMS Microbiol Lett 201(1):1–7.
15. Steller S, Vollenbroich D, Leenders F, Stein T, Conrad B, Hofemeister J, Vater J (2004) Structural and functional organization of the surfactin synthetase multienzyme system. J Biol Chem 279(37):37740–37749.
16. Wen CY, Yen ZG, Wang KX, Chen JG, Shen SS (2011) Structural analysis of surfactin produced by endophytic strain *Bacillus subtilis* EBS05 and its antagonistic activity against *Rhizoctonia cerealis*. Plant Pathol 27(4):342–348
17. DeLong EF (1992) Archaea in coastal marine environments. Proc Natl Acad Sci USA 89(12):5685–5689.
18. Abderrahmani A, Tapi A, Nateche F, Chollet M, Leclère V, Wathelet B, Hacene H, Jacques P (2011) Bioinformatics and molecular approaches to detect NRPS genes involved in the biosynthesis of kurstakin from *Bacillus thuringiensis*. Appl Microbiol Biotechnol 92: 571–581
19. Tapi A, Chollet-Imbert M, Scherens B, Jacques P (2010) A new approach for the detection of non-ribosomal peptide synthetase genes in Bacillus strains by polymerase chain reaction. Appl Microbiol Biotechnol 85:1521–1531
20. Heinz DW, Kleineidam RG, Thomas MG (2015) Structural basis of adenylation domain substrate selectivity in non-ribosomal peptide synthetases. Chem Rev 115(2):1210–1249.
21. Altschul SF, Gish W, Miller W, Myers EW, Lipman DJ (1990) Basic local alignment search tool. J Mol Biology 215(3):403–410.
22. Kumar S, Stecher G, Li M, Knyaz C, Tamura K (2018) MEGA X: Molecular evolutionary genetics analysis across computing platforms. Mol Biol Evol 35(6):1547–1549.
23. Marchler-Bauer A, Derbyshire MK, Gonzales NR, Lu S, Chitsaz F, Geer LY, Bryant SH (2017) CDD/SPARCLE: Functional classification of proteins via subfamily domain architectures. Nucleic Acids Res. 45(D1):D200–D203.

Chapter 27

Flavonoids Genetic Studies on Flavonoid Biosynthesis

Rukhsar Ansari and Sanket Ray

Abstract

The biosynthesis of flavonoids is regulated by the complex web of structural enzymes and transcriptional regulators. This, in turn, affects the chemistry of pigments and defense substances in plants. This chapter presents a full experimental model for the investigation of this pathway using combined genetics, molecular biology, and metabolomic methodologies. Details for using Agrobacterium to create stable transgenic Arabidopsis lines, Gateway-based cloning of regulatory and biosynthetic genes, and lysis of phenolic-rich tissues are provided. Quantitative gene expression quantification with qRT-PCR, targeted metabolite profiling using HPLC–DAD and LC–MS/MS, and the Butanol–HCl assay for proanthocyanidin determination permit accurate measurement of pathway activity. When applied in concert, these tools facilitate investigation of gene function, regulatory logic and steady state metabolic flux through the flavonoid network, and thus provide a solid foundation for studying many other plant taxa.

Keywords Flavonoid biosynthesis, Transcription factors, *Arabidopsis thaliana*, *Zea mays Vitis vinifera*, Transparent testa mutants

1 Introduction

The phenylpropanoid biosynthetic pathway contributes to the formation of many types of flavonoids, such as anthocyanins, flavones, flavanones, flavonols, and proanthocyanidins (PAs), which are tannin-like [1]. Flavonoids are among the most common types of secondary metabolites known within the plant kingdom. They are often found in polyphenolic forms throughout the plant. Flavonoid compounds can be found in leaves, stems, bark, fruit, roots, and flowers of the plant [2].

The anthocyanin pathway is considered the most central pathway in flavonoid metabolism, giving several compounds serving functions in plant defense and color and responding to the environment of the plant [3]. In the anthocyanin pathway, the starting point represents a reaction that involves three molecules of malonyl-CoA that are derived from fatty acid breakdown. Three molecules of coumaroyl-CoA are produced by phenylalanine through its

Dhruti Amin et al. (eds.), *Phytoconstituent Analysis*, Springer Protocols Handbooks,
https://doi.org/10.1007/978-1-0716-5400-2_27,

phenylpropanoid pathway, after which these molecules react with the three molecules of malonyl-CoA to form the basis of the anthocyanin pathway [4].

The synthesis that makes anthocyanins is initiated through a subsequent group of enzymes, which methylate each other, moving down the pathway. It is the action of the first enzyme, chalcone synthase, which initiates the conversion of these precursors into a naringenin chalcone that starts synthesis of precursors to anthocyanin [5]. Acting downstream are enzymes, such as chalcone isomerase, flavanone 3-hydroxylase, dihydroflavonol 4-reductase, and anthocyanidin synthase, that eventually produce the final products. Each enzyme acts one after another to produce the mature form of anthocyanins [6].

The production of anthocyanins is initiated through a group of reactions that produce a diversity of compounds from colorless flavanones to fully colored anthocyanins [1]. These final products give the distinctive red, purple, and blue colors to many fruits and flowers. A key feature of the anthocyanin pathway is that it includes several branching points, particularly the divergent paths that lead to the production of condensed tannins, or proanthocyanidins (PAs). Although PAs lack visible color, they play critical roles in plant survival by strengthening and protecting tissues, especially within seed coats [7].

The key transcriptional activator in this complex is the evolutionarily conserved MYB-bHLH-WD40 (MBW) protein complex. It controls the expression of structural genes in specific tissues and directs metabolite flow into specific flavonoid branches [8, 9].

1.1 Essential Plant Variants for Genetic Studies

Several models and specialized plant systems were previously used in genetic studies of flavonoid biosynthesis (Table 1):

Table 1
Essential plant species used in genetic studies of flavonoid biosynthesis

Plant species	Primary focus of genetic study	Key genetic loci/genes	Citation
Arabidopsis thaliana (Thale Cress)	Flavonols, PAs (seed coat), Anthocyanins (seedlings). Extensive genetic tools (T-DNA lines).	**TT4** (*CHS*), **TT3** (*DFR*), **TT1** (*MYB*), **TT8** (*bHLH*)	[10, 11]
Zea mays (Maize)	Anthocyanin pigmentation in kernels/ tissues. Classic model for transposon tagging and regulatory loci.	**C1** (MYB), **R** (*bHLH*), **P1** (MYB for phlobaphenes)	[12–14]
Vitis vinifera (Grape)	Proanthocyanidins, Anthocyanins in berries (wine quality).	**VvMYBA1/2** (Anthocyanin), **VvMYBPA1** (PA)	[15, 16]
Glycyrrhiza glabra (Licorice)	Isoflavonoid (phytoalexin) biosynthesis, specialized pathway branches.	**GgCHS6**	[17]

2 Materials

This section offers a thorough inventory of the plant materials, specialized equipment, chemical standards, molecular reagents, and analytical tools needed for genetic research on flavonoid biosynthesis.

2.1 Plant Materials and Growth Conditions

The range of the flavonoid study depends on the type of plant material (anthocyanins compared with proanthocyanidins, for example).

1. Model Plants: The common genetic model is *Arabidopsis thaliana* (Col-0). Targeted T-DNA insertion mutants, such as the Salk tt3, tt4, and tt8 lines, can be obtained for loss-of-function studies from carefully selected resources like the SIGnAL T-DNA collection at the Salk Institute [18, 19]. In order to verify pathway activity, these Transparent Testa (TT) mutants are crucial negative controls.
2. Specialized Plants:
   a. *Vitis vinifera* (e.g., Cabernet Sauvignon clone) for specific studies on condensed tannins (PAs) and berry anthocyanins [15].
   b. *Zea mays* (B73 inbred line) for classic studies on pigmentation regulation via the C1/R loci [20].
3. Culture Media: Murashige and Skoog (MS) media [21]. (Phytotechnology grade) adjusted to pH 5.7 ± 0.1. It must be supplemented with 1.0% w/v sucrose and 0.8% w/v agar for preparing solid selection plates.
4. Soil Medium: Standard soil mix (e.g., 3:1 peat moss:vermiculite), sterilized to prevent contamination, used for growing mature plants for floral dip transformation.
5. Environmental Control: Growth chambers must be set to 22 ± 1°C with a 16 h Light/8 h Dark photoperiod.
6. Induction Tools: UV-B lamps (e.g., UV-B 313 nm fluorescent tubes, 1–5 W m^{-2} irradiance) are required to strongly induce the expression of stress-responsive flavonoid genes (like flavonols) in non-pigmented tissues [22]. (For important considerations regarding UV-B treatment and experimental consistency, see Note 1. in section 4)

2.2 Molecular Biology Reagents and Tools

These materials are essential for RNA isolation, cloning, and transformation of the flavonoid genes.

2.2.1 Nucleic Acid Preparation and Synthesis

1. **RNA Isolation Kit:** Total RNA isolation kit (e.g., RNeasy Plant Mini Kit from Qiagen or TRIzol Reagent), specifically chosen for its efficacy in polyphenol-rich plant material.

2. **DNase I:** RNase-free DNase I (Promega or equivalent) for genomic DNA removal, a critical step to ensure qRT-PCR accuracy.
3. **Reverse Transcriptase (RT):** High-capacity cDNA Reverse Transcription Kit (Applied Biosystems or equivalent), utilizing Oligo(dT) or Random Hexamers for first-strand cDNA synthesis.

2.2.2 Cloning and Transformation Reagents

1. PCR Polymerase: High-fidelity, proofreading DNA polymerase (e.g., Phusion High-Fidelity Polymerase) is mandatory for amplifying the CDS to minimize sequence errors prior to cloning.
2. Cloning System: The Gateway™ Cloning System (Invitrogen) reagents, including BP Clonase II™ and LR Clonase II™ enzyme mixes, can be used for efficient, directional cloning into the expression vector [23].
3. Cloning Vectors: For plant transformation, appropriate destination vectors include pK2GW7 or members of the pEarleyGate vector series, which are widely used Gateway®-compatible expression systems designed for *Agrobacterium*-mediated plant transformation [23, 24]. These destination vectors typically incorporate the cauliflower mosaic virus (CaMV) 35S promoter to drive strong constitutive transgene expression in a broad range of plant tissues [25]. These constructs also have selectable marker genes that usually make cells resistant to hygromycin or kanamycin. This helps researchers choose which cells to transform [26].
4. Bacterial Strain: Use *Agrobacterium tumefaciens* strain GV3101 for T-DNA delivery. Strain LBA4404 can be used as an alternative if needed [27].
5. Transformation Additives: To activate the expression of the vir gene, add 200 μM of high-purity acetosyringone to the mixture [28]. During the transformation, add 0.05% (v/v) Silwet L-77 to improve tissue wetting [29].
6. Antibiotics: Rifampicin—to maintain the selection of Agrobacterium cultures, whereas kanamycin and hygromycin—to select transformed plant tissues [30].

2.3 Resources for Gene Expression Measurement Via qRT-PCR

Reliable transcriptional activity quantification is supported by these resources.

1. qPCR Instrument: A real-time quantitative PCR (qPCR) device that can analyze melting curves.

2. Master Mix: SYBR Green Master Mix (2×) or TaqMan Probe Master Mix.
3. Primers: HPLC-purified Oligonucleotides. Requirement: Primers must span an exon exon junction for gDNA exclusion and must have 90%–110% efficiency [31, 32].

2.4 Metabolite Analysis (HPLC–MS) Equipment and Reagents

Analytical-grade solvents and specialized equipment are needed for accurate flavonoid analysis.

2.4.1 Instrumentation and Columns

1. Tissue Grinder: High-speed grinder or homogenizer or liquid nitrogen-chilled mortar and pestle for tissue pulverization.
2. Analytical Column: C18 reverse phase column (150×4.6 mm, 5 μm particle size).
3. Chromatography System: Diode Array Detector (DAD) and High-Performance Liquid Chromatograph (HPLC).
4. Mass Spectrometry: Tandem Mass Spectrometer (LC–MS/MS), preferably Q–TOF, equipped with Electrospray Ionization (ESI) source.
5. Filtration: 0.22 μm PTFE (Polytetrafluoroethylene) syringe filters.

2.4.2 Chemical Standards and Solvents

1. HPLC/LC–MS Grade solvents: water, acetonitrile (ACN), and methanol (MeOH). Analytical techniques should only employ high-purity LC–MS grade solvents.
2. Acid Modifiers (LC–MS Grade): 0.1% *v*/*v*. of either Trifluoroacetic acid (TFA) or Formic Acid (FA).
3. Standards of Reference: Analytical standards for calibration curves with high purity (>95%):
   - Anthocyanin: Cyanidin-3-glucoside.
   - Flavonol: Quercetin-3-glucoside.
   - Flavanone: Naringenin.
4. For the specific Proanthocyanidin quantification assay, use the Proanthocyanidin Reagent (PA Reagent): Butanol-HCl Reagent (95:5 butanol:conc. HCl with ferric ammonium sulfate) [33].

3 Methods

Using the resources mentioned in Sect. 2, this section describes the comprehensive, step-by-step methods for the genetic investigation of flavonoid biosynthesis, ensuring high procedural accuracy and reproducibility.

3.1 High-Quality RNA Extraction and cDNA Synthesis

1. Setting up Tissue: Collect 100 mg of the target plant tissue (such as berry skins of Vitis vinifera or UV-B-induced Arabidopsis seedlings) and flash-freeze it in liquid nitrogen (LN2) instantly. To immediately stop RNase activity, this step is essential [34].
2. Homogenization and Lysis: Use a high-speed Ball Mill Homogenizer (30 Hz for 1 min) or an LN_2-chilled mortar and pestle to grind the frozen tissue into a fine powder.
3. Add 1 mL of the guanidinium thiocyanate-containing lysis buffer from the RNA isolation kit to the powdered tissue immediately [34].
4. Use a column-based kit to extract total RNA, being careful to incorporate chloroform extraction or the suggested wash procedures to efficiently eliminate polyphenols and polysaccharides that might hinder subsequent uses [35]. Elute the final RNA in 50 μL of RNase-free water.(For troubleshooting RNA isolation from polyphenol-rich tissues, see Note 2 in section 4)
5. To eliminate any contaminating genomic DNA, treat 1 μg of total RNA with 1 unit of RNase-free DNase I at 37°C for 30 min [31].
6. cDNA Synthesis: Using Oligo(dT) primers and a high-capacity reverse transcriptase, create first-strand cDNA from 1 μg of DNase-treated RNA by incubating it at 42°C for 60 min [32].

3.2 Gene Amplification and Gateway Cloning

1. High-Fidelity PCR: Use Phusion or another high-fidelity, proofreading DNA polymerase to amplify the entire coding sequence (CDS). Create primers that have attB sites on either side to facilitate Gateway® recombination. To guarantee precise and error-free amplification of the entire CDS, run about 35 PCR cycles [36].
2. Generate the Entry Clone (BP Reaction): Purify the attB-flanked PCR product using a column-based PCR cleanup kit according to the manufacturer's instructions. Next, set up the BP recombination reaction by combining the purified product with the pDONR™ Entry Vector and BP Clonase™ II. Incubate the reaction at 25°C for 1 h, following Gateway® cloning procedures described by [23].
3. Generation of the Expression Clone (LR Recombination Step): To find colonies carrying the correct construct, start by obtaining the entry clone generated in the previous step and evaluating it in *E. coli*. Make plasmid DNA from these candidates, then use Sanger sequencing to verify the insert's orientation and sequence. After a correct entry clone has been confirmed, use the LR recombination system to recombine it with the chosen Gateway® plant destination vector (such as pK2GW7). The final expression clone that will be utilized in

later plant transformation experiments will be produced by this recombination step [23, 24].

4. Verify the Final T-DNA Vector: Extract plasmid DNA from the colonies that result from transforming the LR reaction product into *E. coli*. Verify the final construct by using Sanger sequencing to confirm the sequence integrity and restriction digestion to confirm the insert size and orientation.

3.3 Agrobacterium-Mediated Transformation of Arabidopsis thaliana

This protocol ensures stable integration of the T-DNA carrying the expression cassette using the *A. tumefaciens* GV3101 strain.

1. Agro-Culture Activation: Transform the T-DNA vector into A. tumefaciens. Grow
2. the culture until OD 600 ≈0.8 and resuspend the pellet in infiltration medium (1/2 MS salts, 5% sucrose, 200 μM AS, and 0.05% Silwet L-77) [29].
3. Floral Dip: Dip the inflorescences of 4-week-old Arabidopsis plants into the suspension for 5–10 s.
4. Tl Seed Selection: Harvest Tl seeds. Select resistant seedlings on MS medium containing the appropriate selective antibiotic (e.g., 50 mg/L Kanamycin) [37].
5. T2/T3 Homozygosity Screening: Screen T2 lines for 3:1 segregation. Confirm T3 homozygous lines (100% resistance on selective media) using genomic PCR with the Left Border primer (LB) [18] (Table 2).(Additional guidance on verification of homozygous T-DNA lines is provided in Note 3 in section 4).

3.4 Gene Expression Analysis (Quantitative Real-Time PCR, qRT-PCR)

This qRT-PCR protocol utilizes the $2^{-\Delta\Delta CT}$ method to quantify the effect of the transgene on structural gene expression [38].

1. Homogeneous Tissue Harvest: Harvest 100 mg of T3 homozygous line tissue and wild-type controls ($n \geq 3$). Prepare 1:10 diluted cDNA (Sect. 3.1.).

Table 2
Expected results of transgenic analysis and their interpretations in studies of the flavonoid and anthocyanin pathways

Expected outcome	Interpretation	Citation
T2 segregates 3:1 (Resistant:Susceptible)	Indicates a single T-DNA insertion locus.	[18]
Transgenic line shows deeper purple coloration	Confirms ectopic activation of the anthocyanin pathway.	[22]

2. Primer Quality Control: Use primers designed to span exon–exon junctions. Verify that the primer set has efficiency (E) between 90% and 110% via a standard curve analysis [32].
3. qRT-PCR Reaction Setup: Reactions should be set up in technical triplicates, with a 10–20 μL final volume, using 2× SYBR Green and 250 nM final concentration of primers. A No-Template Control (NTC) should be included.
4. For the amplification stage, perform 40 cycles of PCR. Then, do a melting-curve analysis between 60°C and 95°C to ensure the product is specific [31].
5. For relative expression, the $2^{-\Delta\Delta CT}$ method should be employed, using the CT values of the target gene against those of a stable housekeeping gene such as Actin or UBQ10 (Table 3).

3.5 Flavonoid Quantification and Identification (HPLC-DAD and LC-MS/MS)

It describes a methodology based on HPLC/DAD for accurate quantification of the metabolites, coupled with Liquid Chromatography–Tandem Mass Spectrometry (LC–MS/MS) for reliable structural identification.

3.5.1 Extraction of Metabolites

1. Freeze dry 100 mg of T3 homozygous line tissue. Grind the dried sample into a fine powder using a ball mill homogenizer or mortar and pestle chilled with liquid nitrogen. This will enlarge the surface area and prevent heat labile breakdown of sensitive metabolites.
2. Add 1 mL of extraction solvent (80% HPLC-grade methanol with 0.1% formic acid) to the powdered tissue. Methanol effectively extracts phenolics, while formic acid stabilizes anthocyanins and prevents degradation of pigments during extraction [40, 41].
3. Vortex the mixture for 5 min-to break up the sample and improve the interaction of the solvent with the sample-followed by sonication in an ice-cold bath for 30 min; supersonic waves break up cellular structures and improve the release of metabolites, while the cold temperature stops thermal degradation.

Table 3
Expected molecular outcomes of transgene expression and their interpretations in flavonoid biosynthesis studies

Expected outcome	Interpretation	Citation
10–100-fold upregulation of introduced gene.	Confirms stable integration and high expression of the transgene.	[23]
Structural genes (CHS, DFR) show 5–100-fold upregulation.	Provides molecular evidence of the gene's transcriptional activating function.	[8, 39]

4. To separate soluble metabolites from tissue debris and keep the compound stable, centrifuge the extract at 14,000 × *g* for 15 min at 4°C. Carefully move the supernatant and filter it through a 0.22 μm PTFE syringe filter to get rid of particles that could damage or clog HPLC/LC-MS columns. Put the filtered extracts in amber vials and keep them at −20°C (amber protects anthocyanins from light, and freezing stops oxidation and degradation before analysis) [41, 42].(For recommendations on metabolite stability during extraction and storage, see Note 4 in section 4).

3.5.2 Quantification by HPLC-DAD Analysis

1. Fill a C18 reverse-phase column (150×4.6 mm, 5 μm) with 10–20 μL of the filtered extract while keeping the column temperature at 30°C.
2. Use a binary gradient system to separate the compounds at a 1.0 mL/min flow rate. Make acetonitrile with 0.1% formic acid for solvent B and water with 0.1% formic acid for solvent A.
3. A diode-array detector (DAD) with characteristic wavelengths of 280 nm for flavanones, 350 nm for flavonols, and 520 nm for anthocyanins can be employed to monitor the eluent [43].
4. By integrating peak areas and comparing them to a five-point calibration curve created from external high-purity reference standards, one can ascertain the target flavonoid concentrations [15].

3.5.3 LC–MS/MS Analysis (Structural Identification)

1. Using electrospray ionization (ESI), connect the HPLC system to a tandem mass spectrometer. To identify every flavonoid class, conduct the analysis in both positive and negative ion modes in a sequential manner.
2. Conduct focused MS^2 experiments on particular parent ions. To verify the structure of the aglycone and its attached sugar moieties, examine the resulting fragmentation patterns, such as the neutral loss of 162 Da for hexose [44, 45]
3. By comparing each compound's UV spectrum, retention time, and MS^2 spectra with recognized reference standards and spectral libraries, it can determine the structural identity (Table 4). (Additional interpretation of the Butanol-HCl assay and complementary DMACA staining is provided in Note 5, section 4)

3.6 Butanol–HCl Assay for Proanthocyanidin (PA) Quantification

Proanthocyanidins (PAs) are polymeric substances that are challenging to precisely measure with traditional HPLC. A trustworthy technique for determining their total content is the butanol–HCl Assay [33, 47].

Table 4
Analytical outcomes of advanced profiling techniques for activator genes in flavonoid biosynthesis

Technique	Expected outcome for an activator gene (e.g., MYB)	Interpretation	Citation
HPLC–DAD	twofold to tenfold increase in final product concentrations (e.g., Cyanidin-3-glucoside) alongside a decrease in upstream intermediates.	Confirms the gene successfully directs metabolic flux toward the final product.	[40, 41]
LC–MS/MS	Identification of novel flavonoid conjugates (e.g., unique glycosylation or acylation) not present in the wild type.	Suggests the regulator controls specific tailoring enzymes (transferases) late in the pathway.	[46]

Table 5
Analytical assessment of proanthocyanidin (PA) accumulation

Technique	Expected outcome for a PA activator gene (e.g., VvMYBPA1)	Interpretation
Butanol–HCl Assay	A550 value is significantly higher (twofold to fivefold) in the transgenic line compared to the wild-type control.	Confirms successful increase in total Proanthocyanidin content mediated by the regulatory gene.

1. Fill a glass culture tube with a Teflon-lined screw cap with 50 mg of dry, finely powdered tissue (such as grape or Arabidopsis seeds).
2. Make 1 mL of the butanol–HCl reagent (95:5 v/v butanol:-concentrated HCl) with 0.1 g/L of ferric ammonium sulfate [$FeNH_4(SO_4)_2$]. The acid-catalyzed hydrolysis of PA polymers is accelerated by the ferric ion.
3. After properly sealing the tubes, incubate them for 60 min at 95°C in a water bath or heating block. In this stage, red anthocyanidins are produced by depolymerizing the PA polymers.
4. Let the tubes cool until they reach room temperature. On a spectrophotometer, set a reagent blank (butanol–HCl mixture without tissue) to zero. Then, measure the absorbance at 550 nm (A_{550}). Determine the total PA content using the A_{550} reading (Table 5).

4 Notes

Troubleshooting and Methodological Considerations for Flavonoid Genetic Studies:

To get results that are accurate and can be repeated, keep the following in consideration:

1. **Controlling UV-B Variables for Induction Experiments:** Environmental factors, especially, have a great impact on flavonoid biosynthesis. Use identical light conditions for all biological replicates, including UV-B quality, intensity (1–5 W m^2), and duration (for example, 24 h before harvest). This reduces variability due to gene expression and metabolite accumulation that may occur upon changing light conditions [22].
2. **Managing Polyphenol Interference During the Extraction of RNA:** Check the color and purity of RNA that you obtained from high-polyphenolic tissues. Brownish or colored extracts generally indicate the presence of phenolic compounds. This is because polyphenols may co-precipitate with RNA and inhibit reverse transcription, making qPCR results unreliable. If so, perform an additional cleanup step, such as overnight precipitation, with 3 M lithium chloride at 4°C to remove polyphenolic compounds, which also stabilizes RNA. The A_{260}/A_{280} ratio should fall between 1.8 and 2.0 for high-quality cDNA synthesis [35, 48].
3. **Verification of Genotypic Homozygosity in T-DNA Lines:** Triplex PCR method is a necessary verification of homozygous genotypes by screening T-DNA insertion lines. The first primer pair should flank the site of insertion (LP/RP) in order to identify the wild-type allele, while the second primer set includes the T-DNA Left Border primer (LB) and a gene-specific primer (RP), which identifies the insertion. These two steps ensure distinction between wild type, heterozygous, and homozygous plants. For further analysis, use exclusively T3 or T4 homozygous lines in order to avoid segregation artefacts and ensure genetic stability of experimental results [18].
4. **Metabolite Stability (Cold Chain):** Flavonoids, especially anthocyanins, are very sensitive to heat and light; therefore, those compounds easily degrade and oxidize. All the steps of extraction must be performed on ice or in a cold room at 4°C to maintain the cold chain at all times. Extracts should be stored, when possible, in amber vials at −20°C and in the dark.
5. **Butanol-HCl Interpretation:** While the A550 value indicates the overall quantity of PA, the DMACA (*p*-dimethylamino-cinnamaldehyde) staining reagent can directly indicate and

confirm the colorless proanthocyanidins in Arabidopsis seed coats. This reagent detects flavan-3-ols only.

Acknowledgments

I acknowledge the unwavering support and inspiration of Naran Lala College of Professional and Applied Science in shaping my academic journey and contributing to the realization of this chapter.

References

1. Winkel-Shirley B (2001) Flavonoid biosynthesis: a colorful model for genetics, biochemistry, cell biology, and biotechnology. Plant Physiol 126(2):485–493
2. Wang T-y, Li Q, Bi K-s (2018) Bioactive flavonoids in medicinal plants: structure, activity and biological fate. Am. J Pharm Sci 13(1):12–23
3. Koes RE, Quattrocchio F, Mol JNM (1994) The flavonoid biosynthetic pathway in plants: function and evolution. BioEssays 16(2):123–132
4. Petersen M, Hans J, Matern U (2010) Biosynthesis of phenylpropanoids and related compounds. Annu Plant Rev Biochem Plant Sci Mol 40:182–257.
5. Pandith SA et al (2020) Chalcone synthases (CHSs): the symbolic type III polyketide synthases. Planta 251(1):15
6. Liu W et al (2021) The flavonoid biosynthesis network in plants. Plant Cell Environ 22(23):12824
7. Pervaiz T et al (2017) Naturally occurring anthocyanin: structure, functions, and biosynthetic pathway in fruit plants. J Plant Sci 5(2):1–9
8. Hichri I et al (2011) Recent advances in the transcriptional regulation of the flavonoid biosynthetic pathway. J Exp Bot 62(8):2465–2483
9. Zhao L et al (2013) The R2R3-MYB, bHLH, WD40, and related transcription factors in flavonoid biosynthesis. Plant Sci 13(1):75–98
10. Saslowsky DE, Dana CD, Winkel-Shirley B (2000) An allelic series for the chalcone synthase locus in Arabidopsis. Genetics 255(2):127–138
11. Nesi N et al (2001) The Arabidopsis TT2 gene encodes an R2R3 MYB protein determining proanthocyanidin accumulation in seeds. Plant Cell 13(9):2099–2114
12. Kong Q et al (2012) Regulatory switch enforced by bHLH and ACT-domain mediated dimerizations of the maize transcription factor R. Proc Natl Acad Sci U S A 109(30):E2091–E2097.
13. Hartmann U et al (2005) Differential interactions of cis-elements recognized by R2R3-MYB, BZIP, and BHLH factors regulate phenylpropanoid biosynthesis genes. Plant J 57(2):155–171
14. Taylor LP, Briggs WR (1990) Genetic regulation and photocontrol of anthocyanin accumulation in maize seedlings. Plant Cell 2(2):115–127
15. Bogs J et al (2007) The grapevine transcription factor VvMYBPA1 regulates proanthocyanidin synthesis during fruit development. Plant Physiol 143(3):1347–1361
16. Rinaldo AR et al (2015) A grapevine anthocyanin acyltransferase regulated by VvMYBA produces most acylated anthocyanins in grape skins. Plant Physiol 169(3):1897–1916
17. Zhu Y et al (2025) MeJA elicitation on flavonoid biosynthesis and gene expression in hairy roots of *Glycyrrhiza glabra* L. Plants 16(11):1387
18. Alonso JM et al (2003) Genome-wide insertional mutagenesis of *Arabidopsis thaliana*. Science 301(5633):653–657
19. Nesi N et al (2000) The TT8 gene encodes a bHLH protein required for DFR and BAN expression in Arabidopsis siliques. Plant Cell 12(10):1863–1878
20. Agarwal TR (2016) Syntelogs of MYB31 and MYB42 exhibit divergent roles in phenylpropanoid pathway regulation in maize, sorghum, and rice (Doctoral dissertation, University of Toledo)
21. Murashige T, Skoog F (1962) A revised medium for rapid growth and bio assays with tobacco tissue cultures. Physiol Plant 15(3):473
22. Borevitz JO et al (2000) Activation tagging identifies a conserved MYB regulator of phenylpropanoid biosynthesis. Plant Cell 12(12):2383–2393
23. Karimi M, Inzé D, Depicker A (2002) GATEWAY™ vectors for Agrobacterium-mediated plant transformation. Trends Plant Sci 7(5):193–195

24. Earley KW et al (2006) Gateway-compatible vectors for plant functional genomics and proteomics. Plant J 45(4):616–629
25. Odell JT, Nagy F, Chua N-H (1985) DNA sequences required for activity of the CaMV 35S promoter. Nature 313(6005):810–812
26. Hiei Y et al (1994) Efficient Agrobacterium-mediated transformation of rice (Oryza sativa L.). Plant J 6(2):271–282
27. Hellens RP et al (2000) pGreen: a versatile binary Ti vector for Agrobacterium-mediated plant transformation. Plant Mol Biol 42(6):819–832
28. Stachel SE et al (1985) Signal molecules from wounded plant cells that activate T-DNA transfer in *Agrobacterium tumefaciens.* Nature 318(6047):624–629
29. Clough SJ, Bent AF (1998) Floral dip: a simplified method for Agrobacterium-mediated transformation of *Arabidopsis thaliana*. Plant J 16(6):735–743
30. Koncz C, Schell J, MGG G (1986) The promoter of TL-DNA gene 5 directs tissue-specific expression in new Agrobacterium binary vectors. Mol Gen Genet 204(3):383–396
31. Bustin SA et al (2009) The MIQE guidelines: minimum information for publication of qPCR experiments. Clin Chem 55:4
32. Taylor S et al (2010) Practical approach to RT-qPCR: publishing data conforming to MIQE guidelines. Clin Chem 50(4):S1–S5.
33. Porter LJ, Hrstich LN, Chan BG (1985) Conversion of procyanidins and prodelphinidins to cyanidin and delphinidin. Phytochemistry 25(1):223–230
34. Chomczynski P, Sacchi N (1987) Single-step RNA isolation by acid guanidinium thiocyanate-phenol-chloroform extraction. Anal Biochem 162(1):156–159
35. Ghawana S et al. (2011) An RNA isolation system for plant tissues rich in secondary metabolites. Plant Methods 4(1):85
36. Xu R, Qingshun LQ (2008) Streamlined cloning of genes into binary vectors via the Gateway® TOPO system. Plant Methods 4(1):4
37. Barghchi M (1995) High-frequency Agrobacterium-mediated transformation of *Arabidopsis thaliana* ecotypes "C24" and "Landsberg erecta" using Agrobacterium tumefaciens. In Agrobacterium Protocols. Springer, 135–147
38. Livak KJ, Schmittgen TD (2001) Analysis of relative gene expression using RT-qPCR and the $2^{-\Delta\Delta CT}$ method. Methods 25(4):402–408
39. Koes R, Verweij W, Quattrocchio F (2005) Flavonoids: a colorful model for regulation and evolution of biochemical pathways. Trends Plant Sci 10(5):236–242
40. Fuleki T, Francis FJ (1968) Quantitative methods for anthocyanins: extraction and determination in cranberries. J Food Sci 33(1):72–77
41. Lee J et al (2005) Determination of total monomeric anthocyanin content by the pH differential method: collaborative study. J AOAC 88(5):1269–1278
42. Giusti MM, Wrolstad RE (2001) Characterization and measurement of anthocyanins by UV-visible spectroscopy. Curr Protoc Food Anal Chem 1:F1.2.1–F1.2.13.
43. Qin X et al (2018) Systematic chemical analysis reveals superior antioxidant capacity via synergistic flavonoid effects in red tissues. Plant Sci 6:9
44. Hagazy K, Mulata HN (2016) Structural characterization of flavonoids using mass spectrometry. Res Rev J Pharmacol 6(3):1–19.
45. Wiczkowski W, Szawara-Nowak D, Topolska J (2013) Red cabbage anthocyanins: profile, isolation, identification, and antioxidant activity. Food Res Int 51(1):303–309
46. Kuppusamy P et al (2018) Quantification of phenolic and flavonoid markers in *Lolium multiflorum* by HPLC-DAD. Anal Methods 28:282–288
47. Gullickson ER et al (2019) Development of a cranberry standard for quantifying insoluble proanthocyanidins. J Agric Food Chem 68(10):2900–2905
48. Buendía B et al (2010) HPLC-MS analysis of proanthocyanidin oligomers and phenolics in 15 strawberry cultivars. J Agric Food Chem 58(7):3916–3926

Chapter 28

Saponins: Molecular Engineering for Saponin Production

Juhi Barot and Nafisa Patel

Abstract

Saponins are secondary compounds that contain a glycosidic bond and exhibit a series of biological properties. They currently find numerous applications in medicines, health foods, and agricultural products. However, biotechnological methods are necessary to increase the yield and diversity of these compounds, as their natural levels are too low. Golden Gate cloning is a novel, modular DNA assembler that utilizes Type IIS restriction enzymes. Recently, it has been widely applied for recreating biosynthetic pathways for diverse secondary compounds, including saponins. The approach is helpful for the easy assembly of multiple genes encoding biosynthetic enzymes, including P450 enzymes, glycosyltransferases, and enzymes that provide precursors for these enzymes. Thus, Golden Gate cloning technology significantly accelerates the development of novel biocandidates with improved saponin yield and diversity. In combination with CRISPR/Cas systems and genome-centered regulation, it becomes a versatile and effective tool for sustainable production and development of novel saponin compounds with pre-set biological properties.

Keywords Saponin, Golden Gate cloning, Type IIS restriction enzymes, Heterologous expression

1 Introduction

Saponins are glycosylated triterpenoids that possess amphiphilic properties and exhibit a wide range of biological activities, including potent anti-inflammatory, antifungal, antiviral, immunomodulatory, and anticancer effects [1]. They are naturally produced by many plant genera listed in Table 1. Although pharmacologically valuable, extracting saponins from plants is tedious, inefficient, and yields low amounts due to seasonal changes, slow plant growth, and complex downstream processing [2].

To address these issues, microbial production platforms have been developed as environmentally friendly alternatives for synthesizing plant-derived natural products. *Escherichia coli*, despite its widespread use in metabolic engineering, faces several challenges in producing saponin. Being a prokaryote, *E. coli* lacks the necessary eukaryotic post-translational machinery required for proper folding,

Dhruti Amin et al. (eds.), *Phytoconstituent Analysis*, Springer Protocols Handbooks,
https://doi.org/10.1007/978-1-0716-5400-2_28,

Table 1

Plants, their saponin types, and the genes responsible for saponin production

Plant name (scientific)	Common name	Category/use	Saponin type	Notes/ applications	Saponin source in the plant	Key biosynthetic genes	References
Aesculus hippocastanum	Horse chestnut	Medicinal plant	Aescin (triterpenoid)	Used for vascular issues	Seed	β-amyrin synthase, CYPs	[11]
Agave spp.	Agave	Ornamental/ industrial	Steroidal	Detergent and steroid precursor	Leaves and sap	β-amyrin synthase, sterol side-chain cleavage enzymes	[12]
Avena sativa	Oats	Food plant	Avenacosides (steroidal saponin)	Contains avenacosides	Whole grain (bran layer)	Avenacoside pathway genes, β-amyrin synthase	[13]
Capsicum spp.	Chili peppers	Ornamental	Low saponins	Low-level saponins	Fruit	Unknown or low-expression genes	[14]
Chenopodium quinoa	Quinoa	Food Plant	Triterpenoid	The seed coat has high saponins and needs rinsing	Seed coat	β-amyrin synthase, UGTs, CYP716A	[15]
Cicer arietinum	Chickpea	Food plant	Triterpenoid	Moderate saponins	Seed	β-amyrin synthase, CYP450 family	[16]
Dioscorea spp.	Yam	Ornamental/ medicinal	Steroidal (diosgenin)	Steroid precursor	Tuber	Diosgenin synthase, CYP94D108	[17]
Glycine max	Soybean	Food plant	Triterpenoid	Health benefits, food use	Seed	β-amyrin synthase, UGT73P2, CYP93E1	[18]
Glycyrrhiza glabra	Licorice	Medicinal plant	Glycyrrhizin (triterpenoid saponin)	Sweetener, anti-inflammatory	Root	β-amyrin synthase, CYP88D6, UGTs	[19]

(Continued)

Table 1
continued

Plant name (scientific)	Common name	Category/use	Saponin type	Notes/ applications	Saponin source in the plant	Key biosynthetic genes	References
Gypsophila spp.	Baby's breath	Industrial/ biotech	Triterpenoid	Foaming agent in labs	Roots	β-amyrin synthase, CYPs	[20]
Panax ginseng/ P. quinquefolius	Ginseng	Medicinal plant	Ginsenosides (triterpenoid saponins)	Adaptogen, used in herbal medicine	Root, stem, leaves	Dammarenediol-II synthase, CYP716A47, UGT71A27	[21]
Phaseolus spp.	Beans	Food plant	Triterpenoid	Common legumes with saponins	Seed	β-amyrin synthase, CYP93E1	[22]
Primula spp.	Primrose	Ornamental	Triterpenoid	Contains triterpenoid saponins	Leaves and roots	β-amyrin synthase	[23]
Quillaja saponaria	Soapbark tree	Industrial/ biotech	Quil A (triterpenoid)	Foaming agent in beverages/ vaccines	Bark	β-amyrin synthase, CYP716A, UGT73	[24]
Sapindus spp.	Soapberry	Ornamental/ detergent	Triterpenoid	Natural soap (Reetha)	Fruit (pericarp)	β-amyrin synthase, Sap1, Sap2	[25]
Saponaria officinalis	Soapwort	Medicinal plant	Triterpenoid	Soap/shampoo use	Leaves and roots	β-amyrin synthase	[26]
Spinacia oleracea	Spinach	Food plant	Low/trace saponins	Trace saponins	Leaves	Minimal; saponin genes not well studied	[27]
Trigonella foenum-graecum	Fenugreek	Medicinal plant	Steroidal	Used in herbal medicine for diabetes	Seed	β-amyrin synthase, CYPs (CYP93E1)	[28]
Yucca schidigera	Yucca	Medicinal plant	Steroidal	Used in feed and detergents	Stem and root	β-amyrin synthase, CYPs, UGTs	[29]

modification, and activity of cytochrome monooxygenases (P450s) and glycosyltransferases (UGTs)—enzymes essential for saponin biosynthesis. Moreover, the hydrophobic and cytotoxic nature of triterpenoids, which can disrupt membranes, may harm bacterial cells, thereby reducing their viability and productivity [3, 4].

Plants, as the natural producers of saponins, are inherently capable of synthesizing these complex compounds. However, genetic modification of plant hosts remains challenging due to their long life cycles, genetic redundancy, low transformation efficiency, and regulatory hurdles related to genetically modified organisms (GMOs). These factors collectively hinder the scalability and reproducibility of saponin production in plant systems.

In contrast, *Saccharomyces cerevisiae* is the most suitable host for heterologous saponin production. Its eukaryotic nature allows for intrinsic post-translational modifications, including glycosylation and membrane-associated activities, which are necessary for the functional CYP450s. Its amenability to metabolic engineering stems from well-characterized genetics, strong promoters, and the capacity for integrating complex biosynthetic pathways [5]. Additionally, *S. cerevisiae* tolerates high levels of hydrophobic compounds and can be cultivated easily on an industrial scale. Central to saponin biosynthesis in yeast is enhancing precursor supply via the mevalonate (MVA) pathway, which naturally produces isoprenoid building blocks such as IPP and DMAPP from acetyl-CoA, leading to the formation of the triterpenoid skeleton. By overexpressing a deregulated form of HMG-CoA reductase (HMGR), the rate-limiting enzyme in the pathway, flux can be significantly increased toward squalene and subsequently 2,3-oxidosqualene, the immediate precursor to β -amyrin and other triterpenoids [6, 7].

To further improve production efficiency, this work proposes a comprehensive modular design for engineering yeast cells that integrates multiple strategies:

Enhanced NADPH regeneration for the reduction reactions catalyzed by CYP450, using enzymes such as malic enzyme or NADP-dependent glyceraldehyde-3-phosphate dehydrogenase. Cytosolic acetyl-CoA enhancement via heterologous expression of ATP citrate lyase (ACL) and the pyruvate dehydrogenase bypass. Boosting β-alanine metabolism to maintain nitrogen balance and redox homeostasis. Downregulation of competing pathways, such as lanosterol synthase (ERG7) and oxidosqualene cyclases (OSCs), through CRISPR interference (CRISPRi), to redirect metabolic flux from native sterol synthesis toward triterpenoid production [8]. Co-expression of β-amyrin synthase, various cytochrome P450 enzymes, and UGTs, for stoichiometric control and pathway completeness [9].

The key innovation in this system is the use of Golden Gate cloning, a modular assembly technique based on Type IIS restriction enzymes, which enables the fast, scarless, one-pot construction of multigene pathways [10]. Unlike traditional cloning methods, which are slow and limited by the availability of restriction sites, Golden Gate enables the parallel and directional assembly of multiple genetic parts (e.g., promoters, genes, terminators) into a single plasmid or genomic locus. This approach facilitates rapid testing and optimization of metabolic pathways.

By integrating these modular strategies into a unified plasmid-based system, this work offers a systematic and scalable approach to optimizing saponin biosynthesis in *S. cerevisiae*. Developing such a microbial cell factory not only bypasses the limitations of plant extraction and prokaryotic systems but also sets the stage for large-scale production of high-value therapeutic triterpenoids.

2 Materials

1. Yeast Host: *Saccharomyces cerevisiae* strain BY4741 or CEN. PK2-1C is required for cloning.
2. Growth Media:
   a. YPD: 10 g/L yeast extract, 20 g/L peptone, 20 g/L glucose for the cultivation of the host strain.
   b. Synthetic complete dropout media (SC–URA or SC–LEU) for plasmid selection.
   c. Galactose-based SC media for inducible promoter activation.
3. Level-0 Assembly: Basic Parts Cloning
   a. It involves the amplification and assembly of an individual gene of interest, including promoters, conserved coding regions, and terminators.
   b. As an entry vector, the pYTK001 series vector has been used.
   c. Forward and reverse primers need to be designed appropriately, complementary to amplify the gene of interest and have an overhang for the type II restriction enzyme (BsaI-HFv2).
   d. Q5 High-Fidelity DNA Polymerase for amplification purposes.
   e. T4 DNA ligase, CutSmart buffer, and ligation buffer to perform the cloning of the gene of interest and insert it into the entry vector.
   f. Promoters: TEF1 [30], PGK1 [31], SNR52 [32].
   g. CDS: tHMG1 [33], ZWF1 [34], ACS2, β-AS [35], CYP716A47, UGT71A27, dCas9 [36], INO2, POS5.

   h. Terminators: CYC1t, ADH1t [37], PGK1t [38], SUP4t [39].

4. Level-1 Assembly: Transcription Unit Construction
   a. Transcription UnitDesign has been done by creating a vector having a promoter–gene–terminator region.
   b. TU1: Precursor supply: TEF1–tHMG1–CYC1tPGK1–ACS2–ADH1tTDH3–ZWF1–PGK1t.
   c. TU2: Core saponin pathway: GAL1–β-AS–CYC1tTEF1–CYP716A47–ADH1tPGK1–UGT71A27–PGK1t.
   d. TU3: Regulatory and support modules: TEF1–dCas9–CYC1tSNR52–gRNA_ERG1–SUP4tPGK1–INO2–ADH1tTDH3–POS5–CYC1t.

5. Level-2 Assembly: Multi-TU Pathway Construction
   a. This level assembles multiple Level 1 TUs in a final multigene expression plasmid.
   b. pYTK096 Level 2 destination vector (2μ origin, URA3 selection) used for final assembly.
   c. BpiI (Type IIS enzyme) specifically used for TU assembly.
   d. T4 DNA ligase.

6. Yeast Transformation Material
   a. Lithium acetate (0.1 M) used to permeabilize yeast cells.
   b. PEG 3350 facilitates DNA uptake into the host cell.
   c. Salmon sperm DNA (denatured) used as carrier DNA.
   d. Final Level 2 plasmid.
   e. SC–URA/SC–LEU plates used for the selection of post-transformed cells.
   f. Sterile water, TE buffer for washing and dilution.
   g. Pre-induction: SC–URA + 2% glucose.
   h. Induction media: SC–URA + 2% galactose.
   i. Conditions: 30°C, 200 rpm, 48–72 h.

3 Methods

3.1 Level-0 Assembly: Construction of Basic Genetic Parts (Table 2)

1. Level-0 assembly was performed to produce individual standardized biological parts, including promoters, CDS, and transcriptional terminators. Genetic fragments of interest were PCR-amplified using Q5 High-Fidelity Polymerase to ensure accurate replication without unwanted mutations. Primers

Table 2
Components for the construction of Level-0 assembly

Component	Volume
Insert DNA	50–100 ng
pYTK001 backbone	50 ng
BsaI-HFv2	1 μL
T4 Ligase	1 μL
Buffer	2 μL
Nuclease-free water	To 20 μL

were designed with BsaI Type IIS restriction sites and appropriate 4-bp overhangs to facilitate direction-specific Golden Gate cloning.

2. PCR condition: 98°C for 30 s, 30 cycles: 98°C 10 s, 60°C 20 s, 72°C 30 s/kb.
3. The PCR-amplified fragments were then purified and inserted by BsaI-mediated Golden Gate assembly into MoClo-compatible Level-0 entry plasmids (pYTK001 series). Reactions contained BsaI and T4 DNA ligase to allow for simultaneous digestion and ligation, yielding seamless parts flanked by standardized MoClo fusion sites. Recombinant plasmids were transformed into yeast and verified by colony PCR and sequencing.
4. This step ensured that the design resulted in well-defined promoters (TEF1, PGK1, TDH3, GAL1, SNR52), CDS modules (tHMG1, ZWF1, ACS2, etc.), and terminators (CYC1t, ADH1t, PGK1t, SUP4t) for modular assembly (*see* **Note 1**).

3.2 Level-1 Assembly: Construction of Transcription Units (TUs)

1. Validated Level-0 modules were assembled into complete transcription units (TUs) through BsaI-mediated Golden Gate assembly into the Level-1 acceptor plasmid pYTK073. Each TU consisted of a single promoter–gene–terminator expression cassette, designed according to its metabolic function.
2. The fidelity of the cloning was verified by restriction digestion and Sanger sequencing. Positive constructs were propagated under ampicillin selection.
3. This stage yielded self-contained biosynthetic modules enabling controlled metabolic enhancement in yeast (*see* **Note 2**).

3.3 Level-2 Assembly: Construction of Multigene Pathway Plasmid (Fig. 1)

1. The fully validated Level-1 units were assembled into a single Level-2 destination vector pYTK096 via BpiI-mediated Golden

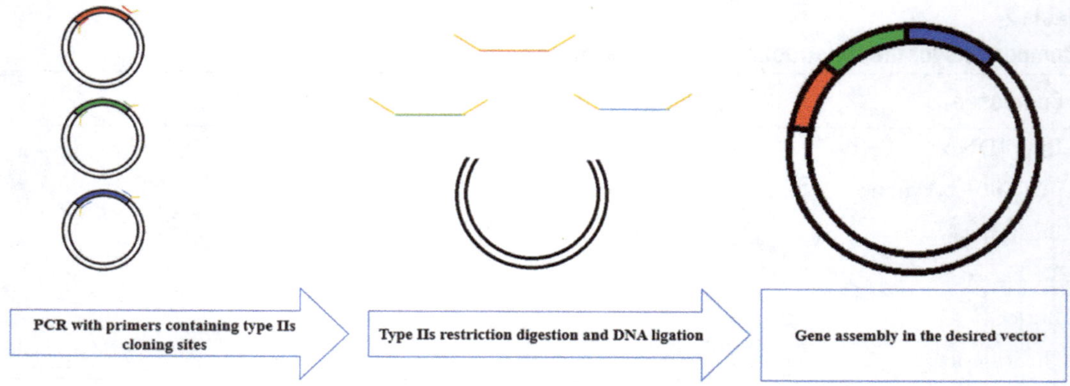

Fig. 1 Assembly of Type IIs insert and vector cloning

Gate assembly. This plasmid includes a 2μ origin for high-copy maintenance in yeast and URA3 for auxotrophic selection.

2. Incubation has been carried out at 16°C.
3. The final construct of ~12–14 kb was confirmed by multiplex PCR and diagnostic digestions. Successful yeast transformation also validated functional integrity (*see* **Note 3**).

3.4 Transformation of Yeast

1. The final Level-2 plasmid was transformed into *S. cerevisiae* BY4741 using a lithium acetate–PEG–carrier DNA protocol. The cells were heat-shocked in the presence of denatured salmon sperm DNA to allow for efficient uptake of the multigene construct.
2. Transformants were selected on SC–URA plates, using the auxotrophic marker provided by pYTK096. Positive colonies were confirmed using plasmid rescue and expression analysis (*see* **Note 4**).
3. Induction and fermentation for producing saponin
4. Yeast cultures harboring the engineered plasmid were initially grown in SC–URA + 2% glucose to establish biomass without induction. Later, cells were shifted to SC–URA + 2% galactose, a condition that activates the GAL1-regulated biosynthetic pathway, TU2, thereby triggering metabolic flux toward ethanol production.
5. Fermentation was carried out at 30°C for 48–72 h (*see* **Note 5**).

3.5 Expression Validation

1. Colony PCR confirmation, qPCR for gene expression, HPLC/GC or LC/MS for saponin and metabolite quantification (*see* **Note 6**).

4 Notes

1. **Level-0 Assembly (Basic Parts Cloning)**
   - Primer design: Check for overhangs and restriction sites (BsaI recognition sequence + correct 4-bp overhangs).
   - PCR fidelity: Q5 DNA polymerase must be used carefully; do not overcycle to minimize non-specific bands.
   - Gel purification: It is essential to purify the DNA fragments to avoid any inhibitors of the ligation.
   - Golden Gate reaction: Proportions of enzyme and buffer should be accurate; do not subject ligase and restriction enzymes to freeze-thaw cycles.
   - Verification: Verify the DNA sequence for the encoded green fluorescent protein by colony PCR and DNA sequencing.
2. **Level-1 Assembly (Transcription Units)**
   - Module compatibility: The promoters, CDSs, and terminators should be MoClo compatible with the appropriate fusion sites.
   - Plasmid backbone: Employ pYTK073 vector; old DNA lowers efficiency.
   - Sequencing junctions: It is essential to verify seamless assembly in junctions.
   - Antibiotic selection: Correct concentration of ampicillin needs to be there; old plates are not effective.
3. **Level-2 Assembly (Multigene Pathway)**
   - Choice of enzyme: BpiI (and not BsaI) is to be used for TU assembling. Enzyme activity should be checked through test digestions.
   - In general, Golden Gate reactions greatly benefit from cycling between 37°C (digestion) and 16°C (ligation).
   - Plasmid size: Larger constructs (~12–14 kb) are more challenging to assemble; confirm using multiplex PCR and diagnostic digests.
   - Destination vector: The insert pYTK096 should be intact; it contains the 2μ origin and URA3 marker.
4. **Yeast Transformation**
   - Cell preparation: Fresh log-phase yeast cultures are required for optimal transformation efficiency.
   - Lithium acetate/PEG cocktail: Make fresh; age-dependent viscosity of PEG.

- Heat shock: Optimize the time of temperature. A longer time will reduce viability; a shorter time will reduce uptake.
- Selection plates: Let SC–URA/SC–LEU plates dry before use to avoid smeared colonies.

5. **Induction and Fermentation**
   - Pre-induction: Verify that glucose is consumed prior to shifting to galactose media.
   - Induction media: Must have a precise concentration of galactose (2%); addition of glucose inhibits GAL1 promoter.
   - Fermentation conditions: Subsequent shaking (200 rpm) and temperatures (30°C).
   - Duration of the culture must be observed at OD600, avoiding overgrowth, which will affect the yield of products if the culture exceeds 72 h.

6. **Expression Validation**
   - Colony PCR: Fresh colonies must be used; older colonies may provide low amplification.
   - qPCR: Normalize to housekeeping genes (e.g., ACT1) for expression analysis.
   - Metabolite analysis: In the analysis of saponins via HPLC, GC, and LC-MS.
   - Replicates: Always do replicates to verify the reproducibility of the data.

References

1. Kundu P, Thorat SA, Kiran A, Goel K, Zinta G (2025) Plant-derived triterpenoid saponins: multifaceted roles and bioengineering prospects. Plant Cell Rep 44(8):172
2. Kaur R, Mishra V, Gupta S, Sharma S, Vaishnav A, Singh SV (2024) Industrial and environmental applications of plant-derived saponins: an overview and future prospective. J Plant Growth Regul 43(9):3012–3026
3. Niazi SK, Magoola M (2023) Advances in Escherichia coli-based therapeutic protein expression: mammalian conversion, continuous manufacturing, and cell-free production. Biologics 3(4):380–401
4. Zhao Yjia, Li C (2018) Biosynthesis of plant triterpenoid saponins in microbial cell factories. J Agric Food Chem 66(46):12155–12165
5. Zhao M, Ma J, Zhang L, Qi H (2024) Engineering strategies for enhanced heterologous protein production by Saccharomyces cerevisiae. Microb Cell Fact 23(1):32
6. Ignea C, Raadam MH, Motawia MS, Makris AM, Vickers CE, Kampranis SC (2019) Orthogonal monoterpenoid biosynthesis in yeast constructed on an isomeric substrate. Nat Commun 10(1):3799
7. Ma T, Shi B, Ye Z, Li X, Liu M, Chen Y et al (2019) Lipid engineering combined with systematic metabolic engineering of Saccharomyces cerevisiae for high-yield production of lycopene. Metab Eng 52:134–142
8. Billerbeck S, Prins RC, Marquardt M (2022) A modular cloning toolbox including CRISPRi for the engineering of the human fungal pathogen and biotechnology host *Candida glabrata* [Internet]. Microbiology. https://biorxiv.org/lookup/, https://doi.org/10.1101/2022.10.17.512482. Accessed 5 Aug 2025
9. Chai L, Che J, Qi Q, Hou J (2024) Metabolic engineering for squalene production: advances and perspectives. J Agric Food Chem 72(50):27715–27725

10. Mukherjee M, Caroll E, Wang ZQ (2021) Rapid assembly of multi-gene constructs using modular Golden Gate cloning. J Vis Exp (168):1–24
11. Dahash SL, Abass OK, Abdul-Razaq MM, Al-Kuraishy HM, Al-Gareeb AI (2021) Aesculus hippocastanum-derived extract β-aescin and in vitro antibacterial activity. J Microsc Ultrastruct 9(1):26–30
12. Niculae M, Hanganu D, Oniga I, Burcă SA, Tiperciuc B, Ielciu I et al (2025) Agave amica (Medik.) Thiede & Govaerts (Asparagaceae)—insights into its valuable phenolic profile and in vitro antimicrobial, antibiofilm, antioxidative, and antiproliferative properties. Antibiotics. 14(7):638
13. Hu J, Hu C, Zhao Y, Lee PS, Sang S (2025) Germination and false germination increase the levels of bioactive steroidal saponins in oats. J Agric Food Chem 73(3):1940–1951
14. Damiri N, Fenton AR, Irsan C, Oktaviani O, Fadly R, Pratama R et al (2025) Utilization of medicinal plant extracts to reduce the mosaic disease complex of chili plant. Biodiversitas J Biol Divers [Internet] 26(3). https://smujo.id/biodiv/article/view/20709. Accessed 30 Jul 2025
15. El Hazzam K, Hafsa J, Sobeh M, Mhada M, Taourirte M, El Kacimi K et al (2020) An insight into saponins from quinoa (Chenopodium quinoa Willd): a review. Molecules 25(5):1059
16. Pérez-Ramírez IF, Escobedo-Alvarez DE, Mendoza-Sánchez M, Rocha-Guzmán NE, Reynoso-Camacho R, Acosta-Gallegos JA et al (2023) Phytochemical profile and composition of chickpea (Cicer arietinum L.): varietal differences and effect of germination under elicited conditions. Plants 12(17):3093
17. Uduak E, Aniefiok O, Ndem E, Reagan A, Lasbrey E, Chineke O et al (2025) Unveiling the evolutionary history and structural dynamics of Dioscorea bulbifera L. and congeners through matK gene in silico analysis. Trop J Nat Prod Res 9(5):2220
18. Li S, Chen J, Hao X, Ji X, Zhu Y, Chen X et al (2024) A systematic review of black soybean (*Glycine max* (L.) Merr.): nutritional composition, bioactive compounds, health benefits, and processing to application. Food Front 5(3):1188–1211
19. Rasool A, Dar TA (2025) Glycyrrhizin and its derivatives: an emerging secondary metabolite arsenal of Glycyrrhiza glabra. Med Chem Res 34(4):745–763
20. Shakeel A, Noor JJ, Jan U, Gul A, Handoo Z, Ashraf N (2025) Saponins, the unexplored secondary metabolites in plant defense: opportunities in integrated pest management. Plants 14(6):861
21. Xing JJ, Hou JG, Liu Y, Zhang RB, Jiang S, Ren S et al (2019) Supplementation of saponins from leaves of Panax quinquefolius mitigates cisplatin-evoked cardiotoxicity via inhibiting oxidative stress-associated inflammation and apoptosis in mice. Antioxidants 8(9):347
22. Cid-Gallegos MS, Sánchez-Chino XM, Corzo-Ríos LJ, Gómez Y, Gómez YDLM, Moguel-Concha DDR, Borges-Martínez JE et al (2025) Phaseolus vulgaris as a source of peptides and biofunctional compounds. In: Improving health and nutrition through bioactive compounds [Internet]. Elsevier, pp 79–91. https://linkinghub.elsevier.com/retrieve/pii/B9780443218736000051. Accessed 30 Jul 2025
23. Trendafilova A, Raykova D, Ivanova V, Novakovic M, Nedialkov P, Paunova-Krasteva T et al (2025) Phytochemical characterization and anti-biofilm activity of Primula veris L. roots. Molecules 30(8):1702
24. Lazo J, Tapia J, Guerra FP (2025) Cadmium and copper stress responses in soapbark tree (Quillaja saponaria): effects on growth, metal accumulation, saponin concentration, and gene expression. Plants 14(5):709
25. Gotur SP, Wadhwan V, Sharma P, Gupta R, Kamarthi N, Tyagi S (2025) Reetha (Sapindus mukorossi) extract as a biosafe surrogate for xylene in hematoxylin and eosin staining procedure. Indian J Pathol Microbiol 68(2):344–348
26. Ashurova LN, Khurramov AR, Bobakulov KM, Sasmakov SA, Ashirov ON, Umarova MR et al (2025) A new withanolide from *Saponaria officinalis* and its antimicrobial and cytotoxic activities. Nat Prod Res 40(9):2392–2401
27. Naseem A, Akhtar S, Ismail T, Qamar M, Sattar D, Saeed W et al (2023) Effect of growth stages and lactic acid fermentation on anti-nutrients and nutritional attributes of spinach (Spinacia oleracea). Microorganisms 11(9):2343
28. Al Raish SM, Almasri RS, Bedir AS (2025) Ancient remedies, modern medicine: a review of antidiabetic, cardioprotective, and antimicrobial activities of date palm (*Phoenix dactylifera*), tomato (*Solanum lycopersicum*), fenugreek (*Trigonella foenum-graecum*), and ashwagandha (*Withania somnifera*). Biology 14(6):695
29. Paray BA, El-Basuini MF, Alagawany M, Albeshr MF, Farah MA, Dawood MAO (2021) Yucca schidigera usage for healthy aquatic animals: potential roles for sustainability. Animals 11(1):93
30. Da Silva NA, Srikrishnan S (2012) Introduction and expression of genes for metabolic engineering applications in Saccharomyces cerevisiae. FEMS Yeast Res 12(2):197–214

31. Partow S, Siewers V, Bjørn S, Nielsen J, Maury J (2010) Characterization of different promoters for designing a new expression vector in Saccharomyces cerevisiae. Yeast 27(11):955–64. https://doi.org/10.1002/yea.1806
32. DiCarlo JE, Norville JE, Mali P, Rios X, Aach J, Church GM (2013) Genome engineering in Saccharomyces cerevisiae using CRISPR-Cas systems. Nucleic Acids Res 41(7):4336–4343
33. Hampton RY (2002) Proteolysis and sterol regulation. Annu Rev Cell Dev Biol 18(1):345–378
34. Gietz RD, Schiestl RH (2007) High-efficiency yeast transformation using the LiAc/SS carrier DNA/PEG method. Nat Protoc 2(1):31–34
35. Moses T, Pollier J, Almagro L, Buyst D, Van Montagu M, Pedreño MA et al (2014) Combinatorial biosynthesis of sapogenins and saponins in *Saccharomyces cerevisiae* using a C-16α hydroxylase from *Bupleurum falcatum.* Proc Natl Acad Sci 111(4):1634–1639
36. Gilbert LA, Larson MH, Morsut L, Liu Z, Brar GA, Torres SE et al (2013) CRISPR-mediated modular RNA-guided regulation of transcription in eukaryotes. Cell. 154(2):442–451
37. Ding N, Zhou S, Deng Y (2021) Transcription-factor-based biosensor engineering for applications in synthetic biology. ACS Synth Biol 10(5):911–922
38. Mumberg D, Müller R, Funk M (1995) Yeast vectors for the controlled expression of heterologous proteins in different genetic backgrounds. Gene 156(1):119–122
39. Koleske AJ, Young RA (1995) The RNA polymerase II holoenzyme and its implications for gene regulation. Trends Biochem Sci 20(3):113–116

Check for updates

Chapter 29

Molecular Mechanisms of Tannin Biosynthesis

Hetvi Naik, Abdullah Khan and Natarajan Amaresan

Abstract

Tannins are abundant plant secondary metabolites that are categorized into two major groups: condensed tannins (proanthocyanidins) (PAs) and hydrolysable tannins (HTs). This chapter explores the molecular mechanisms underlying tannin biosynthesis with an emphasis on enzymatic functions. Condensed tannin biosynthesis originates from the phenylpropanoid pathway, progressing from flavonoid intermediates to flavan-3-ol formation and polymerization. Synthesis of hydrolysable tannins begins with the shikimate pathway leading to gallic acid production and subsequent galloylation reactions with pentagalloylglucose serving as a critical intermediate. This chapter presents detailed protocols for tannin extraction, purification, quantification, and degradation analysis along with enzyme assays for key biosynthetic enzymes including phenylalanine ammonia-lyase, anthocyanidin reductase, leucoanthocyanidin reductase, and cinnamate-4-hydroxylase. Additionally, methods for gene expression analysis and subcellular localization studies provide further insight into the spatial organization of tannin biosynthesis.

Keywords Biosynthesis, qPCR, Condensed tannins, Enzymatic assay, Hydrolysable tannins

1 Introduction

Tannins are abundant plant secondary metabolites, such as condensed tannins (proanthocyanidins, PAs) and hydrolysable tannins (HTs) [1]. PAs are oligomeric or polymeric flavan-3-ols linked through C4→C8 or C4→C6 bonds, whereas HTs consist of a glucose core esterified with gallic acid derivatives [2, 3]. Both serve critical ecological functions in plant defense and significantly affect human health through their antioxidant, anti-inflammatory, and antimicrobial properties.

The biosynthesis of PAs begins with the phenylpropanoid pathway, proceeds through the flavonoid pathway, and culminates in flavan-3-ol synthesis and polymerization [4]. HTs biosynthesis originates from the shikimate pathway intermediate 5-dehydroshikimate, which leads to gallic acid formation and subsequent galloylation reactions with pentagalloylglucose (PGG) serving as a key intermediate [5]. While the early biosynthetic steps are well characterized, the final polymerization steps and regulatory mechanisms remain

Dhruti Amin et al. (eds.), *Phytoconstituent Analysis*, Springer Protocols Handbooks,
https://doi.org/10.1007/978-1-0716-5400-2_29,

poorly understood. This chapter explores the molecular mechanisms of tannin biosynthesis and focuses on enzyme function.

2 Materials

Prepare all solutions using ultrapure water (prepared by purifying deionized water to attain a sensitivity of 18 MΩ-cm at 25°C) and analytical grade reagents. Prepare and store all reagents at room temperature unless otherwise indicated. Diligently follow all waste disposal regulations when disposing of waste materials.

2.1 Buffers and Solutions

1. Extraction buffer: 100 mM Tris–HCl (pH 7.5), 2 mM EDTA, 5 mM DTT, 10% (v/v) glycerol, and 1% (w/v) polyvinylpolypyrrolidone (PVPP). Add 1 mM phenylmethylsulphonyl fluoride (PMSF) immediately before use (*see* **Note 1**).
2. Phosphate-buffered saline (PBS, 10×): 1.37 M NaCl, 27 mM KCl, 100 mM Na_2HPO_4, 18 mM KH_2PO_4, and pH 7.4.
3. 4-Dimethylaminocinnamaldehyde (DMACA) reagent: Dissolve 0.1 g DMACA in 100 mL of butanol (70:30, v/v) (*see* **Note 2**).
4. Folin–Ciocalteu reagent: Dissolve 2 g of phosphomolybdic acid and 1 g of phosphotungstic acid in 100 mL of distilled water, then add 30 g of sodium carbonate to another 100 mL of distilled water and stir both solutions until fully dissolved. Gradually combine the two solutions and dilute with distilled water to a final volume of 1 L.
5. Acetic acid (5%).
6. Tris–HCl buffer.
7. NADPH.
8. Na_2CO_3.
9. Liquid nitrogen.
10. Acetone (HPLC grade).
11. Methanol (HPLC grade).
12. Butanol (HPLC grade).
13. Hydrochloric acid (analytical grade).
14. TRIzol reagent.
15. RNase-free DNase.
16. Reverse transcription kit.
17. qPCR master mix.
18. Gene-specific primers for key enzymes (see Tables 1 and 2).
19. PCR thermal cycler.
20. Real-time PCR system.

Table 1
Gene-specific primers for qPCR analysis of key enzymes in the condensed tannin biosynthetic pathway

Gene	Forward primer (5′-3′)	Reverse primer (5′-3′)
PAL	GCTGACTGGTGAAAAGGTGC	TGTTCGGGATAGCCGATGTT
CHS	ACGTGGAGGAGTATCGCAAG	CGATCAAAGCGACGACGTTT
CHI	TTCGAGACCTTCGCCTACAA	GTCGAAGTTGAGCCCGAAGA
F3H	CACCAAGTTCCTGTCGTCCT	CTCGGGTCGTCGATCAACTT
DFR	CTCGACGCCGAGTTTGAGAT	CGTAGGCCTTGAGGATGTCG
ANS	GCGTTGAAGGCTCGTACTCC	CGCAGTAGGAGCTGGTTTGT
ANR	GCCATGTCCGACTTGCAGTA	AGCCCAACTTGAGACCCTTG
LAR	TGGCTTGCCTGATATCGCTG	GCTCCTCAAGCGCCATCAAT

Table 2
Gene-specific primers for qPCR analysis of key enzymes in the hydrolysable tannin biosynthetic pathway

Gene	Forward primer (5′-3′)	Reverse primer (5′-3′)
PAL	GCTGACTGGTGAAAAGGTGC	TGTTCGGGATAGCCGATGTT
C4H	ATAGAATGGGGCATAGCAGAA	TTGGAGGTAGGGGAGTTTGT
SCPL1	GCTCATCTTCCTCCTAGTGC	GGATTCCCTTCCGACTCTA
DAHPS	TGTCTCAACCTCCGCTCTTT	CGCTGAGATGGGTTTGAGTG

2.2 Condensed Tannins

2.2.1 Enzyme Assays for Early Biosynthetic Steps

1. Phenylalanine ammonia-lyase (PAL) assay buffer: 100 mM Tris–HCl buffer (pH 8.5) containing 5 mM β-mercaptoethanol.
2. PAL substrate solution: 20 mM L-phenylalanine in PAL assay buffer.
3. Chalcone synthase (CHS) assay buffer: 100 mM potassium phosphate buffer (pH 7.0) containing 1 mM EDTA and 10 mM KCN.
4. CHS substrate solutions:
   - 50 μM Malonyl-CoA.
   - 20 μM *p*-Coumaroyl-CoA.

5. Dihydroflavonol reductase (DFR) assay buffer: 100 mM Tris–HCl (pH 7.0) containing 1 mM EDTA.
6. DFR substrate solutions:
   - 100 μM dihydroquercetin.
   - 2 mM NADPH.
7. Anthocyanidin reductase (ANR) assay buffer: 100 mM sodium phosphate buffer (pH 6.5) containing 1 mM EDTA.
8. ANR substrate solutions:
   - 100 μM Cyanidin.
   - 2 mM NADPH.

2.2.2 Analysis of Condensed Tannins

1. HPLCHPLC analysis buffer A: 0.1% (v/v) formic acid in water.
2. HPLC analysis buffer B: 0.1% (v/v) formic acid in acetonitrile.
3. Phloroglucinol reagent: 50 mg/mL phloroglucinol and 10 mg/mL ascorbic acid in methanol containing 0.1 N HCl (*see* **Note 3**).
4. Sephadex LH-20 column chromatography material.
5. C18 reverse-phase HPLC column (4.6 × 250 mm, 5 μm).
6. Mass spectrometry system (LC–MS/MS).
7. Chromatography equipment (HPLC system with diode array detector and fluorescence detector).

2.2.3 Gene Expression Analysis

2.3 Hydrolysable Tannins

2.3.1 Enzyme Assays for Initial Biosynthetic Steps

1. Phenylalanine ammonia-lyase (PAL) assay buffer: 100 mM Tris–HCl buffer (pH 8.5) containing 2 mM β-mercaptoethanol/DTT, 1 mM EDTA, and 10% glycerol (optional).
2. PAL substrate solution: 20 mM L-phenylalanine in PAL assay buffer.
3. Cinnamate-4-Hydroxylase (C4H) assay buffer: 100 mM potassium phosphate buffer (pH 7.5), 0.5 mM cinnamic acid, 0.5 mM NADPH, and $MgCl_2$.
4. 3-Deoxy-D-arabino-heptulosonate-7-phosphate Synthase (DAHP Synthase) assay buffer: 50 mM Tris–HCl (pH 7.5), phosphoenolpyruvate (PEP), erythrose-4-phosphate (E4P), and $MgCl_2$.

5. Gallic Acid Synthesis Enzyme Assay buffer: 50 mM Tris–HCl (pH 7.0), 1 mM 3-dehydroshikimate (3-DHS), and 5 mM $MgCl_2$.
6. Polyketide Synthase Assay (PKS) buffer: 100 mM Tris–HCl buffer (pH 8.0) containing 1–2 mM *β*-mercaptoethanol or DTT, 1 mM EDTA, and 10%–20% glycerol.
7. Substrates for PKS reaction buffer: starter CoA molecule (e.g., acetyl-CoA or galloyl CoA) and 100 mM malonyl-CoA.

2.3.2 Analysis of Hydrolysable Tannins

1. HPLCHPLC analysis buffer A: 0.1% aqueous phosphoric acid (H_3PO_4)
2. HPLC analysis buffer B: Acetonitrile.
3. Dried plant material rich in geraniin (e.g., *Geranium carolinianum* L.).
4. Superose 12 10/300 GL column.
5. C18 reverse-phase HPLC column (4.6×250 mm, 5 μm).
6. Mass spectrometry system (LC–MS/MS).
7. Chromatography equipment (HPLC systemHPLC system with UV detector—set at 274 nm).

2.3.3 Gene Expression Analysis

2.4 Subcellular Localization Studies

1. Binary vectors containing GFP fusion constructs.
2. *Agrobacterium tumefaciens* strain.
3. Infiltration medium: 10 mM MES (pH 5.6), 10 mM $MgCl_2$, 100 μM acetosyringone.
4. Confocal laser scanning microscope.
5. DAPI (4′,6-diamidino-2-phenylindole) staining solution: 1 μg/mL in PBS.
6. ER-Tracker™ Red dye.
7. Anti-GFP antibody.
8. Fluorescence microscope with appropriate filter sets.

3 Methods

Unless otherwise noted, perform all procedures at room temperature.

3.1 Extraction of Tannins

1. Grind 1 g of plant tissue to a fine powder in liquid nitrogen using a pre-chilled mortar and pestle (*see* **Note 4**).
2. Transfer the powder to a 15 mL centrifuge tube and add 10 mL of cold acetone (7:3, v/v) containing 0.1% ascorbic acid.

3. Homogenize the mixture for 1 min using a vortex mixer at maximum speed.
4. Sonicate the mixture in an ice-water bath for 15 min (*see* **Note 5**).
5. Centrifuge at 5000 × *g* for 15 min at 4°C.
6. Collect the supernatant and repeat the extraction twice with the residue.
7. Pool the supernatants and concentrate under reduced pressure at 35°C to remove acetone.
8. Adjust the volume to 5 mL with water and freeze-dry if long-term storage is needed (*see* **Note 6**).

3.2 Condensed Tannins

3.2.1 Purification of Condensed Tannins

1. Prepare a Sephadex LH-20 column (2 × 20 cm) and equilibrate with 50% methanol.
2. Dissolve 300 mg of the crude extract in 2 mL of 50% methanol and load onto the column.
3. Elute sequentially with 100 mL of 50% methanol (to remove low-molecular-weight phenolics) followed by 100 mL of 70% acetone (to elute condensed tannins) (*see* **Note 7**).
4. Collect the 70% acetone fraction, concentrate under reduced pressure at 35°C to remove acetone, and freeze-dry the aqueous solution.
5. Store the purified condensed tannin fraction at −20°C until further analysis.

3.2.2 Spectrophotometric Quantification of Condensed Tannins

1. Dissolve 10 mg of the purified condensed tannin fraction in 10 mL of methanol.
2. In a 1.5 mL microcentrifuge tube, mix 100 μL of the sample solution with 900 μL of DMACA reagent.
3. Incubate the mixture in the dark for 20 min at room temperature.
4. Measure the absorbance at 640 nm against a reagent blank (*see* **Note 8**).
5. Calculate the content of condensed tannins using a standard curve prepared with catechin (5–100 μg/mL) and express the results as mg catechin equivalents per g of sample (*see* **Note 9**).

3.2.3 Phloroglucinol Degradation Analysis of Condensed Tannins

1. Dissolve 5 mg of the purified condensed tannin fraction in 1 mL of methanol.
2. Add 1 mL of phloroglucinol reagent to the sample solution.

3. Heat the mixture at 50°C for 20 min.
4. Cool to room temperature and add 1 mL of 40 mM sodium acetate to stop the reaction.
5. Analyze the degradation products by HPLC under the following conditions:
   - Column: C18 reverse-phase (4.6×250 mm, 5 μm).
   - Mobile phase: A (0.1% formic acid in water) and B (0.1% formic acid in acetonitrile).
   - Gradient: 5% B (0 min), 15% B (15 min), 40% B (30 min), 90% B (35 min), 90% B (38 min), 5% B (40 min), 5% B (45 min).
   - Flow rate: 1 mL/min.
   - Detection: 280 nm.
   - Temperature: 30°C.
6. Calculate the mean degree of polymerization (mDP) by dividing the total peak area of all flavan-3-ol units (extension units + terminal units) by the peak area of terminal units (*see* **Note 10**).

3.2.4 Enzyme Extraction and Assays

Preparation of Enzyme Extract

1. Grind 5 g of plant tissue to a fine powder in liquid nitrogen using a pre-chilled mortar and pestle.
2. Transfer the powder to a 50 mL centrifuge tube and add 15 mL of cold extraction buffer.
3. Homogenize the mixture for 2 min using a vortex mixer at maximum speed.
4. Filter the homogenate through four layers of cheesecloth.
5. Centrifuge the filtrate at 15,000 × *g* for 20 min at 4°C.
6. Collect the supernatant and use it for enzyme assays (*see* **Note 11**).

Anthocyanidin Reductase (ANR) Assay

1. Prepare the reaction mixture in a 1.5 mL microcentrifuge tube containing 50 μL of ANR assay buffer, 10 μL of cyanidin (100 μM), 10 μL of NADPH (2 mM), and 30 μL of enzyme extract.
2. Incubate the reaction mixture at 30°C for 30 min.
3. Terminate the reaction by adding 100 μL of ethyl acetate.
4. Vortex the mixture for 30 s and centrifuge at 10,000 × *g* for 5 min.
5. Collect the upper organic phase and evaporate to dryness under a stream of nitrogen.

6. Reconstitute the residue in 100 μL of methanol for HPLC analysis.
7. Analyze the reaction products by HPLC under the conditions described in Sect. 3.2.3, Step 5.
8. Identify epicatechin by comparing retention time and UV spectrum with authentic standards.

Leucoanthocyanidin Reductase (LAR) Assay

1. Prepare the reaction mixture in a 1.5 mL microcentrifuge tube containing 50 μL of ANR assay buffer (use the same buffer as for ANR), 10 μL of leucocyanidin (100 μM), 10 μL of NADPH (2 mM), and 30 μL of enzyme extract [6].
2. Proceed as described in Section: Anthocyanidin Reductase (ANR) Assay, Steps 2–7.
3. Identify catechin by comparing retention time and UV spectrum with authentic standards.

3.3 Hydrolysable Tannins

3.3.1 Purification of Hydrolysable Tannins

1. Prepare Superose 12 10/300 GL column equilibrated with mobile phase (5% ethanol + 5% acetic acid). Meanwhile, mobile phase B, consisting of 30% v/v ethanol and 30% v/v acetic acid is degassed using ultrasonication.
2. Filter the crude extract using 0.22 μm nylon membrane and inject 0.4 mL of sample at 0.4 mL/min flow rate.
3. Perform gradient elution gradually increasing solvent B from 0% to 40% up to 480 mL, and then switching to 100% B.
4. Monitor fractions at 280 nm, collect based on peaks, and concentrate under reduced pressure [7].
5. Store the purified hydrolysable tannin fraction at −20°C until further analysis.

3.3.2 Spectrophotometric Quantification of Hydrolysable Tannins

1. Dissolve 25 mg of the purified hydrolysable tannin fraction in 50 mL of distilled water.
2. Mix the above extract solution with 0.5 mL of Folin–Ciocalteu reagent and 2.0 mL of 100 g/L Na_2CO_3 and 5.0 mL of distilled water (*see* **Note 12**).
3. Incubate the mixture in the dark for 30–40 min at room temperature.
4. Measure the absorbance at 765 nm against a reagent blank.
5. Calculate the content of hydrolysable tannins using a standard curve prepared with gallic acid (15–100 mg/L) and express the results as mg gallic acid equivalents (GAE) per gram of tannin extract.

3.3.3 Geraniin Degradation Analysis of Hydrolysable Tannins

Sample Preparation

1. Grind the dried plant material into a fine powder and gently heat 1–2 g of the powdered sample in 50–100 mL of distilled water using a reflux system for 90 min.
2. Allow the solution to cool, and then filter it using a 0.45 μm membrane or syringe filter to obtain a clear extract.
3. Analyze the degradation products by HPLC under the following conditions:
   Column: C18 reversed-phase column.
   Mobile Phase: Employ a gradient system using:
   Solvent A: 0.1% aqueous phosphoric acid
   Solvent B: Acetonitrile Gradient Duration: Run over approximately 65 min, adjusting as needed for optimal separation. Flow Rate: Maintain a flow of 1.0 mL per min. Detection: Set the UV detector to 274 nm. Temperature: Keep the column at 25°C. Injection Volume: Inject between 10 and 20 microliters per run.
4. Identify and analyze the chromatographic peaks representing geraniin degradation products, such as gallic acid, ellagic acid, and corilagin. Match the retention times and UV spectra with standard references and calculate concentrations using calibration curves based on known standards [8].

3.3.4 Enzyme Extraction and Assays

Preparation of Enzyme Extract

1. Grind 5 g of plant tissue to a fine powder in liquid nitrogen using a pre-chilled mortar and pestle.
2. Transfer the powder to a 50 mL centrifuge tube and add 15 mL of cold extraction buffer.
3. Homogenize the mixture for at least 2 min using a vortex mixer at maximum speed.
4. Filter the homogenate through four layers of cheesecloth or muslin cloth.
5. Centrifuge the filtrate at 15,000 × *g* for 20 min at 4°C.
6. Collect the supernatant and use it for enzyme assays.

Phenylalanine Ammonia Lyase Activity Assay

1. Approximately 0.5–1.0 g of fresh plant tissue is ground in liquid nitrogen and homogenize in ice-cold extraction.
2. Centrifuge the homogenate at 12,000 × *g* for 20 min at 4°C, and collect the supernatant as the crude enzyme extract. Vortex the mixture for 30 s and centrifuge at 10,000 × *g* for 5 min.

3. Assay is conducted in a 1 mL reaction containing L-phenylalanine, hydrolysable tannin, and enzyme extract in borate buffer (pH 8.8). Controls omit tannins, and blanks exclude the enzyme.
4. Incubate the mixtures at 30°C for 60 min with gentle shaking.
5. Analyze the reaction products by HPLC under the conditions described in Section: Sample Preparation Step 3
6. Use standard solutions of trans-cinnamic acid to construct a calibration curve over a concentration range of 1–100 μM.
7. Identify the sample peaks by retention time comparison with standards and quantify by peak area integration.

Cinnamate-4-hydroxylase (C4H) Assay

1. Repeat Steps 1 and 2 (Sect. 3.5.2) for C4H extract. In a 1 mL reaction, combine 50 μL of 10 mM trans-cinnamic acid (500 μM final) in 50 mM Tris–HCl (pH 7.5).
2. Add 10 μL of hydrolysable tannin solution (10–200 μM) and 100 μL of C_4H enzyme extract.
3. Include 50 μL of NADPH (1 mM final) and adjust the volume with Tris–HCl buffer (50 mM, pH 7.5) (*see* **Note 13**).
4. Incubate the reaction mixtures at 30°C for 30–60 min with gentle shaking to allow the enzyme to convert trans-cinnamic acid to *p*-coumaric acid.
5. Add 50 μL of 1 M HCl to stop the reaction.
6. Vortex the reaction mixture with 1 mL of methanol for 1 min, and then centrifuge at 12,000 × *g* for 10 min to remove insoluble materials. Collect the supernatant for HPLC analysis.
7. Quantify *p*-coumaric acid production by comparing peak areas to a standard curve (1–100 μM). Evaluate the impact of hydrolysable tannins on C_4H activity by comparing reaction rates with and without tannins.

3.4 Analysis of Enzyme Gene Expression

3.4.1 RNA Extraction and cDNA Synthesis

1. Extract total RNA from 100 mg of plant tissue using TRIzol reagent according to the manufacturer's instructions.
2. Treat the RNA with RNase-free DNase to remove genomic DNA contamination.
3. Determine the RNA concentration and purity using a spectrophotometer.
4. Synthesize first-strand cDNA using a reverse transcription kit with oligo(dT) primers according to the manufacturer's instructions.

3.4.2 Quantitative Real-Time PCR (qPCR)

1. Prepare the qPCR reaction mixture with 10 μL of qPCR master mix, 0.5 μL of forward primer (10 μM), 0.5 μL of reverse primer (10 μM), 2 μL of cDNA template, and 7 μL of nuclease-free water.
2. Run the qPCR reactions under the following conditions:
   - Initial denaturation: 95°C for 3 min
   - 40 cycles of: 95°C for 15 s, 60°C for 30 s, and 72°C for 30 s
   - Melting curve analysis: 65°C to 95 °C with 0.5°C increments
3. Add the proper reference genes for normalization, such as actin, ubiquitin, and GAPDH.
4. Determine the target genes' relative expression levels by applying the $2^{-\Delta\Delta Ct}$ technique.

3.5 Subcellular Localization of Biosynthetic Enzymes

3.5.1 Construction of GFP Fusion Proteins

1. Amplify the full-length coding sequences of genes encoding key enzymes (ANR and LAR) without the stop codon using PCR.
2. Clone the PCR products into an appropriate binary vector containing a GFP reporter gene to create C-terminal GFP fusion constructs.
3. Transform the constructs into *Agrobacterium tumefaciens* strain using the freeze–thaw method.

3.5.2 Transient Expression in Nicotiana benthamiana

1. Grow *A. tumefaciens* transformants containing the GFP fusion constructs in LB medium with appropriate antibiotics at 28°C for 24 h.
2. Centrifuge the bacterial culture at 3000 × *g* for 10 min and resuspend the pellet in infiltration medium to an OD600 of 0.5.
3. Incubate the bacterial suspension at room temperature for 2 h.
4. Infiltrate the abaxial side of *N. benthamiana* leaves with the bacterial suspension using a needleless syringe.
5. Allow plants to grow under normal conditions for 48–72 h.

3.5.3 Confocal Microscopy Analysis

1. Cut 1 cm² sections from infiltrated leaf areas.
2. Mount the leaf sections on glass slides with a drop of water.
3. Observe GFP fluorescence using a confocal laser scanning microscope with excitation at 488 nm and emission detection at 500–530 nm.

4. For co-localization studies, stain the leaf sections with appropriate organelle markers (e.g., DAPI for nucleus and ER-Tracker™ Red for endoplasmic reticulum) according to the manufacturer's instructions.
5. Capture images and analyze the subcellular localization of fluorescent signals.

4 Notes

1. PMSF has a short half-life in aqueous solutions. Therefore, it should be added to the extraction buffer immediately prior to use. Alternatively, a protease inhibitor cocktail can be used.
2. The DMACA reagent should be prepared fresh before use and stored in the dark. This reagent reacts specifically with flavan-3-ols and condensed tannins to produce a blue chromophore.
3. Phloroglucinol reagent should be freshly prepared before use. The presence of ascorbic acid helps to prevent oxidation during the reaction.
4. It is important to keep the plant tissue frozen during grinding to prevent enzymatic degradation of condensed tannins. Pre-chill the mortar and pestle with liquid nitrogen before grinding.
5. During sonication, keep the sample in an ice-water bath to prevent heat-induced degradation of condensed tannins.
6. The aqueous extract can be stored at −20°C for short-term (1–2 weeks) or freeze-dried for long-term storage (several months).
7. The elution volumes and solvent compositions may need to be optimized depending on the plant species and tissue type. Monitor the elution by spotting small aliquots of fractions on filter paper and spraying with DMACA reagent. Positive fractions will develop a blue color.
8. The blue color formed by the reaction of DMACA with condensed tannins is time-dependent and light-sensitive. Therefore, it is important to maintain consistent reaction time (20 min) and protect the samples from light during incubation.
9. The DMACA method is specific for flavan-3-ols and condensed tannins with free meta-hydroxyl groups on the A-ring. The reaction produces a blue chromophore with absorption maximum at 640 nm.

10. The mean degree of polymerization (mDP) is calculated using the formula: mDP = (sum of all flavan-3-ol units)/(sum of terminal units). This provides information about the average chain length of condensed tannin polymers.
11. For optimal enzyme activity, all extraction and enzyme assay procedures should be performed at 4°C or on ice unless otherwise specified. The enzyme extract should be used immediately or stored at −80°C in small aliquots to avoid repeated freeze–thaw cycles.
12. A blue complex, which can be measured at 760–765 nm, is created when the Folin–Ciocalteu reagent reacts with the phenolic groups in tannins in an alkaline environment. Na_2CO_3 is added last to avoid precipitation and preserve the soluble ionic form of HT.
13. Because NADPH is oxidation-prone and light-sensitive, it is best to prepare it fresh and store it on ice while using it.

References

1. Smeriglio A, Barreca D, Bellocco E, Trombetta D (2017) Proanthocyanidins and hydrolysable tannins: occurrence, dietary intake and pharmacological effects. Br J Pharmacol 174:1244–1262
2. He F, Pan Q-H, Shi Y, Chang-Qing Duan (2008) Biosynthesis and genetic regulation of proanthocyanidins in plants. Molecules 13:2674–2703
3. Cai Y, Zhang J, Chen NG, Shi Z, Qiu J, He C, Chen M (2017) Recent advances in anticancer activities and drug delivery systems of tannins. Med Res Rev 37:665–701
4. Tanner GJ, Francki KT, Abrahams S, Watson JM, Larkin PJ, Ashton AR (2003) Proanthocyanidin biosynthesis in plants. J Biol Chem 278:31647–31656
5. Niemetz R, Gross GG (2005) Enzymology of gallotannin and ellagitannin biosynthesis. Phytochemistry 66:2001–2011
6. Paolocci F, Robbins MP, Madeo L et al (2007) Ectopic expression of a basic helix-loop-helix gene transactivates parallel pathways of proanthocyanidin biosynthesis: structure, expression analysis, and genetic control of leucoanthocyanidin 4-reductase and anthocyanidin reductase genes in *Lotus corniculatus*. Plant Physiol 143:504–516
7. Guo L, Qiang T, Ma Y et al (2021) Purification and characterization of hydrolysable tannins extracted from *Coriaria nepalensis* bark using macroporous resin and their application in gallic acid production. Ind Crops Prod 162:113302
8. Wu Q-Y, Zhou Y, Jin X et al (2011) Chromatographic fingerprint and the simultaneous determination of five bioactive components of *Geranium carolinianum* L. water extract by high performance liquid chromatography. Int J Mol Sci 12:8740–8749

Part IV

Artificial Intelligence (AI) in Phytoconstituent Analysis

Check for updates

Chapter 30

AI-Powered Prediction of Secondary Metabolite Biosynthesis

Preeti Bhatt, Rakesh Savant, Kajal Patil, Anand Patel and Bhavin Patel

Abstract

Secondary metabolites are small bioactive compounds produced by plants that support defense, stress tolerance, and ecological interaction. Predicting the biosynthesis of these metabolites is complex because it involves multiple genes, enzymes, and environmental factors. Traditional laboratory-based prediction is slow and costly. Integrating Artificial Intelligence (AI) into metabolite synthesis prediction enables researchers to analyse large biological datasets, such as genomics datasets. AI techniques, such as Machine Learning (ML), Deep Learning (DL), and Transfer Learning, can detect hidden patterns, identify key enzymes, classify metabolites, and predict biosynthetic pathways with improved accuracy. This protocol provides a step-by-step method for developing and evaluating AI-based models to predict secondary metabolite biosynthesis from multiomics data. The procedure covers data collection, preprocessing, feature extraction, model training, validation, and deployment. It is suitable for applications in biotechnology, agriculture, plant science, and metabolite discovery.

Keywords Phytoconstituents, Biosynthetic gene clusters, Genome mining, Predictive modeling

1 Introduction

A metabolite is a small molecule involved in the biochemical processes inside living organisms. Plant metabolites are broadly classified as primary and secondary metabolites; both groups are essential for growth and survival, but secondary metabolites have specialized roles in defense, signaling, and ecological interactions [1]. Primary metabolites, such as carbohydrates, proteins, lipids, nucleic acids, and amino acids, support core processes like photosynthesis, respiration, and cell division, and supply precursors for secondary metabolite biosynthesis [2]. Secondary metabolites (for example, phenolics, terpenoids, alkaloids, and flavonoids) influence plant color, flavor, scent, and defense responses and are strongly affected by biotic and abiotic stresses.

The biosynthesis of secondary metabolites is complex because it involves the coordinated action of genes, enzymes, regulatory transcription factors, and environmental cues. Elicitation (the detection

Dhruti Amin et al. (eds.), *Phytoconstituent Analysis*, Springer Protocols Handbooks,
https://doi.org/10.1007/978-1-0716-5400-2_30,

of a stress or signal) often upregulates defense-related transcription factors and shifts metabolic flux toward specific secondary pathways. Experimental characterization of these pathways requires extensive wet-lab work and can be time-consuming and costly. Artificial Intelligence (AI) offers a complementary route to speed up discovery by analyzing large multiomics datasets (genomic, transcriptomic/proteomic, and metabolomic) to detect patterns that indicate which enzymes and pathways are active and which metabolites are likely to be produced under given conditions. Integrating omics data improves prediction power because genomic data show potential (genes, BGCs), proteomic data indicate expressed enzymes, and metabolomic data provide direct evidence of metabolite presence and abundance [3–5]. AI methods (machine learning and deep learning) can therefore prioritize candidate metabolites, suggest biosynthetic routes, and guide targeted experiments.

Data quality and preprocessing are crucial for reliable AI predictions. Biological datasets often contain noise, batch effects, and missing values; therefore, cleaning, normalization, feature extraction (e.g., k-mer counts, sequence embeddings, protein domain features, spectral peaks), and dimensionality reduction are necessary steps before model training [6–9]. Model selection (classical ML vs deep learning), validation (k-fold cross-validation external validation), and biological interpretation (feature importance, pathway consistency) together determine whether predictions are trustworthy and actionable [8, 10–14].

This protocol provides a concise, step-by-step workflow for developing AI models to predict plant secondary metabolites biosynthesis. It covers data collection, preprocessing, feature extraction, model training, validation, and deployment, with emphasis on reproducibility and biological relevance. The approach aims to accelerate metabolite discovery and support applications in biotechnology, agriculture, and natural-product research.

2 Materials

2.1 Software and Tools

1. Python 3.x: Primary programming language for data analysis and AI model development.
2. Jupyter Notebook/Google Colab: For writing and running Python scripts.

2.2 Python Libraries

1. NumPy—numerical computation.
2. Pandas—data handling and preprocessing.
3. Scikit-learn—machine learning models (SVM, Random Forest, k-NN, etc.).

4. TensorFlow/Keras or PyTorch—deep learning model development.
5. Matplotlib/Seaborn – data visualization.
6. RDKit—chemical structure processing (optional, if using metabolite fingerprints).

2.3 Bioinformatics Tools (If Required)

1. BLAST—sequence similarity search.
2. PlantTFDB/KEGG/MetaCyc—for pathway, gene, and enzyme reference.
3. NCBI datasets—for genomic and proteomic downloads.

2.4 Datasets

Prepare at least one of the following:

1. Genomic Data: Raw DNA sequences, biosynthetic gene clusters (BGCs), or sequences related to metabolite pathways.
2. Promote Data: Protein expression profiles obtained from LC–MS/MS or public proteomics databases.
3. Metabolomic Data: LC-MS, GC-MS, or NMR spectra for metabolite identification and quantification.
4. Metadata Files: Species/tissue information, Stress conditions, Experimental batch details.

All datasets are saved in formats like FASTA, CSV, TSV, mzML, or raw spectral formats, depending on the source.

2.5 Hardware

1. A standard laptop or workstation with at least: (≥8 GB RAM).
2. GPU is optional for deep learning.

3 Methods

3.1 Data Collection

1. Collect genomic data from plant samples, such as DNA sequences or biosynthetic gene clusters involved in metabolite biosynthesis [15].
2. Collect proteomic data using mass spectrometry or related techniques to identify enzyme expression and protein abundance.
3. Collect metabolomic data directly from plant tissues using LC–MS, GC–MS, or NMR to measure metabolite levels.
4. Organize sample metadata (species, tissue, growth condition, stress treatment) along with raw genomic, proteomic, and metabolomic files.
5. Convert all datasets into standard formats, such as FASTA, CSV, or spectral files, to ensure compatibility with AI/ML tools.

3.2 Data Preprocessing and Feature Extraction

1. Clean raw data to remove duplicated entries, missing values, and noise produced during biological experiments [6, 8, 9].
2. Normalize the datasets using appropriate scaling techniques (e.g., Min–Max scaling or StandardScaler) to reduce batch effects and ensure consistent model input [7].
3. Extract meaningful features from each omics dataset:
   a. Genomics: k-mer frequency, GC content, sequence patterns, and gene expression signatures.
   b. Proteomics: protein abundance, enzyme class distribution, and interaction patterns.
   c. Metabolomics: spectral peaks, peak intensities, retention time, and metabolite concentrations.
4. Use dimensionality-reduction methods, such as Principal Component Analysis (PCA), to simplify complex datasets and remove redundant variables [6, 8, 9].
5. Combine all selected features into a structured dataset suitable for machine learning (ML) or deep learning (DL) model training.

3.3 Model Training

3.3.1 Select Appropriate ML or DL Techniques Based on Data Type and Prediction Needs [16]. Common Models Include

1. Machine Learning:
   - Support Vector Machine (SVM),
   - Random Forest,
   - k-nearest neighbor (k-NN),
   - Linear Regression.
   - Decision Tree:
2. Deep Learning:
   - CNNs for spectral or image-like data [17],
   - RNNs for sequence-based genomic data [8, 11–14].
3. Split the dataset into training, validation, and test sets using random or stratified sampling.
4. Train the selected model by feeding extracted features and allowing the system to learn patterns associated with specific metabolites.
5. Use gradient descent and backpropagation (for DL models) to optimize model weights.
6. Perform hyperparameter tuning using Grid Search or Bayesian optimization to improve model performance and reduce error.
7. Save the final trained model for later validation and deployment.

3.4 Model Validation and Accuracy Assessment

Validate model performance using k-fold cross-validation leave-one-out validation, or external datasets not used during training [10]. For classification tasks, compute evaluation metrics such as:

1. **Accuracy**
   - Accuracy tells how many predictions made by the model are correct.
   - It is useful when the dataset is balanced (a similar number of samples in each class).
   - High accuracy means the model correctly identifies most metabolites or classes.

2. **Precision**
   - Precision shows how many of the metabolites predicted by the model were actually correct.
   - It reduces false positives.
   - High precision means the model does not wrongly predict metabolites that are not present.

3. **Recall**
   - Recall measures how many actual metabolites the model was able to detect correctly.
   - It reduces false negatives.
   - High recall indicates that the model successfully identifies most of the actual metabolites in the sample.

4. **F1 score**
   - F1 score is the harmonic mean of precision and recall.
   - It is useful when the dataset is imbalanced.
   - A high F1 score indicates that the model strikes a good balance between accurately detecting most metabolites (recall) and predicting them correctly (precision).

5. **For regression-based predictions, use:**
   - RMSE (Root Mean Squared Error).
   - MAE (Mean Absolute Error).
   - R^2 (Coefficient of determination).

     1. Assess biological relevance by checking whether predicted metabolites match known plant pathways or published metabolic responses.
     2. Select the best-performing model based on both statistical metrics and biological consistency.

3.5 Model Deployment

1. Export the final trained model using formats such as pickle, ONNX, or TensorFlow SavedModel.
2. Deploy the model using one of the following:
   - Cloud-based platforms (AWS, Google Cloud, Microsoft Azure) for scalable and high-performance prediction [18].
3. Web or desktop applications that allow users to upload new omics datasets and receive predictions [19].
4. APIs to integrate model predictions directly into existing laboratory workflows.
5. Connect metabolomic instruments (LC–MS/NMR) or genomic data pipelines to the deployed system for automated or semi-automated predictions.
6. Allow researchers to analyze new plant samples without retraining the model, ensuring fast and reproducible results.
7. Periodically update or retrain the deployed model as new datasets become available .

4 Notes

1. Data quality is critical: Noise, missing values, and batch effects must be removed for reliable predictions.
2. Normalization is required. Omics data vary widely, so scaling to a common range is necessary for model performance.
3. Avoid data leakage. Ensure that training and testing datasets do not share samples.
4. Choose models based on dataset size. Classical ML) works better for small datasets; deep learning requires larger data.
5. Cross-validation is recommended. It gives a more accurate estimate of model performance.
6. Hyperparameter tuning improves accuracy. Use Grid Search or similar optimization methods.
7. Biological interpretation is important. Predictions should match known pathways or published plant studies.
8. Document all preprocessing steps. Ensures reproducibility for future users.

References

1. Jeyasri R et al (2023) Methyl jasmonate and salicylic acid as powerful elicitors for enhancing the production of secondary metabolites in medicinal plants: an updated review. Plant Cell Tissue Organ Cult 153(3):447–458
2. Tariq A et al (2023) Regulation of metabolites by nutrients in plants, Plant ionomics: sensing, signaling, and regulation. John Wiley & Sons, Hoboken, NJ, USA, pp. 1–18
3. Zhao J, Davis LC, Verpoorte R (2005) Elicitor signal transduction leading to production of plant secondary metabolites. Biotechnol Adv 23(4):283–333
4. Breitling R et al (2013) Metabolomics for secondary metabolite research. Metabolites 3(4):1076–1083
5. Emwas A-H et al (2019) NMR spectroscopy for metabolomics research. Metabolites 9(7):123
6. Ahmad T, Zakarya F (2020) Supervised learning methods for skin segmentation classification. Cent-Eur J New Technol Res Educ Pract 3(1):40–56
7. Misra BB, van der Hooft JJ (2016) Updates in metabolomics tools and resources: 2014–2015. Electrophoresis 37(1):86–110. p
8. Reddy JL et al (2019) Hand gesture recognition using PCA, KNN and SVM. Int J Sci Res Comput Sci Eng Inf Technol 5(2):547–550
9. Tripathi K, Nandi NBG (2015) Continuous Indian sign language gesture recognition and sentence formation. Procedia Comput Sci 54:523–531. p
10. Broadhurst DI, Kell DB (2006) Statistical strategies for avoiding false discoveries in metabolomics and related experiments. Metabolomics 2(4):171–196
11. Kim Y, Kim GB, Lee SY (2021) Machine learning applications in genome-scale metabolic modeling. Curr Opin Syst Biol 25:42–49
12. Prasad CR, Kishore D, Rao MSS (2014) Recognizing Gestures of Sign Language Using Feed Forward Neural Networks. Int J Syst Technol 7(1):13–21.
13. Rajaganapathy S et al (2015) Conversation of sign language to speech with human gestures. Procedia Comput Sci 50:10–15
14. Rao GA et al (2018) Deep convolutional neural networks for sign language recognition, In: 2018 conference on signal processing and communication engineering systems (SPACES). IEEE, Piscataway, NJ, USA
15. Wishart DS (2019) Metabolomics for investigating physiological and pathophysiological processes. Physiol. Rev 99(4):1819–1875
16. Feehan R, Montezano D, Slusky JS (2021) Machine learning for enzyme engineering, selection and design. Protein Eng Des Sel 34:gzab019
17. Poltorak V et al (2014) Spec2Class: accurate prediction of plant secondary metabolite class using deep learning. bioRxiv 2024.03.17.585408
18. Mendez KM, Broadhurst DI, Reinke SN (2019) The application of artificial neural networks in metabolomics: a historical perspective. Metabolomics 15(11):142
19. Chi J et al (2024) Artificial intelligence in metabolomics: a current review. Trends Analyt Chem 178:117852

Check for updates

Chapter 31

Machine Learning-Driven Profiling of Legume Phenolics for Diabetes Management

Hitarth Kansara, Vishwam Singh and Krutika Abhyankar

Abstract

Phenolic compounds are increasingly recognized for their essential role in the management of diabetes mellitus, primarily through mechanisms involving the reduction of blood glucose levels, enhancement of insulin secretion, improvement of pancreatic beta cell function, mitigation of oxidative stress, and inhibition of key carbohydrate-metabolizing enzymes, such as α-amylase, α-glucosidase, and dipeptidyl peptidase IV (DPP-IV). Among plant-based sources, legumes constitute a significant reservoir of phenolic compounds, many of which have been reported as potent inhibitors of α-glucosidase activity. In this study, the dataset concerning phenolic profiles of selected legume species is curated based on metabolomics data drawn from existing literature. The chapter outlines a methodological framework employing a machine learning (ML) approach—specifically, random forest classification—to predict the α-glucosidase inhibitory activity of legume-derived phenolic compounds. Feature selection and model interpretability analyses indicate that CYP450 1A2 inhibitor status, Log P value, binding affinity, molecular weight, and the number of hydroxyl groups constitute the most influential descriptors in predicting inhibitory potential. This protocol provides a computational strategy for screening phenolic candidates from legumes as prospective α-glucosidase inhibitors, facilitating further research and development in diabetes management.

Keywords Alpha-glucosidase inhibitor, Classification analysis, Metabolomics, Machine learning, Random forest

1 Introduction

Plant metabolites are reported to exhibit profound implications for human health, agriculture, and environmental sustainability. These metabolites are reported from metabolomics-based studies. It uses hyphenated techniques as chromatographic separation (UHPLC and/or GC) coupled to mass spectrometry (MS) for comprehensive metabolic profiling of vegetal matrices [1, 2]. The base peaks so obtained from the profiling of samples (plant cultivars, animal tissue samples, etc) help to characterize the metabolites. These metabolites, based on existing literature, are classified into different categories. Specifically considering legume seeds, a wide variety of phenolic

Dhruti Amin et al. (eds.), *Phytoconstituent Analysis*, Springer Protocols Handbooks,
https://doi.org/10.1007/978-1-0716-5400-2_31,

compounds are reported and searched for their bioactive properties [1, 3]. Plant-derived bioactive compounds are observed to inhibit alpha-glucosidase, and it emphasizes the most promising therapeutic candidates for Type 2 Diabetes mellitus (T2DM) management via targeting glucosidase enzyme [4]. Alpha-glucosidase is present at the brush border of enterocytes and mediates the last stage in glucose metabolism, particularly in carbohydrate digestion. Therefore, the inhibition of alpha-glucosidase enzyme results in a glucose production delay, which contributes to its therapeutic role in T2DM. The relationship between alpha-glucosidases' catalytic characteristics, particularly substrate selectivity, and their structures have been the subject of much research in the past two decades [4]. At this point, machine learning (ML), as a subfield of artificial intelligence, can be beneficial in overcoming experimental limitations by achieving a more comprehensive understanding of the effects of various phytochemicals under different conditions. ML is widely used to learn from the experiences hidden in large data sets; it uses statistics and some algorithms that can help to see the patterns in the data, make predictions, or develop heuristic rules to guide experimental or clinical studies in the future [5]. The data set can be constructed through extensive online research. The data set should include details about the phytochemical name, IC_{50} values against alpha-glucosidase enzyme, log P value, rotatable bonds, binding affinity values, ADMET properties, etc. The model development can be based on R, Java, Python, etc. The ML approach can be useful in identifying the most effective components among phytochemicals whose efficacy has already been proven by experiments. The Random Forest method (RFM) can be used for regression and classification problems. The RFM integrates a multitude of decision trees; predictions from each tree contribute to the final prediction, which is computed by averaging the predictions of all trees [6]. This protocol highlights effective phytochemicals as inhibitors of alpha-glucosidase based on previous in vitro studies. Thus, this classification analysis methodology is provided in the chapter. It deals with the steps required to search for plant phenols as alpha-glucosidase inhibitors using ML for management of diabetes.

2 Materials

1. Tools and libraries include scikit-learn' (Random Forest), RDKit (molecular descriptors), matplotlib, "seaborn" (visualization).
2. For molecular docking, consider following tools Autodock vina, MgI tool (ligand and protein preparation), Discovery studio (visualization) and Open Babel (format conversion), ADMET studies using admetsar1 (https://lmmd.ecust.edu.cn/admetsar1/home/).

3 Methods

3.1 Preparing Dataset

1. Consider web-based literature on the legume seed metabolomics profile (*see* **Note 1**); state their IC_{50} (mg/ml) values against alpha-glucosidase enzyme in vitro system (*see* **Note 2**).
2. Mention add-on features to the phenolic compounds (molecular weight, H-bond acceptors/donors, rotatable bonds, hydroxyl groups, heavy atoms, and log P value) reported to exhibit IC_{50} values against alpha-glucosidase enzyme (*see* **Note 3**). Some feature labels for the given phenolic compounds can be noted from PubChem and RDKit.
3. Add features addressing binding energy values for respective phenolic compounds based on the molecular docking carried against alpha-glucosidase enzyme (PDB ID—5NN3) as shown in (Fig. 1) (*see* **Note 4**) [7].
4. Note the reported legume phenolic compounds for their in-silico toxicity and ADMET studies using (admetsar1: https://lmmd.ecust.edu.cn/admetsar1/home/).

This step helps to filter the candidates/test compounds with optimal distribution, metabolism, excretion, and low toxicity.

3.2 Random Forest analysis

1. Implement random forest algorithms in Python's Jupyter Notebook and adhere to the workflow for building a predictive model (Fig. 2) (*see* **Note 5**).
2. Load the phenolic compounds dataset with all the required labels remove the compounds name column and separate the features i.e., input variables stated as "x" from the target variable i.e., antidiabetic potency stated as "y" (*see* **Note 6**).
3. Split the data into 80% training and 20% test sets using train_test_split is the command using Random state 41 (*see* **Note 7**).
4. Carry out data preprocessing by converting the categorical target classes (high/low/moderate potency) to numeric values using LabelEncoder. Further apply StandardScaler to normalize all features to similar scales (*see* **Note 8**).
5. For model training, use 80% dataset from the data split as mentioned before. Further, for optimization, use Randomized Search CV with eightfold cross-validation to find optimal parameters. Cross-validation cross-validation will evaluate the model performance on unseen data by testing it multiple times on different data splits and will generate the best CV score with random state 41 (*see* **Note 9**).
6. Based on the split test dataset and the model accuracy (>70%) using random forest, (≥60%) reflects a good prediction model and considers further steps for the analysis (*see* **Note 10**).

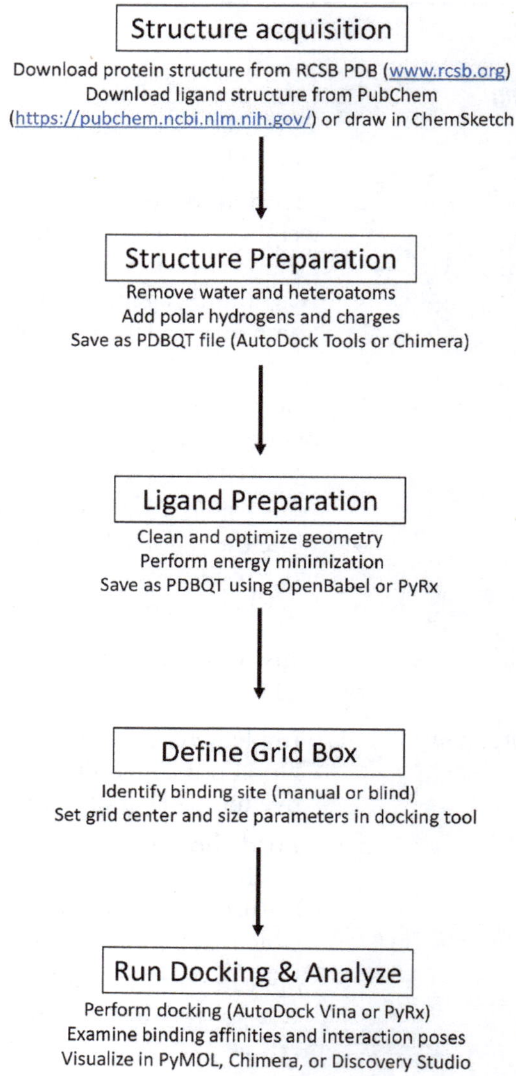

Fig. 1 Flowchart depicting the methodology employed for docking the target protein to the ligand

7. Evaluate the models by selecting metrics for classification by assessing weighted averages and macro averages. The average values are calculated based on efficacy related to low, moderate, and high anti-diabetic potentials of phenolic compounds.
8. The model will generate a weighted average with specific values for accuracy and precision, Recall, F1 score, such as (0.7000). Additionally, macro average values are also generated for accuracy and precision, Recall and F1 score, such as (0.5167).

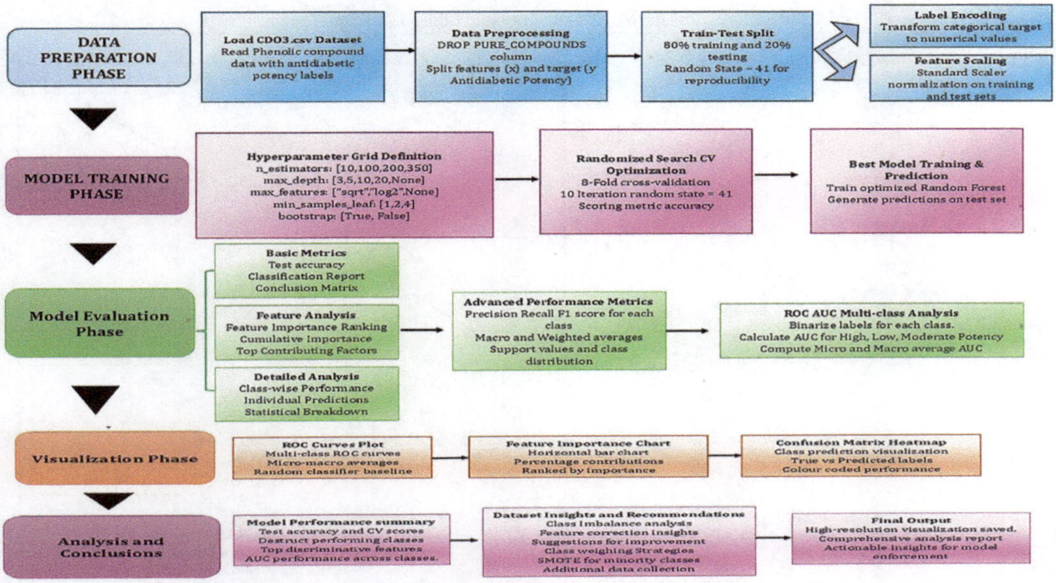

Fig. 2 Random forest phenolic compound classification pipeline

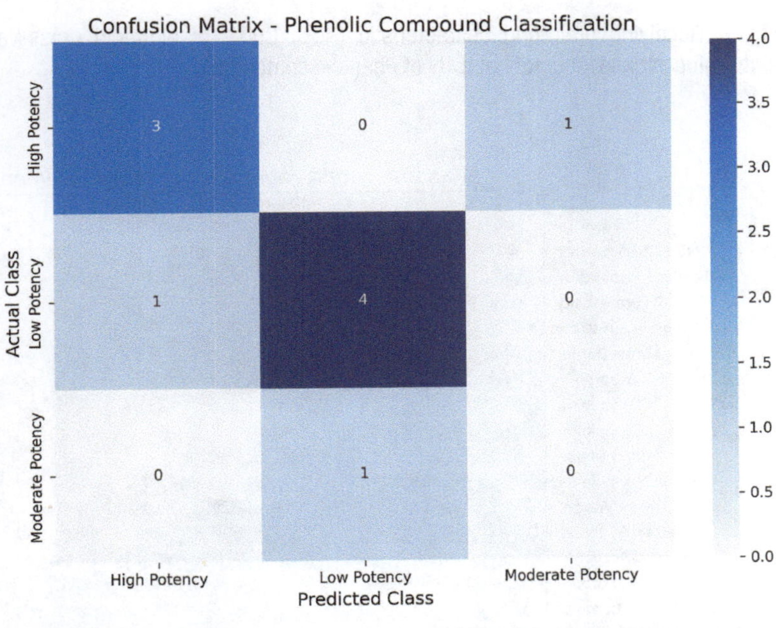

Fig. 3 Confusion matrix plot to compare the predictions made by the model with actual results

9. Further, construct the confusion matrix plot and ROC–AUC curve for the developed Random Forest model using matplotlib and seaborn as shown in (Fig. 3) and (Fig. 4).

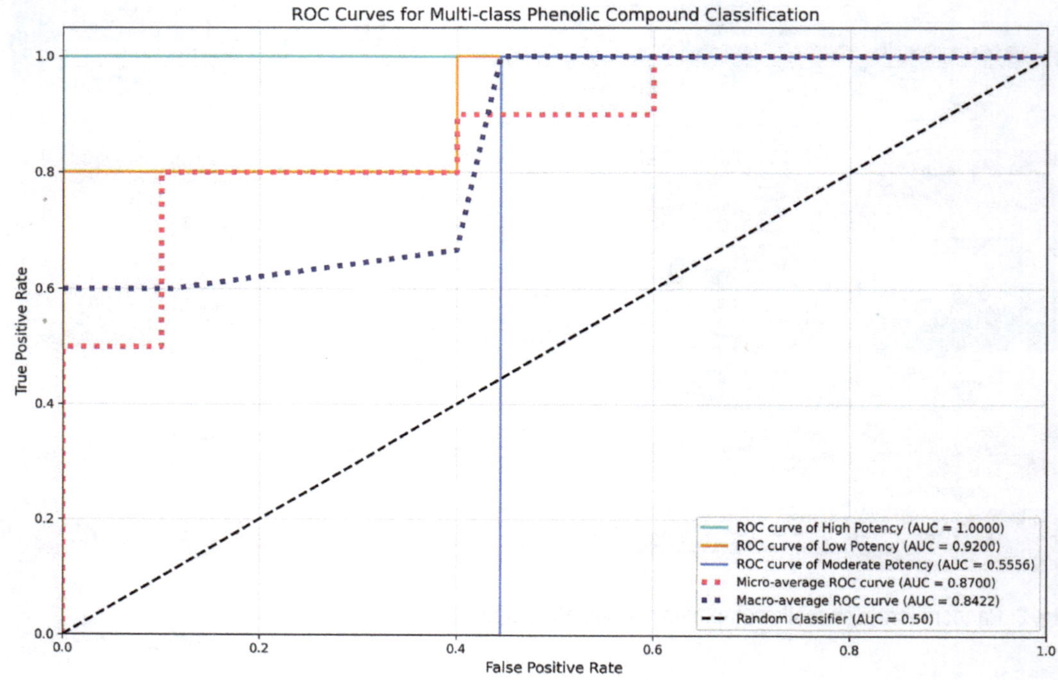

Fig. 4 Receiving operating characteristic (ROC) Curve of eightfold cross-validation on the prediction of alpha-glucosidase inhibitory activity of Phenolic compounds

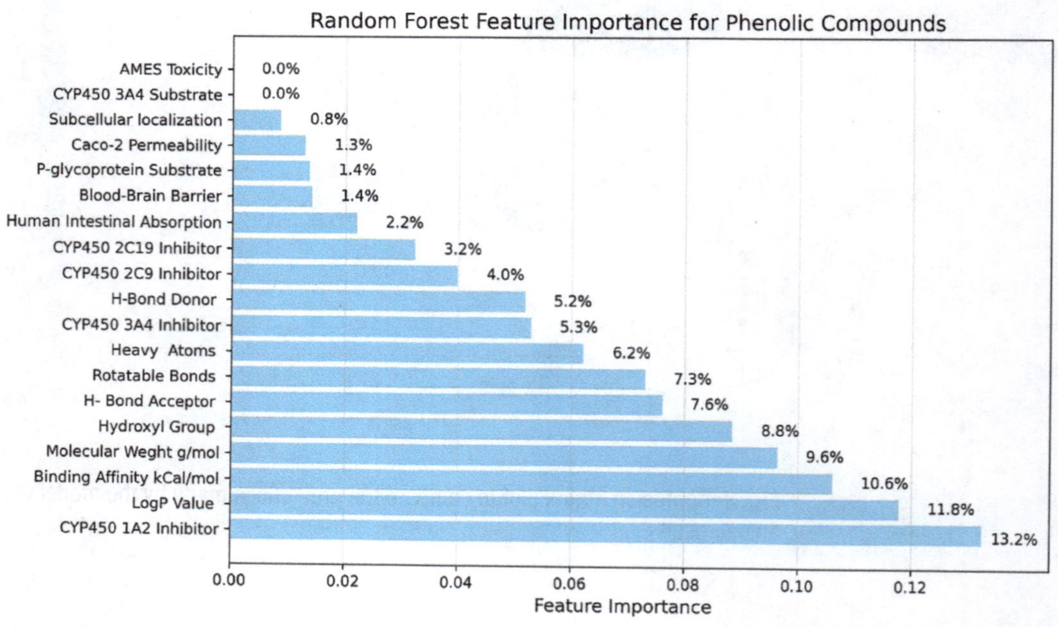

Fig. 5 Random forest feature importance analysis for alpha-glucosidase inhibitory activity of the phenolic compounds

10. Plot Random Forest feature importance for phenolic compounds for model interpretation and validation, correlating important features' IC_{50} values with chemical properties, docking values, and ADMET features (Fig. 5). Wherein highest relevance will be shown for specific features, such as CYP450 1A2 Inhibitor, followed by log P value, Binding Affinity kCal/mol and hydroxyl group, etc (Fig. 5).
11. Compare predictions with experimental results to validate model reliability.

4 Notes

1. The data set should be constructed by considering academic literature search engines, such as Google scholar, Web of Science, and PubMed. Few keywords, such as "Legumes", "Phenolic compounds", "Metabolomics" "alpha-glucosidase inhibitors", and "Anti-diabetic", can be considered as suggestions. Additionally, the searched studies considered for literature should possess a specific timeline, such as 2015–2025.
2. The IC_{50} values against alpha-glucosidase inhibitory activity are separated as follows:

   a. High Potency: $IC_{50} < 0.01$ mg/mL (highly effective at low concentrations).
   b. Moderate Potency: IC_{50} 0.01–0.1 mg/mL (moderately effective).
   c. Low Potency: $IC_{50} > 0.1$ mg/mL (weakly effective, requiring higher concentrations).

3. The phenolic compounds features are available from PubChem and additional details can be obtained using RDKit which is an open—source cheminformatics software.
4. The protein preparation step is very crucial while processing alpha-glucosidase obtained from PDB ID—5NN3. The structure procured from PDB Database contains six identical chains (A–F). For docking, only one chain, i.e., Chain A, will be required. In this chain, the standard amino acid residues have 872 amino acids only. From residue 873–952 are non-peptide heteroatoms like glycans, such as NAG(N-acetylglucosamine). There is also one oxidized cysteine, i.e., cysteine sulfenic acid, which can be converted to cysteine. Further clean the protein by removing other non-protein molecules like water, sulfate ions ($SO4^{2-}$), chloride (Cl^-), PEG (Polyethylene Glycol), EDO (Ethylene Glycol), etc.

5. In Python, all input values should be numeric. Import the required libraries for data processing, machine learning, and visualization as follows:

```
import pandas as pd
import numpy as np
from sklearn.model_selection import train_test_
split, RandomizedSearchCV, GridSearchCV
from sklearn.preprocessing import LabelEncoder,
StandardScaler, label_binarize
from sklearn.ensemble import
RandomForestClassifier
from sklearn.metrics import accuracy_score,
classification_report, confusion_matrix, roc_
curve, auc
import matplotlib.pyplot as plt
import seaborn as sns
from itertools import cycle
```

6. Split data into training (80%) and testing (20%) sets with a fixed random state for reproducibility using:

```
X_train, X_test, y_train, y_test = train_test_
split(X, y, test_size=0.2, random_state=41)
```

7. Separate features (X) and target variable (y) (low, moderate, high potency) using

```
X = compound_descriptor.loc[:, (compound_
descriptor.columns!= 'Antidiabetic_potency')]
```

```
y = compound_descriptor['Antidiabetic_potency']
```

Encode categorical target labels to numerical format for machine learning by applying

```
label_encoder = LabelEncoder()
y_train = label_encoder.fit_transform(y_train)
y_test = label_encoder.transform(y_test)
```

8. Standardize feature values to have mean = 0 and standard deviation = 1 for better model performance by using `scaler = StandardScaler()`.

```
Fit the scaler on the training data with X_
train = scaler.fit_transform(X_train) and
transform the test data with X_test = scaler.
transform(X_test).
```

9. Initialize a Random Forest classifier with a fixed random state using

```
rf_model =RandomForestClassifier(random_state=41).
```

Perform randomized search cross-validation to find the optimal hyperparameters by executing

```
grid_search = RandomizedSearchCV(rf_model, par-
am_grid, cv=8, scoring='accuracy', n_jobs=-1,
random_state=41, n_iter=10)
```

```
Then fit the grid search on the training data
with grid_search.fit(X_train, y_train).
Get the best performing model from the hyperpa-
rameter search using
best_rf = grid_search.best_estimator_
Generate predictions on the test set with
y_pred = best_rf.predict(X_test)
```

10. Calculate and display the model's performance by computing the test accuracy:

```
test_accuracy = accuracy_score(y_test, y_pred)
print(f"Test Accuracy: {test_accuracy:.4f}")
```

References

1. Abu-Reidah IM, Arráez-Román D, Warad I, Fernández-Gutiérrez A, Segura-Carretero A (2017) UHPLC/MS2-based approach for the comprehensive metabolite profiling of bean (*Vicia faba* L.) by-products: a promising source of bioactive constituents. Food Res Int 93:87–96
2. Ammar S, del Mar Contreras M, Belguith-Hadrich O, Bouaziz M, Segura-Carretero A (2015) New insights into the qualitative phenolic profile of *Ficus carica* L. fruits and leaves from Tunisia using ultra-high-performance liquid chromatography coupled to quadrupole-time-of-flight mass spectrometry and their antioxidant activity. RSC Adv 5(26):20035–20050
3. Solanki K, Abhyankar KS (2024) Metabolomic profiling of germinated and non-germinated *Lablab purpureus* seeds: antioxidant properties and α-amylase inhibitory activities for diabetes management. Food Prod Process Nutr 6(1):99
4. Kashtoh H, Baek KH (2022) Recent updates on phytoconstituent alpha-glucosidase inhibitors: an approach towards the treatment of type two diabetes. Plants 11(20):2722
5. Alzubi J, Nayyar A, Kumar A (2018) Machine learning from theory to algorithms: an overview. J Phys: Conf Ser 1142:012012
6. Patil SS, Deshannavar UB, Gadekar-Shinde SN, Gadagi AH, Kadapure SA (2023) Optimization studies on batch extraction of phenolic compounds from *Azadirachta indica* using genetic algorithm and machine learning techniques. Heliyon 9(11):e21991
7. Fadimu GJ, Farahnaky A, Gill H, Olalere OA, Gan CY, Truong T (2022) In-silico analysis and antidiabetic effect of α-amylase and α-glucosidase inhibitory peptides from lupin protein hydrolysate: enzyme-peptide interaction study using molecular docking approach. Foods 11(21):3375

Index

Dhruti Amin et al. (eds.), *Phytoconstituent Analysis*, Springer Protocols Handbooks, https://doi.org/10.1007/978-1-0716-5400-2,

FSC
www.fsc.org
MIX
Papier | Fördert gute Waldnutzung
FSC® C083411

Kolibri 360 GmbH
Ferdinand-Jühlke-Straße 7
99095 Erfurt, Deutschland
produktsicherheit@kolibri360.de